Wireless World

Guide to Broadcasting Stations

Newnes Technical Books

The Butterworth Group

United Kingdom	Butterworth & Co (Publishers) Ltd London: 88 Kingsway, WC2B 6AB
Australia	Butterworths Pty Ltd Sydney: 586 Pacific Highway, Chatswood, NSW 2067 Also at Melbourne, Brisbane, Adelaide and Perth
Canada	Butterworth & Co (Canada) Ltd Toronto: 2265 Midland Avenue, Scarborough, Ontario, M1P 4S1
New Zealand	Butterworths of New Zealand Ltd Wellington: T & W Young Building, 77–85 Customhouse Quay, 1, CPO Box 472
South Africa	Butterworth & Co (South Africa) (Pty) Ltd Durban: 152–154 Gale Street
USA	Butterworth (Publishers) Inc Boston: 10 Tower Office Park, Woburn, Mass. 01801

First published 1946
Eighteenth edition 1980 by Newnes Technical Books

British Library Cataloguing in Publication Data

Guide to broadcasting stations. – 18th ed.
1. Radio stations – Directories
I. 'Wireless world' guide to broadcasting stations
621.3841'6'025 TK6555

ISBN 0 408 00467 3

Typeset by Scribe Design, Gillingham, Kent
Printed in England by Redwood Burn Ltd., Trowbridge and Esher

CONTENTS

ACKNOWLEDGEMENT

Thanks are due to the BBC for the lists of broadcasting stations, which were prepared by the BBC Receiving Station, Caversham Park, Reading.

A GUIDE TO LISTENING

AERIALS

For medium- and long-wave reception most receivers have an internal ferrite-rod aerial, which enables them to receive the local stations and the stronger of the more distant stations. These aerials are directional and give very poor results when the rod points in the direction of the transmitter, so it is worthwhile checking whether the aerial is favourably oriented. Some portable receivers have a turntable built into the base to enable them to be rotated conveniently, and larger receivers sometimes have a control which rotates the aerial within the case. In searching the wavebands, it is easily possible to miss signals from transmitters in line with the aerial, and it is a good plan, therefore, to repeat the search with the aerial at right angles to its former position. Ferrite-rod aerials are not used for short-wave reception and these directional effects are not present.

Many receivers have aerial and earth sockets and it is possible to effect a great improvement in reception by using an external aerial. Suitable forms of aerial are discussed later. When an external aerial is used the effect on reception of rotating the ferrite rod is much less marked and may even be absent altogether.

Short-wave receivers often have telescopic aerials which can be extended to a metre or so in length and can sometimes be tilted. These, too, can provide satisfactory reception of the stronger signals.

Improved reception is often possible using an aerial external to the receiver, supported, for example, on the wall of a room or in the roof-space. Results from indoor aerials are, however, often disappointing because the aerial is screened from the wanted signals by the walls and/or roof of the building and is near the electrical wiring and domestic electrical equipment. While it may be easy to suppress noise and interference from your own washing machine and light dimmer, it is less easy to suppress your neighbour's, which in flats and terraces may be even nearer than your own. Indoor aerials are thus liable to pick up a high level of electrical interference.

For best results an outdoor aerial is essential and, if electrical interference is a problem, the aerial should be located in an interference-free area and special precautions taken to ensure that the cable connecting the aerial to the receiver does not pick up interference from the electrical system of the house.

An inverted-L aerial, Fig. 1a, is quite suitable for long- and medium-wave reception. Results improve as the length of the horizontal section and the height above the ground are increased. The horizontal section should be insulated from the supporting wires or ropes by several small porcelain insulators at each end. The downlead should be a continuous length of wire with the aerial and not joined separately because soldered and other kinds of joints are likely to deteriorate with weathering and eventually cause crackles and other effects in the receiver. The lead-in should be arranged to drop from the aerial well away from the building to avoid contact with gutters and to minimize pick-up of noise from the domestic electrical supply. If a tree is used to support the far end of the

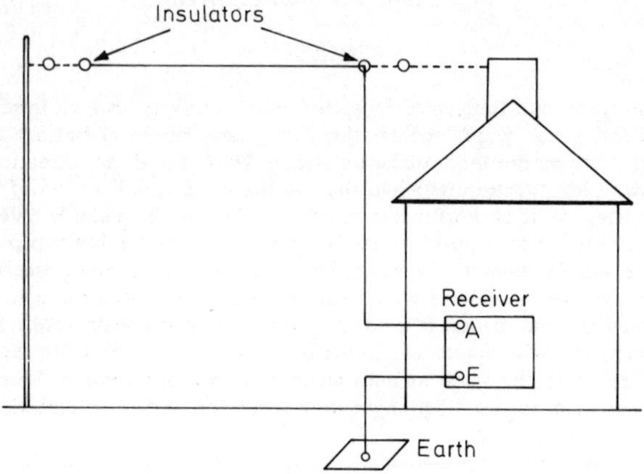

Fig. I(a). Inverted-L aerial . . .

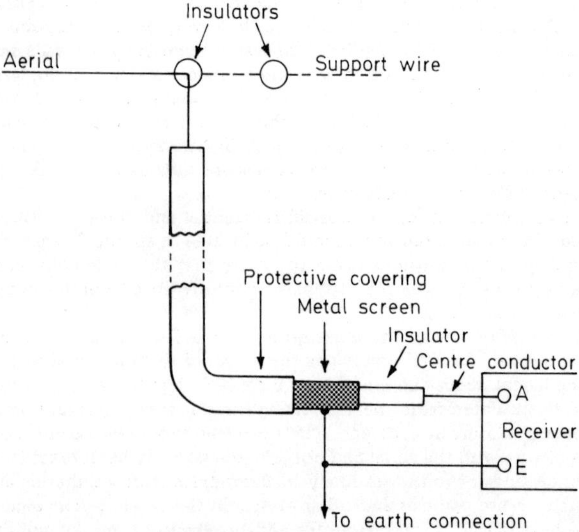

. . . (b) and screened down-lead

aerial, allowance must be made for the movement of the tree under windy conditions. The terminating wire or rope should be passed over a pulley and terminated with a suitable weight. In this way the tension in the aerial wire can be maintained independent of movement of the tree.

Sometimes it is convenient to take the downlead from the centre point of the horizontal section. The resulting aerial is known as a T-aerial and its performance is very similar to that of the inverted-L.

As a precaution against electrical interference the downlead can take the form of a coaxial cable, the inner conductor providing the connection to the receiver and the outer conductor being earthed as shown in Fig. 1b. By this means the downlead is screened so that only signals picked up by the horizontal wire are conveyed to the receiver.

Where there is insufficient space for an inverted-L or T-aerial or where electrical interference is a serious problem, a vertical rod say 5 m long may be used. This should be mounted in an area where interference is a minimum (a chimney top is often a suitable place) and connected to the receiver by a screened lead as shown in Fig. 2. Aerial manufacturers market kits containing all the parts for such an installation including matching transformers for use at the aerial base and receiver input.

An inverted-L, T-aerial or vertical rod aerial is suitable for short-wave reception but where space permits there are more efficient types which can be used: these are directional aerials which should therefore

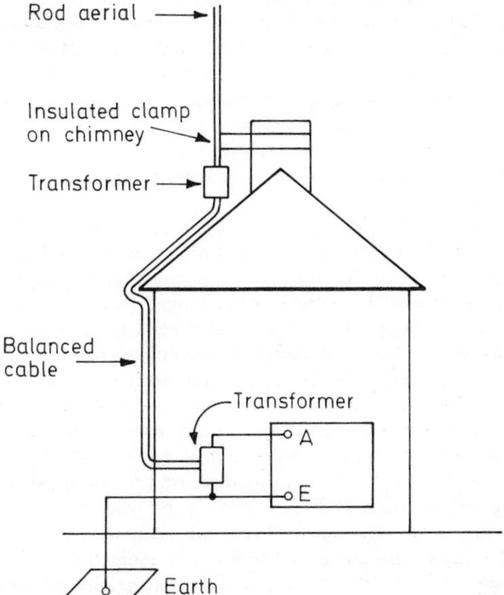

Fig. 2. Vertical rod aerial

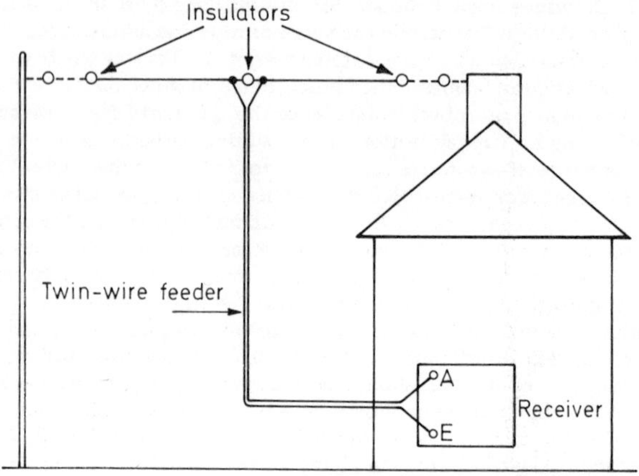

Fig. 3. Simple dipole aerial

be positioned to favour the direction of the transmitters it is desired to receive.

One suitable aerial is the half-wave dipole illustrated in Fig. 3. It consists of two horizontal arms connected to the receiver by a balanced feeder. The dipole should be mounted as high as possible but 10 m is probably the maximum height which is convenient for most domestic situations. The length of each of the two horizontal arms should be chosen to suit the wavelength of the signals it is desired to pick up and varies between 13 m for the 49-m band to 3 m for the 11-m band. The aerial has maximum response to signals travelling at right angles to its length and has minimum response to transmissions arriving in line with the aerial.

A disadvantage of the simple dipole is that it is less effective on wavebands other than those for which it has been designed. If, however, the two leads of the feeder are connected together and to the receiver aerial terminal, the earth terminal being connected to ground, the aerial then becomes a T type which can be used for long- and medium-wave reception as well as for short waves. A two-pole change-over switch can be used to convert the aerial from the dipole to the T form.

A better form of directional short-wave aerial is the inverted-V, Fig. 4. This provides a greater signal to the receiver than the simple dipole and by using the dimensions shown it can be effective over all the short-wave bands. It requires only a single support pole, one end of the aerial being earthed via a 400-ohm terminating resistor, the other being connected to the receiver input. This aerial has maximum sensitivity to signals travelling in the plane of the aerial as indicated in the diagram.

The Beverage aerial demands length but not height and consists of a length of wire supported by a series of short poles, say 2 or 3 m high and spaced sufficiently close to prevent undue sag. Each should be surmounted with an insulator to which the wire is bound, not looped, the aerial being terminated at the far end by a 600-ohm resistor. Wire length is not critical but it should not be less than about 50 m and the lead-in should be direct to the receiver without significant deviation from the general line; if this can be achieved an r.f. transformer and coaxial line are not required to connect the aerial to the receiver. This aerial favours the reception of signals travelling in line with the aerial from the terminating resistor end, and is used professionally with wire lengths up to 1000 m.

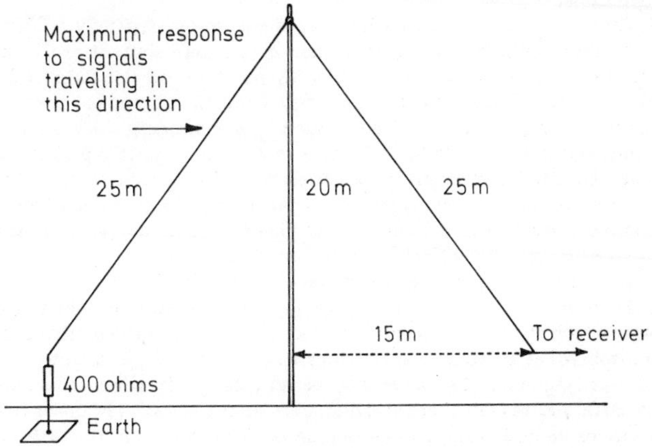

Fig. 4. Inverted-V aerial

When a receiver is supplied from a 3-pole main socket there is a natural temptation to use the earthed pole of the socket as an earth connection for the receiver. Such a connection is likely to be unsatisfactory because the physical connection of the main earth to ground is often at a considerable distance from the mains socket. Consequently the earth path may have appreciable resistance and can carry signals capable of causing interference to radio reception.

Where a receiver is provided with a signal earth terminal, local interference may be reduced by connecting the terminal by a short lead to a copper plate or earth rod buried in the ground. A similar connection is also required for inverted-V and some other aerials. A connection to a gas pipe is usually an unsatisfactory earth and may be extremely dangerous; most underground metal gas pipes are being replaced by plastics pipes. A connection to a metal water pipe is satisfactory only if the pipe is connected directly to an underground water main: in many modern housing estates the metal pipes within the house are connected to buried polythene pipes and do not provide a satisfactory earth connection.

Propagation of radio waves is a complex subject and in this brief chapter we can give only a general description of those aspects which may interest those whose hobby is listening to broadcasts generally and who may be sufficiently enthusiastic to extend their listening to more distant and difficult signals.

A knowledge of the basic facts will ensure that listening is carried out at the right time of day for a given frequency and will certainly provide more enjoyment by enabling the listener to anticipate good reception conditions and eliminate fruitless searching when propagation is poor. Awareness of the trends in propagation will leave the listener in no doubt as to causes of changes in reception and will enable selection of the most favourable periods for searching for the weaker and seldom-heard signal.

There are good reasons why a particular broadcast may within a short period improve to a degree when programme content can be appreciated or conversely may virtually disappear. It can also happen that strong signals from a given area may suddenly disappear within a minute or two, yet are received at their former strength thirty minutes or more later. Normal fading of signals may become more rapid, accompanied by a fall in strength and a corresponding increase in noise. These are some of the effects which the listener will observe and which, if carefully considered, will enable assessment of some of the changes in the ionosphere which affect reception conditions.

The basic facts governing short-wave propagation can be summarized in the following way. Short-wave radio communication is achieved by waves which strike the ionosphere (electrified layers in the earth's upper atmosphere) at an oblique angle and are reflected back to earth to cover the receiving area. The waves may be reflected again when they strike the earth and reach other receiving areas after successive bounces from the ionosphere. However in certain areas, for example in the area between the transmitter and the first earth-reflection point, the transmission may be very difficult to receive: this is a so-called skip zone.

For satisfactory short-wave communication the frequency must be chosen with care. If it is too high, the waves penetrate the ionosphere and are lost in space: if it is too low the waves are attenuated by absorption in the lower regions of the ionosphere. Best results are achieved by using the highest frequency which does not penetrate the ionosphere and the value of this, the highest probable frequency (HPF), depends on the degree of ionization of the gases in the ionosphere. This in turn depends largely on the extent to which the ionosphere over the chosen path is illuminated by the sun. Thus the HPF varies with the time of day and with the time of year.

Any changes in the degree of ionization of the reflecting layer can affect long-distance reception and such changes can be produced by increased radiation from the sun, e.g. from blemishes on its surface such as sunspots and invisible areas called M regions. As seen from the earth, the sun takes 27 days to rotate on its axis and some effects on reception, particularly those due to long-lived M regions, tend to have a 27-day periodicity. Moreover the incidence of sunspots follows an 11-year cycle; this in turn causes an 11-year periodicity in short-wave reception conditions.

At any particular time, a survey of all the broadcast bands will indicate that some are very active (many stations being receivable, possibly with a fair amount of interference), while other bands may appear to be

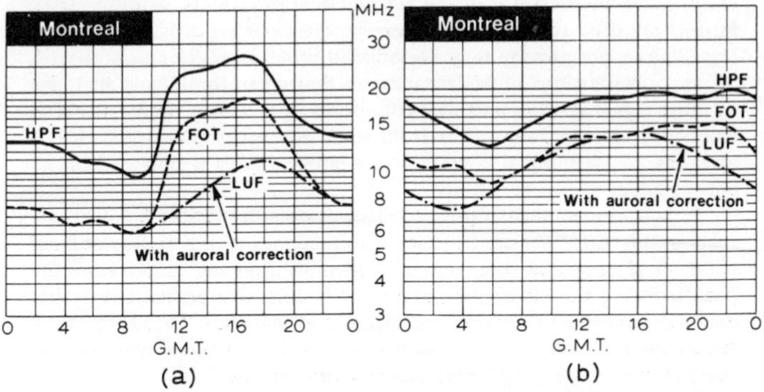

Fig. 5. Examples of HF prediction curves for the U.K.-Montreal path for January (a) and July (b). The highest probable frequency (HPF) is the median usable frequency exceeded on 10% of the days. The LUF (lowest usable frequency) curves are for commercial telegraphy and assume the use of high-power transmitters and rhombic aerials. The path to Montreal passes through the Northern Auroral Zone and waves are subject to additional absorption: a correction is made for this in calculating the LUF. The term optimum traffic frequency (FOT) is self explanatory

practically devoid of signals, apart from weak scattered radiation from stations some few hundred kilometres from the receiving site. These situations arise because transmissions are so arranged that programmes can be received at maximum signal strength in a desired area at local peak listening time. The choice is governed largely by HPF applicable to the required ionospheric path at that time, but the precise frequency may be somewhat lower to ensure that day-to-day variations in HPF do not seriously affect reception throughout the period of the programme or of the transmission schedule, which may be required to continue without alteration for a number of months. Two examples of prediction curves are given in Fig. 5. The upper curve represents the HPF and, in general, frequencies above this value are heard infrequently. The lower curve indicates the frequency below which the signal-to-noise ratio of the received signal becomes unacceptable. If frequencies between these two boundary curves are used the transmitted wave normally propagates over the particular path and provides a service in the target zone. Frequencies which approach the HPF produce the stronger signal but their propagation is more likely to be affected by ionospheric disturbances. It is impossible to predict with accuracy the variations to which signals are likely to be subjected, although short-term predictions based on daily observation of signals received can provide fair accuracy.

It is not good practice to make frequent changes of frequency in a broadcast schedule because the listener expects to find the programme at the same spot on the tuning scale. Thus to offset the variations of

MUF and make best use of the transmission paths, two or more transmitters are used to radiate the same programmes on different frequencies. Thus a programme may be radiated simultaneously on say the 17, 15 and possibly the 11 MHz bands, so that when the HPF is high the 17 MHz signal is good and well supported by 15 MHz, whilst the low-frequency channel may suffer from some absorption. When the HPF is low, the 17 MHz signal is weak and a better service is obtained on 15 and 11 MHz.

Announcements made prior to close-down and radiated by all broadcasts in the same network mention the frequency of the broadcast band which is closing and that which is opening. For any target zone the peak listening time is evening and the schedules of tranmissions to that area are arranged to provide programmes at that time. Frequency separation on the short-wave bands is only 5 kHz and there may be difficulty in receiving a programme clear of interference.

The broadcast bands and their frequency limits are shown elsewhere in this book, and in general transmissions must, by international agreement, be confined to these bands. Other services are similarly restricted to certain frequencies. The highest allotted frequency used in short-wave broadcasting is 26.100 MHz: thus when the HPF exceeds that figure, maximum use of propagation conditions cannot be obtained. However, most domestic receivers have an upper tuning limit as low as 21 or even 17 MHz.

Comparison of Fig. 5a and b shows that under summer-time conditions the HPF curves flatten considerably, day-time frequencies being lower and night-time frequencies higher than in winter-time. In the summer more transmissions are crowded into fewer bands and interference problems increase.

At periods of minimum solar activity HPF's are generally lower throughout the year and the reduced spectrum available for broadcasting causes increased interference.

Sunspot maximum conditions occur in 1979 and there will be a gradual decrease in the HPFs until sunspot minimum conditions are reached in 1985, after which the HPFs will increase toward the next maximum.

The ionosphere is subject to disturbances which can affect radio reception. The disturbances are usually caused by sunspots and their effect is to make the reception of certain of the short-wave broadcast bands difficult or even impossible. Thus, under certain conditions, signals in the high-frequency bands may be weak although the low-frequency bands are normal. Alternatively, the high-frequency bands may be normal and the low-frequency bands weak. Under more exceptional circumstances all the broadcast bands may be inaudible.

Thus, if short-wave reception is found to be very poor, the most likely cause is a disturbance in the ionosphere and it is unlikely to last more than a few days. Most of the disturbances last only a few hours.

SIGNAL IDENTIFICATION

Tuning scales of receivers are often marked with a wealth of station names, but it does not follow that all these stations can be received, even with a good external aerial. Equally, it should not be assumed that stations, even if they can be received, will be picked up at precisely the point indicated by the name on the scale. The calibration of a receiver is not always exact, even when it is new, and it tends to drift as the receiver gets older. Calibration can be checked by tuning in certain stations which maintain their allotted frequencies with great accuracy. Most transmitters have a reasonably good frequency stability but those on 200 KHz, 5, 10, 15 and 20 MHz are particularly accurate. For further details of standard frequencies and time codes consult the Wireless World Diary.

Signal identification involves a knowledge of broadcasting organizations and their programmes, transmission schedules and target areas, rather than merely a knowledge of transmitting stations. Interval signals, clock chimes, times of operation, types of programme and signal strength also aid identification.

The large number of languages used in short-wave broadcasting would be beyond the ability of one person to learn, but consistent listening to broadcasts from known countries, many radiating similar versions of the current world news, gives good practice in recognizing languages. The sound pattern of an unrecognized language can be compared with other broadcasts of languages which appear similar, remembering that a dialect may be used. Knowledge of the normal occupants of a waveband in terms of broadcasters and their programme schedules is also useful in language recognition.

Interval signals, or particular tunes, are often used to preface the start of transmissions or programmes, typical examples being the use of Bow Bells, Greenwich Time Signal and Big Ben by the BBC, the Canadian National Anthem by Sackville. the Kremlin Bells by Moscow and the Kookaburra by Melbourne. Eastern European stations often use the first few bars of a well-known melody, which may have been written by an eminent composer.

If these signals can be recorded on tape, a library of interval signals can be built up. Each recording can be annotated with the details of reception, to increase its usefulness as a reference guide.

The make-up and timing of broadcasts can often prove useful in identification. If a continuous programme is well balanced between music, speech, drama and other items, it is probably intended for home consumption and the opening and closing times of the transmission will give some idea of the time of day in the country of origin. A programme consisting of short items, with a preponderance of speech, starting or finishing at odd times, is likely to be a service for listeners outside the country. Clock chimes may narrow the choice, by fixing the time zone, and they often precede an announcement or news bulletin. Don't forget that some countries have summer or daylight saving time. The relaying of programmes can produce difficulties; for instance, London's Big Ben is heard from stations all over the world. Nevertheless, continued listening may provide a clue, which can be a change of atmosphere at the conclusion of a relay, or an announcement that follows.

Most broadcasts begin with a period of tone for technical alignment purposes, followed by an interval signal and announcement, then possibly a time check, and finally the programme. The frequency of the line-up tone differs from one organization to another; thus the BBC uses 1 kHz, Federal Germany 900 Hz, and some authorities use 440 Hz.

The close-down of a transmission is also important, because of the probability of announcements, and perhaps a national anthem or clock chime.

The type of programme may yield evidence of the nationality of the broadcasting organization and of the intended zone of reception. Domestic services can generally be recognized by the parochial nature of the news, the coverage of world events being small. Programmes for a country's nationals abroad are often a blend of domestic and world news, with commentaries in the national language; a typical example is the BBC World Service. Frequent news bulletins, almost exclusively concerned with world events and given in many languages, strongly suggest a service intended for foreign listeners.

When a programme whose source is unknown is sufficiently intelligible to be followed to a limited extent and a guess made at the language, a search for the identical programme on different frequencies may help identification. A second receiver is useful for this, because it can be tuned to known stations operating services in the supposed language. If another transmission carrying the programme is found, it may be assumed that both originate from the same source, though not necessarily from co-sited transmitters. One transmission may be a relay, and if so the quality of the unknown transmission may not be as good as the known.

It may still be difficult to determine the location of the unknown station, though listening at times of programme change for local or regional announcements can help in reaching a conclusion. At such times there may be changes in fading characteristics and background noise, indicating the conclusion of a relay and suggesting that the signal has been affected twice by ionospheric conditions. A typical example of relays is provided by the BBC World Service broadcast from the UK and relayed by bases in the Middle East, Far East and South Atlantic; other examples are provided by Deutsche Welle in Germany and its relay base in Africa, by Paris and Brazzaville, and by the Voice of America at Greenville and its overseas stations at Tangier, Munich, Monrovia and elsewhere.

The stronger of two signals carrying the same programme may not necessarily be that of the nearer transmitter. The receiving location may be in the skip zone of this transmitter and thus obtains a weaker signal. A better signal may also be obtained from the more distant transmitters if this is beamed toward the receiver site.

Programmes which are broadcast simultaneously on a fair number of frequencies can be generally quickly identified as belonging to the same country or programme network. Even if foreign languages cause difficulty, the sound pattern of any language may indicate that the programme is originating from the same source irrespective of the number of transmitter outlets it may be heard on. With some experience, it becomes possible to identify languages without understanding them; thus, if Cairo broadcasting in Arabic is positively identified, it is then feasible to recognize Arabic programmes in the external service of another country.

If a simultaneous broadcast cannot be found, but the programme pattern can be established, a search of programme schedules issued by the various countries may show details which conform closely to those of the unknown station.

A tape recorder is useful to aid identification, to give positive proof of reception, and to provide a tape library of announcements and call signs, and the interval signals and jingles which characterize so many programmes and broadcast services. The tape machine should be close to the receiver and available for immediate use with its input connected to the receiver output, the mains supply switched on and a tape ready to record.

Any announcement heard which is not readily identifiable may be recorded and later played back repeatedly to help in identifying the language or recognizing some feature. Microphone facilities are useful to enable details of the time, date and approximate frequency or wavelength to be added to the recorded announcement. Such recordings could well form the beginning of an index of station announcements, which might later be arranged in country or geographical order to facilitate further research.

Tape recordings can be made of the signature tunes which most stations use either prior to their opening announcement or before particular programmes. Signature tunes are usually repeated for some minutes before the scheduled opening time, and as indicated previously, they may consist of a well-known melody characteristic of the country, of a few tones, or of bells or clock chimes. These tunes, when memorized, can provide an instant means of identification, but while some are distinctive, others are not, and a tape recording is often useful for comparison.

RECEPTION REPORTS

Reports on reception are always welcomed by broadcasting organizations, whether the listener is located in the target area or not. Such reports can provide useful information on transmissions, and help the broadcaster to assess the accuracy of the assessments on which his schedule was based and the effectiveness of the service.

Reception reports should be concise and accurate and should follow established form. This is preferable to a letter, which takes time to read and assess, and may require the extraction and tabulation of detail by qualified engineers to make it suitable for comparison with other similar reports. The assimilation of reports in a large broadcasting organization must follow a procedure requiring minimum effort, and this is possible only if listeners set out their reports in a standard manner. The information given can then be quickly and accurately assessed by junior staff, who may be trained to present the results in a form suitable for analysis by computer.

The detail which can be provided in a reception report is, however, quite large, and is of great importance when it is based on a test transmission. Information on every aspect of such transmissions is required, and each reception report is studied in detail. Where broadcasts follow a pattern or schedule of long standing, much detail can be omitted and the report can be shortened. The analysis of abbreviated reports of daily reception conditions supplies the transmission schedule engineer with a constant flow of information on signal strength, interference and overall merit. Thus any deviation from normal reception is easily detectable and can be investigated. Possibly the ionospheric path may have changed and a different frequency or aerial array may be needed; perhaps new interference has appeared and steps must be taken to eliminate or avoid it.

SINPFEMO, SINPO AND SIO

The generally recognized form for reports is based on the SINPFEMO code. Each letter signifies a particular aspect of reception and is followed by a rating figure (1 to 5) the significance of which is indicated in the table.

SINPFEMO code

Symbol and Meaning	1	2	3	4	5
S Signal strength	barely audible	poor	fair	good	excellent
I Interference	extreme	severe	moderate	slight	nil
N Noise	extreme	severe	moderate	slight	nil
P Propagation disturbance	extreme	severe	moderate	slight	nil
F Frequency of fading	very fast	fast	moderate	slow	nil
E Modulation quality	very poor	poor	fair	good	excellent
M Modulation depth	over-mod.	poor/nil	fair	good	maximum
O Overall merit	unusable	poor	fair	good	excellent

Restricted forms of this code are now more commonly used, for example, SINPO, in which no indication is given of the frequency of fading or the quality and depth of modulation. An even simpler code is SIO, which embraces only three criteria, namely signal strength, interference and overall merit. The number of rating figures has also been reduced: this is possible because if a signal is classified as 1 reception is unusable, and the difference bwteen 4 and 5 is so small in short-wave reception that the higher of these can be ignored. Where signals are poor enough to justify a rating of less than 2, or where interference is non-existent, 0 may be used.

Reception report forms are available from most broadcasting organizations on request from listeners who indicate their willingness to provide reports on a continuing basis, and some notes on the compilation of a SINPO report are given below. A full SINPFEMO report could be provided merely by adding the F, E and M criteria.

The use of the code is simple if care is taken in assessing the value of the signal. Few broadcasts other than those from a local transmitter qualify for rating of S5 or O5, but with these exceptions all other ratings are feasible. Enthusiasm should not be allowed to distort the report and the signal should be analysed with some precision for each aspect of the SINPO code.

The strength of the signal reported on can be compared with that of well-known broadcasts and the assessment is even simpler if the receiver has a tuning meter indicating signal strength. Such meters are often calibrated in dB above one microvolt, but the calibration is frequently incorrect and should not be accepted unless means are available of checking it.

The assessment of interference depends on the type and character of the interfering signal. This signal is often a whistle or heterodyne note, caused by reception of two signals with a carrier-frequency difference less than the bandwidth of the receiver. Thus, if the receiver bandwidth is 8 or 9 kHz, and the interfering signal is say 3 or 4 kHz from the wanted broadcast, a heterodyne whistle of this frequency is audible. The interference is, however, more troublesome if the frequency difference is only 1 to 2 kHz, because the ear is more sensitive at these lower frequencies. Even though the strength of two interfering signals may be the same, if one is displaced 4 kHz and the other 1 kHz from the wanted signal, a rating of 14 may apply to one and of 13 to the other. Similarly, a weak background of programme is less disturbing than a whistle or steady tone. Thus the rating to be entered is a measure of the intelligibility of the wanted signal.

Atmospheric noise is seldom worse than N3, except during periods of ionospheric disturbance, summer static or the precipitation of electrified rain. Good conditions, and the use of the higher frequency bands, do not normally produce ratings better than N4. Care should be taken to ignore noise introduced by the receiver, especially when the signal is weak and the receiver is operating at full gain.

Propagation disturbance may be more difficult to assess: it is related to the intensity of atmospheric noise and the degree of fading of the received signal. If noise is high and fading rapid, but the programme can be followed, a rating of P3 is justified, but rapid fading to a depth causing programme mutilation qualifies for P1. If little or no noise is apparent and the fades are shallow and do not exceed about one second, being well held by a.g.c., the rating should be P4 or P5.

Overall merit is assessed by taking the average of the individual rating figures to the nearest whole number. There is no need to add a plus or minus sign, or to indicate small differences in merit, because each rating in the code is intended to cover a wide range of conditions and if the listener is certain that one rating does not apply the next figure must be correct.

The best report loses its value if the listener fails to give such essential details as his name and address, the date and time of reception, and the approximate frequency or wavelength (the waveband alone is not enough). Any definitely identified interference should be specified, but if this cannot be done, details of the type of programme or other interfering signal should be mentioned.

LONG AND MEDIUM-WAVE EUROPEAN STATIONS

This list includes only those stations which are believed to be active on the frequencies indicated and which may be heard in Europe. Certain stations located outside the Continent of Europe are sometimes heard in Western Europe and these are included in this section, although they are situated outside the 'European Broadcasting Area', as defined in the Copenhagen Plan. This area is bounded on the south by 30° north latitude, that is, by the territories bordering the Mediterranean Sea, excluding those parts of Arabia and Saudi Arabia within this area but including Iraq. On the west it encloses Iceland, Eire and the Azores, and on the east it is bounded by the meridian 40° east of Greenwich.

Stations are listed against the frequency on which they have been heard, which may in some cases be the frequency allocated in the Copenhagen Plan. Wavelength in metres is shown beside the frequency. The power is in kW.

Alternative station names or exact location of transmitters, where known, are shown after the usual station name. In appropriate cases station names have been given the anglicized spelling.

In certain instances, groups of low powered stations are indicated by a numeral following the name of the main station in the group, e.g. Cagliari + 5, the figure being the number of additional stations to that named.

Abbreviations used in the list are as follows: –

AFN	American Forces Network
BBC	British Broadcasting Corporation
BRT	Belgische Radio en Televisie (in Dutch)
COPE	Cadena de Ondas Populares Espanolas
DDR	Deutscher Demokratischer Rundfunk
IBA	Independent Broadcasting Authority
NDR	Norddeutscher Rundfunk
Prog	Programme
R	Radio
RCE	Radiocadena Espanola
RIAS	Rundfunk im Amerikan Sektor von Berlin
RNE	Radio Nacional de Espana
RTBF	Radiodiffusion-Television Belge (in French)
SER	Sociedad Espanola de Radiodifusion
St	Station
VOA	Voice of America

kHz	Metres	Station	Country	Power	Programme
155	1936	Brasov	Roumania	1200	1st Programme
		Donebach	Germany (W)	250	Deutschlandfunk
		Tromsoe	Norway	10	
		Engels	USSR	150	
164	1829	Allouis	France	2000	France-Inter
		Tachkent	USSR	150	
173	1734	Kaliningrad	USSR	1000	
		Lvov	Ukraine	500	
182	1648	Saarlouis-Felsberg	Germany (W)	2000	Europe No. 1
		Ankara	Turkey	1200	2nd Programme
		Oranienburg-Rehmate	Germany (E)	750	Stimme Der DDR
		Alma Ata	USSR	250	
191	1571	Motala	Sweden	300	1st Programme
		Caltanissetta	Italy	10	National Programme
		Tbilissi	USSR	500	
200	1500	Droitwich	UK	400	Radio 4 UK
		Burghead	UK	50	Radio 4 UK
		Westerglen	UK	50	Radio 4 UK
		Warszawa	Poland	200	2nd Programme
		Etimesgut	Turkey	200	1st Programme
		Leningrad	USSR	150	
		Moscow	USSR	100	
209	1436	Azilal	Morocco	800	Prog A (Arabic)
		Kiev	Ukraine	500	
		Reykjavik + 1	Iceland	100	
		Mainflingen	Germany	15	Deutschlandfunk
218	1376	Monte-Carlo	Monaco	1400	
		Oslo	Norway	200	1st Programme
		Baku + 1	USSR	500	
227	1322	Warszawa 1	Poland	2000	1st Programme
236	1271	Junglinster	Luxembourg	2000	
		Kichinev + 2	USSR	1000	
245	1224	Kalundborg	Denmark	200	1st Programme
		Taldy Kurgan + 3	USSR	500	
		Erzurum	Turkey	200	1st Programme
254	1181	Tipaza	Algeria	1500	French Programme
		Lahti	Finland	200	Programme 1
		Duchanbe + 4	USSR	300	
263	1141	Moskva	USSR	2000	
		Burg	Germany (E)	200	
		Irkutsk + 1	USSR	1000	
272	1103	Ceskoslovensko	Czechoslovakia	1500	Hvezda Programme Czech/Slovak
		Novosibirsk	USSR	150	
281	1068	Minsk	Bielorussia	500	
		Achkabad + 1	USSR	150	

kHz	*Metres*	*Station*	*Country*	*Power*	*Programme*
433		Oulu	Finland	10	Programme 1
520	577	Innsbruck Aldrans + 2	Austria	10	
		Hof-Saale + 1	Germany (W)	0.2	Bayerischer Rundfunk
529	567	Ain Beida	Algeria	600	Arabic Programme
531	565	Beromunster	Switzerland	500	German Programme
		Jerusalem	Israel	200	
		Greifswald	Germany (E)	500	Radio DDR 1
		Torshavn	Denmark	5	
		Cheboksary	USSR	30	
540	555.5	Carraroe	Ireland	2	
		Solt	Hungary	2000	1st Programme
		Tripoli	Libya	600	Arabic Programme
		Waver-Overijse	Belgium	150	
		Oulu 1	Finland	10	Programme 1
		Sulaibiyah	Kuwait	1500	
		Orenburg	USSR	100	
549	546	Les Trembles	Algeria	600	Arabic Programme
		Kichinev + 4	USSR	1000	
		Nordkirchen + 1	Germany (W)	100	Deutschlandfunk
		Beli Kriz	Yugoslavia	6	
558	537	Abu-Zaabal	Egypt	40	
		Faro	Portugal	10	1st Programme
		Targu Jiu + 3	Roumania	200	2nd Programme
		Helsinki 1	Finland	100	Programme 1
		Monte-Ceneri Passo	Switzerland	100	Italian Programme
		Maribor	Yugoslavia	100/20	
		Rostock + 1	Germany (E)	20	DDR 1
		Qeslagh	Iran	1000	
567	529	Homs	Syria	300	General Service Arabic
		Valenca do Minho	Portugal	10	1st Programme
		Tullamore	Ireland	500	R Telefis Eireann
		Berlin	Germany (W)	100	Sender Freies Berlin
		Brasov + 1	Roumania	50	
		Bologna + 3	Italy	25	National Programme
		Volgograd	USSR	250	
576	521	Bechar	Algeria	400	Arabic Programme
		Braganca	Portugal	1	1st Programme
		Vidin	Bulgaria	1000	Sofia 2
		Riga + 1	USSR	500	
		Stuttgart	Germany (W)	300	Suddeutscher Rundfunk
		Schwerin	Germany (E)	250	DDR 1
		Tel-Aviv 2	Israel	200	

kHz	Metres	Station	Country	Power	Programme
585	513	Gafsa	Tunisia	350	National Programme Arabic
		Wien-Bisamberg + 3	Austria	600/240	
		Madrid	Spain	200	RNE 1
		Paris + 1	France	10	France-Inter Paris et Provence
		Riyadh	Saudi Arabia	1200	
		Perm	USSR	30	
594	505	Oujda 1	Morocco	100	Prog A (Arabic)
		Frankfurt + 1	Germany (W)	400	Hessischer Rundfunk
		Pleven	Bulgaria	250	Sofia 1
		Ijevsk	USSR	150	
603	497.5	Lyon-Tramoyes	France	300	France Culture
		Oradea + 2	Roumania	50	
		Koenigswuster-hausen	Germany (E)	20	DDR 1
		Nicosia	Cyprus	20	
		Newcastle	UK	2	Radio 4 UK
		Mariehamn	Finland	1	Swedish Programme
		Tiumen	USSR		
612	490	Sarajevo + 2	Yugoslavia	600	
		Sebaa Aioun	Morocco	300	Prog A (Arabic)
		Tallinn + 1	USSR	100	
		Tullamore	Ireland	200	
		Qashr'Shirin	Iran	400	
621	483	Vila Real + 1	Portugal	10	1st Programme
		Batra	Egypt	450	
		Wavre-Overijse	Belgium	150	RTBF 1 (French)
		Ukhta	USSR	150	
		Santa Cruz Tenerife	Canary Is.	100	RNE 1
630	476	Tunis-Djedeida	Tunisia	600	National Programme Arabic
		Miranda do Douro	Portugal	1	1st Programme
		Cukurova	Turkey	300	1st Programme
		Timisoara	Roumania	400	2nd Programme
		Dannenberg	Germany (W)	10	Sender Freies Berlin
		Vigra	Norway	100	1st Programme
		Saratov	USSR		
639	469	Praha	Czechoslovakia	1500	Praha Prog (Czech)
		La Coruna + 3	Spain	100	RNE 1
		Zakaki	Cyprus	100	BBC
		Bonab	Iran	400	
		Omsk	USSR		
648	463	Crowborough	UK	500	External Service

kHz	Metres	Station	Country	Power	Programme
648	463	Orfordness	UK	50	
contd		Simferopol + 1	Ukraine	150	
		Plovdiv	Bulgaria	30	2nd Programme
		Murska Sobota	Yugoslavia	10	
657	457	Tel-Aviv 1	Israel	200	
		Napoli + 4	Italy	120	National Programme
		Murmansk	USSR	150	
		Tchernovtsy	Ukraine	25	
		Neubranden-burg	Germany (E)	20	Berliner Rundfunk
		Tantan	Morocco	50	Prog A (Arabic)
666	450	Damas-Sabboura	Syria	100	General Service Arabic
		Vilnius + 1	USSR	500	
		Bodensee-sender	Germany (W)	300/180	Sudwestfunk
		Lisboa	Portugal	135	1st Programme
		Sombor	Yugoslavia	10	
675	444	Marseille	France	600	France-Inter
		Benghazi	Libya	100	
		Jerusalem	Israel	20	
		Lopik	Netherlands	120	3rd Programme
		Ujgorod + 1	Ukraine	50	
		Bodoe	Norway	10	1st Programme
684	438.5	Beograd	Yugoslavia	1000	
		Sevilla	Spain	250	RNE 1
		Hof-Saale	Germany (E)	100	RIAS
		Mashad	Iran	100	
		Tselinograd	USSR	300	
693	433	Basra	Iraq	1200	
		Viseu	Portugal	10	1st Programme
		Berlin	Germany (E)	250	Berliner - Rundfunk
		Barrow	UK	1	Radio 2
		Bexhill	UK	1	Radio 2
		Brighton	UK	1	Radio 2
		Burghead	UK	50	Radio 2
		Droitwich	UK	150	Radio 2
		Exeter	UK	1	Radio 2
		Folkestone	UK	1	Radio 2
		Plymouth	UK	1	Radio 2
		Postwick	UK	10	Radio 2
		Redmoss	UK	1	Radio 2
		Stagshaw	UK	50	Radio 2
		Nicosia	Cyprus	20	1st Programme
		Ufa	Israel	150	
702	427	Banska Bystrica + &	Czechoslovakia	400	Bratislava (Slovak) + Regional
		Monte-Carlo	Monaco	300	2nd Programme
		Radio Andorra	Andorra	250	R. Andorra

kHz	Metres	Station	Country	Power	Programme
702	427	Umraniye	Turkey	150	2nd Programme
contd		Sebaa-Aioun	Morocco	140	Prog C (Berber)
		Finnmark	Norway	20	1st Programme
		Aachen-Stolberg + 3	Germany (W)	5/1,2	NDR/WDR
		Duchanbe + 1	USSR	50	
711	422	Jerusalem	Israel	10	
		Rennes 1	France	300	France Culture
		Donetsk	Ukraine	150	
		Abu-Zaabal	Egypt	100	
		Tallinn + 3	USSR	50	
		Heidelberg + 4	Germany (W)	5	Suddeutscher Rundfunk
		Nis	Yugoslavia	20	
720	417	Norte 1	Portugal	100	
		Sfax	Tunisia	200	Regional Programme
		Holzkirchen	Germany (W)	150	Radio Free Europe
		Zakaki	Cyprus	100	BBC
		Lisnagarvey	UK	10	Radio 4
		London	UK	500	Radio 4
		Londonderry	UK	0.25	Radio 4
		Bailesti + 10	Roumania	2	
		Taybad	Iran		
729	411.5	Mirandela	Portugal	10	1st Programme
		Athinai	Greece	150	National Programme
		Oviedo + 3	Spain	50	RNE 1
		Puttbus	Germany (E)	10	DDR 1
		Sadiyat	United Arab Emirates	750	
738	406.5	Tel-Aviv 2	Israel	1200	
		Poznan	Poland	300	2nd Programme
		Barcelona	Spain	250	RNE 1
		Tcheliabinsk	USSR	150	
747	402	Sarakeb	Syria	100	General Service (Arabic)
		Petrich	Bulgaria	500	Sofia 2
		Lopik	Netherlands	250	2nd Programme
		Sarajevo + 1	Yugoslavia		
		Ouagadougou	Upper Volta	100	
		Karaganda	USSR		
756	397	Lugoj + 1	Roumania	1000	
		Braunschweig	Germany (W)	800/200	Deutschlandfunk
		Salman Pack	Iraq	300	
		Lisboa	Portugal	135	
		Kuopio	Finland	10	Programme 1
		Redruth	UK	2	Radio 4 S West
		Carlisle	UK	1	
		Radio Malta	Malta	20	
765	392	Sottens	Switzerland	500	French Programme

kHz	Metres	Station	Country	Power	Programme
765	392	Odessa	Ukraine	150	
contd		Medvejiegorsk	USSR	150	
		Ioannina	Greece	20	Armed forces
		Dakar	Senegal	400	
774	388	Abis	Egypt	1000	
		Voronej + 2	USSR	150	
		Caceres + 3	Spain	60	RNE 1
		Stockholm	Sweden	100	1st Programme
		Split + 3	Yugoslavia	50	
		Sofia + 1	Bulgaria	30	Sofia 1
		Klagenfurt + 10	Austria	30	4th Programme
		Leeds	UK	0.5	Local Radio
		Farahabad	Iran	20	
783	383	Tartus	Syria	600	
		Miramar	Portugal	100	3rd Programme
		Burg	Germany (E)	1000	Stimme Der DDR
		Kiev + 2	Ukraine	100	
		Kazan	USSR		
792	379	Kavalla	Greece	500	VOA
		Limoges 1	France	300	France Culture
		Sevilla EAJ.5	Spain	20	SER
		Jesenik + 2	Czechoslovakia	1	
		Astrakhan	USSR	50	
801	374.5	Amman	Jordan	200	
		Leningrad	USSR	1000	
		Munchen-Ismaning + 1	Germany (W)		Bayerischer Rundfunk
		Barnstaple	UK	2	Radio 4 S West
		Ulan-Ude + 3	USSR	1000	
810	370	Skopje	Yugoslavia	1000	
		Burghead	UK	100	
		Dumfries	UK	2	Radio Scotland
		Redmoss	UK	5	
		Westerglen	UK	100	
		Madrid EAJ.7	Spain	20	SER
		Berlin 3	Germany (W)	5	BBC
		Abu Dhabi	United Arab Emirates	50	
		Volgograd	USSR	150	
819	366	Sud Radio	Andorra	600	Sud Radio
		Batra 2	Egypt	450	
		Warszawa	Poland	300	2nd Programme
		Trieste	Italy	25	National Programme
		Rabat 1	Morocco	25	Prog A (Arabic)
828	362	Sebha	Libya	300	
		Oujda 2	Morocco	100	Prog B (French)
		Shumen + 1	Bulgaria	500	Sofia 2
		Hannover + 1	Germany (W)	100	NDR/WDR
		Freiburg	Germany (W)	40	Sudwestfunk

kHz	*Metres*	*Station*	*Country*	*Power*	*Programme*
828 contd	362	Barcelona EAJ.1	Spain	20	
		Castelo Branco	Portugal	1	1st Programme
		Gorkii	USSR	150	
837	358	Beyrouth	Lebanon	100	
		Ponta Delgada	Azores	10	
		Nancy 1	France	200	France Culture
		Kharkov	Ukraine	150	
		Novi Sad + 1	Yugoslavia	50	
		Las Palmas EAK.35	Canary Is.	10	COPE
846	355	Zefat	Israel	5	
		Roma	Italy	540	2nd Programme
		Moskva	USSR	60	
		Ceske Bude-jovice + 1	Czechoslovakia	30	Prague Prog (Czech)
		Elista	USSR	30	
855	351	Bucuresti	Roumania	750	2nd Programme
		Murcia + 2	Spain	125	RNE 1
		Berlin	Germany (W)	100	RIAS
		Plymouth	UK	1	Radio 4 S West
		Blackburn	UK	0.5	Local Radio
		Amman	Jordan	10	
		Tchelhbinsk	USSR	150	
864	347	Ksar es Souk	Morocco	15	Prog A (Arabic)
		Paris	France	300	France Culture
		Plovdiv	Bulgaria	150	Sofia 1
		Erevan	USSR	150	
873	344	Leningrad	USSR	150	
		Frankfurt	Germany (W)	150	AFN
		Abu Zaabal	Egypt	50	
		Zaragoza EAJ.101	Spain	20	SER
		Budapest + 1	Hungary	20	2nd Programme
		Gijon EAJ.34	Spain		
882	340	Titograd	Yugoslavia	100	
		Koenigswuster-hausen	Germany (E)	250	
		Washford	UK	70	Radio Wales
		Penmon	UK	10	Radio Wales
		Tywyn	UK	5	Radio Wales
		Wrexham	UK	2	Radio Wales
		Damman	Saudi Arabia		
		Stavropol	USSR		
891	337	Antalya	Turkey	600	1st Programme
		Ujgorod + 1	Ukraine	150	
		Hengelo-Overijssel	Netherlands	10	Relay Lopik 2 & 3
		Alger	Algeria	200	

kHz	Metres	Station	Country	Power	Programme
900	333	Milano	Italy	600	National Programme
		Brno + 4	Czechoslovakia	30	
		Iochkar-Ola + 2	USSR	50	
909	330	Thourah	Iraq	200	
		Bournemouth	UK	1	Radio 2
		Brookmans Park	UK	140	Radio 2
		Clevedon	UK	50	Radio 2
		Fareham	UK	1	Radio 2
		Lisnagarvey	UK	10	Radio 2
		Londonderry	UK	1	Radio 2
		Moorside Edge	UK	100	Radio 2
		Redruth	UK	2	Radio 2
		Torquay	UK	1	Radio 2
		Westerglen	UK	50	Radio 2
		Whitehaven	UK	1	Radio 2
		Cluj + 2	Roumania	50	2nd Programme
		Iman	USSR	50	
918	327	Paphos	Cyprus	50/2	
		Ljubljana	Yugoslavia	600	
		Madrid EAJ.29	Spain	20	
		Mezen	USSR	100	
927	324	Wolvertem	Belgium	300	BRT 1 (Dutch)
		Izmir	Turkey	200	1st Programme
		Zakynthos	Greece	50	Regional Programme
		Lamego + 1	Portugal	1	3rd Programme
		Nebit Dag	USSR	50	
936	320.5	Cairo	Egypt	100	
		Agadir 1	Morocco	600	Prog A (Arabic)
		Bremen + 1	Germany (W)	100	Radio Bremen
		Lvov	Ukraine	500	
		Djakovica + 1	Yugoslavia	10	
		Rio de Janeiro	Brazil	100	
		Rezaieh	Iran	10	
945	317.5	Toulouse 1	France	300	France Culture
		Rostov/Don + 1	USSR	300	
		Pleven	Bulgaria	30	2nd Programme
		Miercurea Ciuc	Roumania	14	2nd Programme
		Larissa	Greece	10	Armed forces
		Volgograd	USSR		
954	314.5	Brno + 3	Czechoslovakia	750	Prague (Czech) & Regional
		Trabzon	Turkey	300	1st Programme
		Madrid EAJ.2	Spain	20	RCE
		Iraklion	Greece	20	Armed forces
		Al Arish	Qatar	750	
963	311.5	Tunis	Tunisia	100	
		Beograd	Yugoslavia	200	
		Sofia	Bulgaria	150	Sofia 1

kHz	Metres	Station	Country	Power	Programme
963	311.5	Turku	Finland	100	Programme 1
contd		Paris 4	France	8	Radio Sorbonne
		Coimbra + 1	Portugal	1	
		Soba	Sudan	200	
972	309	Marrakech 1	Morocco	1	Prog A (Arabic)
		Nikolaev	Ukraine	500	
		Hamburg + 2	Germany (W)	300	NDR/WDR
		Korca	Albania	15	
981	306	Alger	Algeria	200	
		Megara	Greece	200	Armed forces
		Goeteborg	Sweden	150	1st Programme
		Trieste	Italy	10	Regional Service
		Bor + 2	Yugoslavia	10	
		Cheb + 2	Czechoslovakia	1	
		Shiraz	Iran	400	
990	303	Amchit	Lebanon	100	
		Berlin-West + 1	Germany (W)	300	RIAS
		Barcelona EAJ.15	Spain	10	SER
		Exeter	UK	1	Radio 4 S West
		Wolverhampton	UK	0.3	IBA Beacon Radio
		Shiraz	Iran	400	
999	300	Kukes	Albania	8	
		Kichinevq	USSR	100	
		Hoyerswerda + 2	Germany (E)	20	Berliner Rundfunk
		Rimini + 1	Italy	6	Second Programme
		Delimara	Malta	5	
		Fareham	UK	0.3	Local Radio S West (BBC)
		Nottingham	UK	1.0	IBA Radio Trent
		Tselinograd	USSR	30	
1008	298	Kerkyra	Greece	50	Regional Programme
		Lopik	Netherlands	300	1st Programme
		Aleksinac	Yugoslavia	200	
		Mozyr + 3	Bielorussia	50	
		Asswan	Egypt	1	
		Semman	Iran	20	
		Las Palmas EAJ.50	Canary Is.	10	SER
1017	295	Tetuan	Morocco	20	
		Istanbul	Turkey	1200	1st Programme
		Wolfsheim	Germany (W)	600	Sudwestfunk
		Nitra + 4	Czechoslovakia	30	Hvezda & Regional
		Venezia + 1	Italy	25	Regional Service
		Tripolis	Greece	10	Armed forces
1026	292	Jerusalem	Israel	200	
		Graz-Dobl + 14	Austria	100	

kHz	*Metres*	*Station*	*Country*	*Power*	*Programme*
1026	292	Brest + 2	Bielorussia	5	
contd		Rabat	Morocco	1	Prog B (French)
		Safi	Morocco	1	Prog A (Arabic)
		Vigo EAJ.48	Spain	10	SER
		Madrid	Spain		
		Alicante EAJ.31	Spain		
		Belfast	UK	1	IBA Downtown Radio
		Tabriz	Iran	100	
1035	290	Porto Alto	Portugal	120	
		Babylon	Iraq	2000	
		Tallinn	USSR	500	
		Milano + 9	Italy	50	
		Gillingham	UK	1	Local Radio
		Sheffield	UK	1	Local Radio
		Yazd	Iran	20	
1044	287	Thessaloniki	Greece	50	Regional Programme
		Burg	Germany (E)	1000	DDR 1
		Sebaa Aioun	Morocco	300	Prog B (French)
		Tbilissi	USSR	200	
1053	285	Tripoli	Libya	50	
		Tetuan 2	Morocco	20	Prog C (Berber)
		Iasi	Roumania	1000	
		Barnstaple	UK	1	Radio 1
		Barrow	UK	1	Radio 1
		Bexhill	UK	2	Radio 1
		Brighton	UK	2	Radio 1
		Burghead	UK	20	Radio 1
		Droitwich	UK	150	Radio 1
		Dundee	UK	1	Radio 1
		Folkestone	UK	1	Radio 1
		Hull	UK	1	Radio 1
		Londonderry	UK	1	Radio 1
		Postwick	UK	10	Radio 1
		Stagshaw	UK	50	Radio 1
		Start Point	UK	100	
		Kurgan	USSR	50	
1062	282	Norte	Portugal	100	2nd Programme
		Diyabakir	Turkey	300	1st Programme
		Kalundborg	Denmark	100	3rd Programme
		Cairo	Egypt	50	
		Cagliari + 6	Italy	25	National Programme
		Zagreb + 1	Yugoslavia	10	
		Saransk	USSR	150	
1071	280	Tartus	Syria	60	
		Brest + 4	France	100	France-Inter
		Riga + 3	USSR	60	
		Banja Luka	Yugoslavia	25	

kHz	Metres	Station	Country	Power	Programme
1071 contd	280	Mnich Hradiste + 1	Czechoslovakia	25	
		Dniepro-petrovsk	Ukraine	20	
		Mesolongion	Greece		
		Rajkot + 1	India	1000	
1080	278	Katowice	Poland	1500	
		Koper-Belikriz	Yugoslavia	200	
		Mallorca EAJ 3	Spain	5	SER
		Granada EAJ 16	Spain	2	SER
		Casablanca	Morocco	1	Prog A (Arabic)
		Crestias	Greece	20	Armed forces
		Abadan	Iran	600	
1089	275.5	Krasnodar	USSR	300	
		Brookmans Park	UK	150	Radio 1
		Fareham	UK	1	Radio 1
		Lisnagarvey	UK	10	Radio 1
		Moorside Edge	UK	150	Radio 1
		Redmoss	UK	2	Radio 1
		Redruth	UK	2	Radio 1
		Tywyn	UK	1	Radio 1
		Washford	UK	50	Radio 1
		Westerglen	UK	50	Radio 1
		Whitehaven	UK	1	Radio 1
		Durres	Albania	100	
		Novi-Sad	Yugoslavia	20	
1098	273	Bratislava	Czechoslovakia	400	Bratislava (Slovak) & Reg.
		Bologna	Italy	60	Regional Service
		Alma-Ata + 1	USSR	150	
		Santa Cruz de Palma	Canary Is.	5	
1107	271	Batra	Egypt	600	
		Kaunas + 4	USSR	150	
		Novi-Sad	Yugoslavia	150	
		Munchen + 4	Germany (W)	100	AFN
		Madrid EFE.14	Spain	20	RCE
		Valladolid EFE.1	Spain	2	RCE
1116	269	Ouarzazate + 1	Morocco	15	Prog A (Arabic)
		Bari + 5	Italy	150	2nd Programme
		Kaliningrad	USSR	30	
		Derby	UK	0.5	Local Radio
1125	267	El Beida	Libya	500	
		Stara Zagora	Bulgaria	500	Sofia 2
		Zagreb + 1	Yugoslavia	200	

kHz	Metres	Station	Country	Power	Programme
1125 contd	267	La Louviere	Belgium	150	RTBF 2 (French)
		Leningrad	USSR	20	
1134	264.5	Zagreb/ Tovarnik + 3	Yugoslavia	300	
		Sevilla EAK.2	Spain	10	COPE
		Bilbao EAK.13	Spain	10	COPE
		Zaragoza EAK.6	Spain	10	COPE
		Valencia EAK.5	Spain	10	COPE
		San Sebastian EAK.44	Spain	2	COPE
		Zamora	Spain		
		Calcutta	India	1000	
		Sulaibiyah	Kuwait	750	
1143	262	Kaliningrad	USSR	150	
		Nova Gradiska + 1	Yugoslavia	100	
		Les Trembles	Algeria	40	Prog A (Arabic)
		Stuttgart + 13	Germany (W)	10	AFN
		Messina	Italy	6	2nd Programme
		Duchanbe + 1	USSR	150	
1152	260	Cluj 2 + 1	Roumania	950	2nd Programme
		Birmingham	UK	1	IBA BRMB Radio
		Glasgow	UK	2	IBA R Clyde
		London	UK	5.5	IBA LBC
		Manchester	UK	1	IBA Piccadilly Radio
		Plymouth	UK	0.5	IBA Plymouth Sound
		Tyneside	UK	1	IBA Metro Radio
		Marrakech	Morocco	1	Prog B (French)
		Tabriz	Iran	100	
1161	258	Strasbourg + 2	France	200	France-Inter
		Tanta	Egypt	200	
		Biala Slatina + 1	Bulgaria	150	Sofia 1
		Duchanbe	USSR	1000	
1170	256	Porto	Portugal	10	
		Moghilev	Bielorussia	1000	
		Beli Kriz + 1	Yugoslavia	50	
		Ipswich	UK	0.5	IBA R Orwell
		Portsmouth	UK	0.8	IBA R Victory
		Swansea	UK	0.8	IBA Swansea Sound
		Teeside	UK	1.0	IBA R Tees
		Erfurt + 1	Germany (E)		Berliner Rundfunk
1179	254	Bacau + 2	Roumania	200	
		Thessaloniki	Greece	50	Forces Radio
		Hoerby	Sweden	100	1st Programme
		Murcia	Spain	5	SER

kHz	Metres	Station	Country	Power	Programme
1179 contd	254	Badalona E EAJ.39	Spain	5	SER
1188	252.5	Szolnok + 1	Hungary	135	2nd Programme
		Cork 1	Ireland	10	R Telefis Eireann
		San Remo	Italy	6	
		Kuurne	Belgium	5	BRT 2 (Dutch)
		Casablanca	Morocco	1	Prog B (French)
		Teheran	Iran	100	
1197	251	Portalegre	Portugal	1	1st Programme
		Munchen-Ismaning	Germany (W)	300	VOA
		Minsk + 3	Bielorussia	50	
		Agadir	Morocco	20	Prog C (Berber)
		Alexandria	Egypt	10	
		Cambridge	UK	0.1	Radio 3
		Sana'a	Yemen	10	
1206	249	Bordeaux 1	France	100	France Culture
		Wroclaw + 3	Poland	200	
		Haifa	Israel	10	
1215	247	Durres	Albania	240	
		Brighton	UK	1	Radio 3
		Brookmans Park	UK	50	Radio 3
		Burghead	UK	20	Radio 3
		Droitwich	UK	50	Radio 3
		Fareham	UK	1	Radio 3
		Hull	UK	0.15	Radio 3
		Lisnagarvey	UK	10	Radio 3
		Londonderry	UK	0.25	Radio 3
		Moorside Edge	UK	50	Radio 3
		Newcastle	UK	2	Radio 3
		Plymouth	UK	1	Radio 3
		Postwick	UK	1	Radio 3
		Redmoss	UK	1	Radio 3
		Redruth	UK	2	Radio 3
		Tywyn	UK	0.5	Radio 3
		Washford	UK	60	Radio 3
		Westerglen	UK	40	Radio 3
		Tartu + 2	USSR	50	
		Las Palmas ECS 4	Canary Is.	20	
1224	245	Vidin	Bulgaria	1000	Sofia 2
		Beer-Sheva	Israel	10	Forces Station
		Madrid EAK 1	Spain	20	COPE
		Villarreal EAK 78	Spain	2	COPE
		Rio de Janeiro	Brazil	100	
		Djambul	USSR	150	
1233	243	Praha + 4	Czechoslovakia	750	Hvezda & Regional

kHz	Metres	Station	Country	Power	Programme
1233 contd	243	Cape Greco	Cyprus	600	
		Tanger	Morocco	200	R Tanger
		Liege	Belgium	5	RTBF 2 (French)
1242	241.5	Marseille 2	France	150	France Culture
		Kiev + 3	Ukraine	150	
		Vaasa	Finland	25	Swedish Programme
		Tel Aviv	Israel	10	
		Muscat	Oman	100	
1251	240	Tripoli	Libya	500	Arabic Programme
		Chaves	Portugal	1	1st Programme
		Siofok + 1	Hungary	135	2nd Programme
		Amsterdam	Netherlands	1	Relay Lopik 3
		Dublin	Ireland	20	R Telefis Eireann
1260	238	Rhodos	Greece	500	VOA
		Szczecin	Poland	160	2nd Programme
		San Sebastian EAJ 8	Spain	10	SER
		Valencia EAJ 3	Spain	20	SER
		Alcala Henares	Spain	5	
1269	236	Neumunster	Germany (W)	600	Deutschlandfunk
		Novi-Sad	Yugoslavia	100	
		Radio Paradise	British West Indies	50	
1278	235	Strasbourg	France	300	France Culture
		Odessa	Ukraine	150	
		Florina	Greece	20	Forces Programme
		Turku 2	Finland	4	Swedish Programme
		Bradford	UK	0.1	IBA Pennine Radio
		Assuit	Egypt	10	
		Kabul	Afghanistan	100	
1287	233	Tel-Aviv 1	Israel	100	Forces Station
		Lisboa	Portugal	2.5	
		Ceskoslovensko 2 + 3	Czechoslovakia	300	Hvezda & Regional
		Mytilini	Greece	5	Forces Programme
		Lar	Iran	20	
1296	231	Orfordness	UK	500	External Services
		Loznica + 1	Yugoslavia	10	
		Rabat	Morocco	1	Prog C (Berber)
		Semnan	Iran	10	
		Rio de Janeiro	Brazil		
		Sennar	Sudan	1500	
		Duchanbe + 1	USSR	1000	
1305	230	Constantine 2 + 1	Algeria	40	French Network
		Haifa	Israel	20	
		Rzeszow + 3	Poland	100	
		Marche	Belgium	50	RTBF 2 (French)
1314	228	Shkoder	Albania	10	
		Stavanger	Norway	100	1st Programme

kHz	Metres	Station	Country	Power	Programme
1314	228	Constantza + 2	Roumania	50	1 & 2 Programme
contd		Tripolis	Greece	20	Forces Programme
		Ancona + 3	Italy	6	2nd Programme
		Shkoder	Albania	10	
		Valencia	Spain	18	RCE
		Madrid ECS.11	Spain	20	RCE
		Cadiz EJF.5	Spain	2	RCE
		Skopje + 1	Yugoslavia	20	
		Fort de France	Martinique	50	
		Arbadil	Iran	20	
1323	227	Leipzig	Germany (E)	150	Soviet Programme
		Zyyi	Cyprus	50	BBC
		Targu Mures + 3	Roumania	15	
1332	225	Funchal	Madeira	10	
		Elvas	Portugal	1	1st Programme
		Roma + 3	Italy	300	National Programme
		Parnu + 2	USSR	30	
		Brno Mesto + 4	Czechoslovakia	25	
		Gnjilane + 3	Yugoslavia	2	
		Teheran	Iran	100	
1341	224	Lisnagarvey	UK	100	Radio Ulster
		Londonderry	UK	0.25	Radio Ulster
		Budapest-Lakihegy	Hungary	300	2nd Programme
		Delimara	Malta	20	
		Alma-Ata + 1	USSR	30	
		Alexandria + 1	Egypt	10	
1350	222	Nancy + 1	France	100	France-Inter
		Madona + 1	USSR	50	
		Beograd-Studio B	Yugoslavia	10	
		Kuwait	Kuwait	200	
		Nouakchott	Mauretania	50	
		Sukhumi	USSR	30	
1359	221	Berlin	Germany (E)	250/100	Stimme Der DDR
		Tirane	Albania	120	
		Moskva	USSR	15	
		Bournemouth	UK	0.3	Local Radio
		Kirkuk	Iraq	240	
		Valencia	Spain		
1368	219	Yamit	Israel	20	
		Porto	Portugal	10	1st Programme
		Krakow + 1	Poland	60	2nd Programme
		Venezia + 13	Italy	25	3rd Programme
		Valjevo	Yugoslavia	10	
		Foxdale	UK	2	Manx Radio
1377	218	Lille 1	France	300	France Culture
		Lutsk + 1	Ukraine	50	

kHz	Metres	Station	Country	Power	Programme
1377	218	Kardzali	Bulgaria	30	1st Programme
contd		Kumrovec + 4	Yugoslavia	20	
		St. Pierre & Miquelon	St Pierre & Miquelon	4	
1386	216	Kaunas	USSR	1000	
		Athinai	Greece	50	2nd Programme
		Tuzla + 1	Yugoslavia	2	
		Orense ECS.18	Spain	2	RCE
		La Coruna EFJ.11	Spain	5	RCE
		Ahvaz	Iran	400	
1395	215	Lushnje	Albania	500	
		Malaga + 9	Spain	100	
		Hoogezand	Netherlands	10	Relay Lopik 3
		Leon EFE 5	Spain	2	RCE
		Alicante EFE.8	Spain	2	RCE
		Granada ECS.5	Spain	5	RCE
1404	214	Tripoli	Libya	20	
		Dniepro-petrovsk + 2	Ukraine	30	
		Bastia + 2	France	40	France Culture
		Baia Mare + 1	Roumania	15	
		Varazdin + 1	Yugoslavia	10	
		Helsinki 2	Finland	2	Swedish Programme
		Kipe	Guinea	400	
1413	212	Pristina	Yugoslavia	100	
		Bad Mergent-heim + 2	Germany (W)	3	Suddeutscher Rundfunk
		Sevilla ECS 8	Spain	5	RCE
		Barcelona EFJ 15	Spain	10	RCE
		Granada ECS 5	Spain	5	RCE
		Oviedo EFE 22	Spain	5	RCE
		Zaragoza EFJ 46	Spain	20	RCE
		Serona ECS 14	Spain	2	RCE
		Bejar EFJ 18	Spain		RCE
		Ile Masirah	Oman	1500	
1422	211	Saarbrucken	Germany (W)	1200/600	Saarlandischer Rundfunk
		Valmiera + 2	USSR	50	
		Alger 3	Algeria	40	
1431	210	Montemor-velho + 5	Portugal	100	
		Krivoi Rog	Ukraine	500	
		Skive	Denmark	70	3rd Programme
		Foggia + 2	Italy	2	2nd Programme
		Reading	UK	0.1	Radio 210
		Bernburg + 2	Germany (E)		Berliner Rundfunk
		Och	USSR	50	

kHz	Metres	Station	Country	Power	Programme
1440	208	Marnach	Luxembourg	1200	2nd Programme
		Kraljevo	Yugoslavia	100/10	
1449	207	Misurata	Libya	20	
		Coimbra	Portugal	1	1st Programme
		Kichinev + 1	USSR	50	
		Squinzano + 24	Italy	50	2nd Programme
		Karlovac + 1	Yugoslavia	10	
		Berlin	Germany (W)	5	Sender Freies Berlin
		Redmoss	UK	2	Radio 4 UK
1458	206	Lushnje	Albania	500	
		Torquay	UK	1	Radio 4 S West
		Birmingham	UK	10	Local Radio
		Carlisle	UK	0.5	Local Radio
		London	UK	50	Local Radio
		Manchester	UK	5	Local Radio
		Newcastle	UK	2	Local Radio
		Constanta 2	Roumania	50	
		Svetozarevo + 1	Yugoslavia	10	
1467	204.5	Monte-Carlo-Fontbonne	Monaco	400/200	
		Kiev + 1	Ukraine	300	
		Dededoruk	Turkey	10	
		Zvornik + 3	Yugoslavia	10	
		Isfahan	Iran	100	
		Frunze	USSR	30	
1476	203	Wien-Bisamberg	Austria	600	
		Lvov	Ukraine	120	
		Barcelona	Spain	25	
		Bilbao EFJ 43	Spain	10	RCE
1485	202	Bournemouth	UK	2	Radio 1
		Carlisle	UK	1	Radio 4 UK
		Brighton	UK	1	Local Radio
		Humberside	UK	2	Local Radio
		Merseyside	UK	2	Local Radio
		Oxford	UK	0.5	Local Radio
		Tours	France	1	France-Inter
		Aquila + 11	Italy	1	
		Bielsko Biala + 18	Poland	1	
		Anklam + 11	Germany (E)	1	DDR 1
		Saviese	Switzerland	1	
		Cesme + 4	Turkey	1	
		Orense EAJ 57	Spain		
		Dubai	United Arab Emirates		
1494	201	Rhodos	Greece	5	Regional Programme
		Ajaccio + 1	France	40	France Culture
		Edintsy + 1	USSR	25	
		Hulsberg	Netherlands	10	Relay Lopik 3
		Guarda	Portugal	0.02	

kHz	Metres	Station	Country	Power	Programme
1503	200	Stargard Szczecinski	Poland	300	2nd Programme
		Ulcinj + 1	Yugoslavia	10	
		Stoke-on-Trent	UK	0.5	Local Radio
		Beograd + 2	Yugoslavia	10	
		Burgos EFJ 52	Spain	3	RCE
		Pamplona EFE 57	Spain	2	RCE
		Bilbao	Spain		
		Nicosia	Cyprus	1	British Forces
		Alma-Ata	USSR	10	
1512	198	Chania	Greece	5	
		Beltem	Belgium	20	BRT 2 (Dutch)
		Pristina 2	Yugoslavia	100/20	
		Sotchi + 2	USSR	30	
		Aktiubinsk	USSR	100	
1521	197	Kosice + 6	Czechoslovakia	600	Hvezda & Regional
		Djedeida	Tunisia	100	
		Kalevala + 3	USSR	5	
		Nottingham	UK	0.5	Local Radio
		Oviedo EAJ.19	Spain	5	SER
		Pontevedra EAJ 40	Spain	3	
		Duba	Saudi Arabia	2000	
1530	196	Citta del Vaticano	Vatican	450	Radio Vatican
		Jitomar + 2	Ukraine	5	
		Funchal	Madeira	10	
1539	195	Voice of Peace		50	
		Mainflingen	Germany (W)	700	Deutschlandfunk
		Ialta + 2	Ukraine	25	
		Daugavpils + 2	USSR	5	
		Valladolid EAJ 47	Spain	5	
		Istok	Yugoslavia	2	
1548	194	Vinnitza	Ukraine	50	
		Edinburgh	UK	2	IBA R Forth
		Liverpool	UK	1	IBA R City
		London	UK	27.5	IBA Capital Radio
		Sheffield	UK	0.3	IBA R Hallam
		Bristol	UK	5	BBC Local Radio
		Cleveland	UK	1	BBC Local Radio
1557	193	Cyclops	Malta	600	
		Nice 2	France	300	France Culture
		Kaunas + 4	USSR	75	
		Osijek	Yugoslavia	50	
		Arbadil	Iran	10	
1566	191.5	Amalias	Greece	1.25	

kHz	Metres	Station	Country	Power	Programme
1566	191.5	Covilha	Portugal	1	1st Programme
contd		Sarnen	Switzerland	300	German Programme
		Leningrad	USSR	60	
		Odessa + 2	Ukraine	5	
		Smarje	Yugoslavia	2	
		Sfax	Tunisia	1200	National Programme (Arabic)
		Sandar Abbas	Iran	100	
1575	190.5	Madrid	Spain	5	RNE
		Genova + 17	Italy	50	3rd Programme
		Dresden + 3	Germany (E)		Berliner Rundfunk
		Porto	Portugal	10	3rd Programme
1584	189	Burgos + 28	Spain	1	
		Calvi + 1	France	1	France-Inter
		Branievo + 17	Poland	1	
		Bad Doberan + 8	Germany (E)	1	DDR 1
		Bar + 43	Yugoslavia	1	
		Leicester	UK	0.5	Local Radio
		Reus EAJ 11	Spain	2	SER
		Pamplona EAJ 6	Spain	2	SER
1593	188	Langenberg	Germany (W)	800	Westdeutscher Rundfunk
		Miscolc + 1	Hungary	20	
		Baneasa + 4	Roumania	15	
		Lisboa	Portugal	10	
1602	187	Sabadell EAJ 20	Spain	2	SER
		Bologna + 8	Italy	1	
		Cieszyn + 12	Poland	1	
		Angermuende + 12	Germany (E)	1	
		Alcira EAJ 54	Spain		SER

GEOGRAPHICAL LIST OF LONG AND MEDIUM WAVE EUROPEAN STATIONS

	kHz
AFGHANISTAN	
Kabul	1278
ALBANIA	
Durres	1089
	1215
Korca	972
Kukes	999
Lushnje	1395
	1458
Shkoder	1314
Tirane	1359
ALGERIA	
Algiers	891
	981
	1422
Ain Beida	529
Bechar	576
Constantine	1305
Les Trembles	549
	1143
Tipaza	254
ANDORRA	
Radio Andorra	702
Sud Radio	819
AUSTRIA	
Graz-Dobl	1026
Innsbruck Aldrans	520
Klagenfurt	774
Wien-Bisamberg	585
	1476
AZORES	
Ponta Delgada	837
BELGIUM	
Kuurne	1188
La Louviere	1125
Liege	1233
Marche	1305
Veltem	1512
Waver-Overijse	540
	620
Wolvertem	927

	kHz
BIELORUSSIA	
Brest	1026
Minsk	281
	1197
Moghilev	1170
Mozyr	1008
BRAZIL	
Rio de Janeiro	936
	1224
	1296
BRITISH WEST INDIES	
Radio Paradise	1269
BULGARIA	
Biala Slatina	1161
Kardzali	1377
Petrich	747
Pleven	594
	945
Plovdiv	648
	864
Shumen	828
Sofia	774
	963
Stara Zagora	1125
Vidin	576
	1224
CANARY ISLANDS	
Las Palmas	837
	1008
	1215
Santa Cruz de Palma	1098
Santa Cruz Tenerife	621
CYPRUS	
Cape Greco	1233
Nicosia	603
	693
	1503
Paphos	918
Zakaki	639
	720
Zyyi	1323

CZECHOSLOVAKIA	kHz
Banska Bystrica	702
Bratislava	1098
Brno	900
	954
Brno Mesto	1332
Ceske Budejovice	846
Ceskoslovenko	272
	1287
Cheb	981
Jesenik	792
Kosice	1521
Nitra	1017
Mnich Hradiste	1071
Praha	639
	1233

DENMARK	
Kalundborg	245
	1062
Skive	1431
Torshavn	531

EGYPT	
Abis	774
Abu-Zabaal	558
	711
	873
Alexandria	1197
	1341
Assuit	1278
Asswan	1008
Batra	621
	819
	1107
Cairo	936
	1062
Tanta	1161

FINLAND	
Helsinki	558
	1404
Kuopio	756
Lahti	254
Mariehamn	603
Oulu	433
	540
Turku	963
	1278
Vaasa	1242

FRANCE	kHz
Ajaccio	1494
Allouis	164
Bastia	1404
Bordeaux	1206
Brest	1071
Calvi	1584
Lille	1377
Limoges	792
Lyons-Tramoyes	603
Marseille	675
	1242
Nancy	837
	1350
Nice	1557
Paris	585
	864
	963
Rennes	711
Strasbourg	1161
	1278
Toulouse	945
Tours	

GERMANY (EAST)	
Angermuende	1602
Anklam	1485
Bad Doberan	1584
Berlin	693
	1359
Bernburg	1431
Burg	263
	783
	1044
Dresden	1575
Erfurt	1170
Greifswald	531
Hoyerswerda	999
Koenigswusterhausen	603
	882
Leipzig	1323
Neubrandenburg	657
Oranienburg-Rehmate	182
Puttbus	729
Rostock	558
Schwerin	576

GERMANY (WEST)	
Aachen-Stolberg	702
Bad Mergentheim	1413
Berlin	567

Germany (West) – contd	kHz
Berlin (cont.)	810
	855
	1449
	990
Bodenseesender	666
Braunschweig	756
Bremen	936
Dannenberg	630
Donebach	155
Frankfurt	594
	873
Freiburg	828
Hamburg	972
Hannover	828
Heidelberg	711
Hof-Saale	520
	684
Holzkirchen	720
Langenberg	1593
Mainflingen	209
	1539
Munchen	1107
Munchen-Ismaning	801
	1197
Neumunster	1269
Nordkirchen	549
Saarbrucken	1422
Saarlouis-Felsburg	182
Stuttgart	576
	1143
Wolfsheim	1017
GREECE	
Amalias	1566
Athens	729
	1386
Chania	1512
Florina	1278
Ioannina	765
Iraklion	954
Kavalla	792
Kerkyra	1008
Larissa	945
Megara	981
Mesolongion	1071
Mytilini	1287
Orestias	1080
Rhodos	1260
	1494
Thessaloniki	1044
	1179

	kHz
Tripolis	1071
	1314
Zakynthos	927
GUINEA	
Kipe	1404
HUNGARY	
Budapest	873
Budapest-Lakihegy	1341
Miscolc	1593
Siofok	1251
Solt	540
Szolnok	1188
ICELAND	
Reykjavik	209
INDIA	
Calcutta	1134
Rajkot	1071
IRAN	
Abadan	1080
Ahvaz	1386
Arbadil	1314
	1557
Bonab	639
Farahabad	774
Isfahan	1467
Lar	1287
Mashad	684
Qashr' Shirin	612
Qeslagh	558
Rezaieh	936
Sandar Abbas	1566
Semnan	1008
	1296
Shiraz	981
	990
Tabriz	1026
	1152
Taybad	720
Teheran	1188
	1332
Yazd	1035
IRAQ	
Babylon	1035
Basra	693
Kirkuk	1359

Iraq – contd	kHz
Salman Pack	756
Thourah	909
IRELAND (EIRE)	
Carraroe	540
Cork	1188
Dublin	1251
Tullamore	567
	612
ISRAEL	
Beer-Sheva	1224
Haifa	1206
	1305
Jerusalem	531
	675
	711
	1026
Tel Aviv	576
	657
	738
	1242
	1287
Ufa	693
Yamit	1368
Zefat	846
ITALY	
Ancona	1314
Aquila	1485
Bari	1116
Bologna	567
	1098
	1602
Cagliari	1062
Caltanissetta	191
Foggia	1431
Genova	1575
Messina	1143
Milano	900
	1035
Napoli	657
Rimini	999
Roma	1332
San Remo	1188
Squinzano	1449
Trieste	819
	981
Venezia	1017
	1368

JORDAN	kHz
Amman	801
	855
KUWAIT	
Kuwait	1350
Sulaibiyah	540
	1134
LEBANON	
Amchit	990
Beyrouth	837
LIBYA	
Benghazi	675
El Beida	1125
Misurata	1449
Sebha	828
Tripoli	540
	1053
	1251
	1404
LUXEMBOURG	
Junglinster	236
Marnach	1440
MADEIRA	
Funchal	1332
	1530
MALTA	
Cyclops	1557
Delimara	999
	1341
Radio Malta	756
MARTINIQUE	
Fort de France	1314
MAURETANIA	
Nouakchott	1350
MONACO	
Monte Carlo	218
	702
	1467
MOROCCO	
Agadir	936
	1197
Azilal	209

Morocco – contd	kHz
Casablanca	1080
	1188
Ksar es Souk	864
Marrakech	972
	1152
Ouarzazate	1116
Oujda	594
	828
Rabat	819
	1026
	1296
Safi	1026
Sebaa Aioun	612
	702
	1044
Tanger	1233
Tantan	657
Tetuan	1017
	1053

NETHERLANDS	
Amsterdam	1251
Hengelo-Overijssel	891
Hoogezand	1395
Hulsberg	1494
Lopik	675
	747
	1008

NORWAY	
Bodoe	675
Finmark	702
Oslo	218
Stavanger	1314
Tromsoe	155
Vigra	630

OMAN	
Ile Masirah	1413
Muscat	1242

POLAND	
Bielsko Biala	1485
Branievo	1584
Cieszyn	1602
Katowice	1080
Krakow	1368
Poznan	738
Rzeszow	1305
Stargard Szczecinski	1503

	kHz
Szczecin	1260
Warsaw	200
	227
	819
Wroclaw	1206

PORTUGAL	
Braganca	576
Castelo Branco	828
Chaves	1251
Coimbra	963
	1449
Covilha	1566
Elvas	1332
Faro	558
Guarda	1494
Lamego	927
Lisboa	666
	756
	1287
	1593
Miramar	783
Miranda do Douro	630
Mirandela	729
Montemorvelho	1431
Norte	720
	1062
Portalegre	1197
Porto	1170
	1368
	1575
Porto Alto	1035
Valenca do Minho	567
Villa Real	621
Viseu	693

QATAR	
Al Arish	954

ROUMANIA	
Bacau	1179
Baia Mare	1404
Bailesti	720
Baneasa	1593
Brasov	155
	567
Bucuresti	855
Cluj	909
	1152

Roumania – contd	kHz
Constantza	1314
	1458
Lasi	1053
Lugoj	756
Miercurea Ciuc	945
Oradea	603
Targa Jiu	558
Targa Mures	1323
Timisoara	630

ST PIERRE & MIQUELON	
St Pierre & Miquelon	1377

SAUDI ARABIA	
Damman	882
Duba	1521
Riyadh	585

SENEGAL	
Dakar	765

SPAIN	kHz
Alcala Henares	1260
Alcira	1602
Alicante	1026
	1395
Badalona	1179
Barcelona	738
	828
	990
	1413
	1476
Bejar	1413
Bilbao	1134
	1503
	1476
Burgos	1503
	1584
Caceres	774
Cadiz	1314
Gijon	873
Granada	1080
	1395
	1413
La Coruna	639
	1386
Leon	1395
Madrid	585
	810
	918
Madrid – contd	954
	1026
	1107
	1224
	1314
	1575
Malaga	1395
Mallorca	1080
Murcia	855
	1179
Orense	1386
	1485
Oviedo	729
	1413
	1521
Pamplona	1503
	1584
Pontevedra	1521
Reus	1584
Sabadell	1602
San Sebastian	1260
Serona	1413
Sevilla	684
	792
	1134
	1413
Valencia	1134
	1260
	1314
	1359
Valladolid	1107
	1539
Vigo	1026
Villareal	1224
Zamora	1134
Zaragoza	873
	1134
	1413

SUDAN	
Sennar	1296
Soba	963

SWEDEN	
Goeteborg	981
Hoerby	1179
Motala	191
Stockholm	774

SWITZERLAND	
Beromunster	531

Switzerland – contd	kHz
Monte-Ceneri Passo	558
Sarnen	1566
Saviese	1485
Sottens	765

SYRIA

Homs	567
Damas-Sabbourah	666
Sarakeb	747
Tartus	783
	1071

TUNISIA

Djedeida	1521
Gafsa	585
Sfax	720
	1566
Tunis	630
	963

TURKEY

Ankara	182
Antalya	891
Cesme	1485
Cukurova	630
Dedoruk	1467
Diyabakir	1062
Erzurum	245
Etimesgut	200
Istanbul	1017
Izmir	927
Trabzon	954
Umraniye	702

UKRAINE

	kHz
Dniepropetrovsk	1071
	1404
Donetsk	711
Ialta	1539
Jitomar	1530
Kharkov	837
Kiev	209
	783
	1242
	1467
Krivoi Rog	1431
Lutsk	1377
Lvov	173
	936
	1476
Odessa	765
	1278
	1566
Simferopol	648
Tchernovtsy	657
Ujgorod	675
	891
Vinnitza	1548

UNITED ARAB EMIRATES

Abu Dhabi	810
Dubai	1485
Sadiyat	729

UNITED KINGDOM

Barnstaple	801
	1053
Barrow	693
	1053
Belfast	1026
Bexhill	693
	1053
Birmingham	1152
	1452
Blackburn	855
Bournemouth	909
	1359
	1485
Bradford	1278
Brighton	693
	1053
	1215
	1485
Bristol	1548
Brookmans Park	909
	1089
	1215
Burghead	200
	693
	810
	1053
	1215
Cambridge	1197
Carlisle	756
	1458
	1485
Clevedon	909
Cleveland	1548
Crowborough	648
Derby	1116

United Kingdom – contd	kHz
Droitwich	200
	693
	1053
	1215
Dumfries	810
Dundee	1053
Edinburgh	1548
Exeter	693
	990
Fareham	909
	999
	1089
	1215
Folkestone	693
	1053
Foxdale	1368
Glasgow	1152
Gillingham	1035
Hull	1053
	1215
Humberside	1485
Ipswich	1170
Leeds	774
Leicester	1584
Lisnagarvey	720
	909
	1089
	1215
	1341
Liverpool	1548
	1152
	1458
	1548
Stoke on Trent	1503
Swansea	1170
Teeside	1170
Torquay	909
	1458
Tyneside	1152
Tywyn	882
	1089
	1215
Washford	882
	1089
	1215
Westerglen	200
	810
	909
	1089
	1215

	kHz
Whitehaven	909
	1089
Wolverhampton	990
Wrexham	882
USSR	
Achkabad	281
Aktiubinsk	1512
Alma Ata	182
	1098
	1341
	1503
Astrakahn	792
Baku	218
Cheboksary	531
Daugavpils	1539
Djambul	1224
Duchanbe	254
	702
	1143
	1161
	1296
Edintsy	1494
Elista	846
Engels	155
Erevan	864
Frunze	1467
Gorkii	828
Ijevsk	594
Iman	909
Iochkar-Ola	900
Irkutsk	263
Kalevala	1521
Kaliningrad	173
	1116
	1143
Karaganda	747
Kaunas	1107
	1386
	1557
Kazan	783
Kichinev	236
	549
	999
	1449
Krasnoder	1089
Kurgan	1053
Leningrad	200
	801
	873

USSR – contd	kHz
Leningrad – contd	1125
	1566
Madona	1350
Medvejiegorsk	765
Mezen	918
Moscow	200
	263
	846
	1359
Murmansk	657
Nebit Dag	927
Nikolaev	972
Novosibirsk	272
Och	1431
Omsk	639
Orenburg	540
Parnu	1332
Perm	585
Riga	576
	1071
Rostov	745
Saransk	1062
Saratov	630
Sotchi	1512
Stavropol	882
Sukhumi	1350
Tachkent	164
Taldy Kurgan	245
Tallinn	612
	711
	1035
Tartu	1215
Tbilissi	191
	1044
Tcheliabinsk	738
	855
Tiumen	603
Tselinograd	684
	999
Ukhta	621
Ulan-Ude	801
Valmiera	1422
Vilnius	666
Volgograd	567
	810
	945
Voronej	775

VATICAN	
Vatican City	1530

YEMEN	kHz
Sanaa'a	1197

YUGOSLAVIA	
Aleksinac	1008
Banja Luka	1071
Bar	1584
Beli Kriz	549
	1170
Beograd	684
	963
	1350
	1503
Bor	981
Djakovica	936
Gnjilane	1332
Istok	1539
Karlovac	1449
Koper-Belikriz	1080
Kraljevo	1440
Kumrovec	1377
Loznica	1296
Ljubljana	918
Maribor	558
Murska Sobota	648
Nis	711
Nova Gradiska	1143
Novi Sad	837
	1089
	1107
	1269
Osijek	1557
Pristina	1512
	1413
Sarajevo	612
	747
Skopje	810
	1314
	1566
Smarje	1566
Sombor	666
Split	774
Svetozarevo	1458
Titograd	882
Tuzla	1386
Ulcinj	1503
Valjevo	1368
Varazdin	1404
Zagreb	1062
	1125
	1134
Zvornik	1467

In general, short-wave stations adjust their frequency schedules four times a year, because of different propagation conditions in spring, summer, autumn and winter. Alterations are arranged on an international basis.

Although some stations may use virtually the same channels throughout the year with only minor differences, others use particular frequencies during only one or two of the four periods. The short-wave list therefore has columns marked M, M, S and N which indicate the four plan periods used by the ITU for the HF Tentative Schedules commencing March, May, September and November respectively when channelling is changed. The underline symbol (_) indicates that the station was planned for the period in the previous year's tentative schedule, and the oblique stroke (/) shows that the station was included in the plans for the year starting in March 1979. An X in these columns shows that a station originally allocated for that plan period has, during the period, been deleted from the plan. The list also includes stations, not shown in the Tentative Schedule, which are known to be active or which are expected to operate in the near future. Some frequencies listed may only be audible in favourable reception conditions, and others may only be in use seasonally. Stations that are known to transmit broadcast programmes, using ISB or SSB transmission modes, which are not intended for general reception by the public, are identified by the addition of an asterisk * after the station name or if the site is unknown after the country code. Other changes noted by the listener can be recorded similarly. The columns also indicate the extent of each short-wave band allocated to broadcasting: these indications exclude the out-of-band frequencies which are also occasionally used.

Transmitter power in the short-wave bands is not easily defined, because the majority of stations have a number of senders of varying power, any one of which may be used as required. The powers quoted are therefore the lowest and highest known to operate at a location over 24 hours and should be used only as a rough guide, because it is impossible to cover all the possibilities.

A high-gain aerial, beamed towards the listener, can provide a strong signal from a comparatively low-powered transmitter, although a narrow-beam array, powered with 250 kW but directed away from the receiving site, may be barely audible. Thus power figures merely indicate the *capability* of a station in terms of field strength: the direction of main radiation may or may not favour a listener outside the target zone.

A station name can be that of the large town nearest to the transmitting site, or it can be the capital of the country even, although there may be more than one transmitting site. Occasionally two different place names are given, separated by an oblique stroke; this indicates that the channel is shared. Where the same transmitter operates at different times on adjacent channels, separate entries are made; this accounts for the multiplicity of entries under some place names. In most cases clandestine stations are listed by their slogan only and no country is shown.

This list of stations is compiled from information obtained from broadcasting authorities and the BBC Receiving Station, Caversham Park, Reading covering the period November 1978–March 1979.

A geographical list of short-wave stations will be found on page 179.

MHz	*Metres*	*Station*	*Country*	*kW*
2.500	120.00	Rugby Std Freq	UK	
2.670	112.36	Sariwon	Korea (N)	
2.739	109.53	Riyadh*	Saudi Arabia	
2.745	109.29	Sinuiju	Korea (N)	
2.765	108.50	Pyongyang	Korea (N)	
2.775	108.11	Hamhung	Korea (N)	
2.850	105.26	Pyongyang	Korea (N)	
3.000	100.00	Fukien Front Stn	China Rep	
3.015	99.50	Pyongyang	Korea (N)	120
3.030	99.01	Wonsan	Korea (N)	
3.155	95.09	Peshawar	Pakistan	10
3.195	93.90	Baghdad	Iraq	50
3.200	93.75	Fukien Front Stn	China Rep	
3.205	93.60	Ibadan	Nigeria	10
		Lucknow	India	10
3.210	93.46	Maputo	Mozambique	
3.215	93.31	Rawalpindi	Pakistan	10
3.220	93.17	Peking	China Rep	120
3.222	93.11	Lama Karma	Togo	10
3.223	93.08	Simla	India	2.5
		Mbabne	Swaziland	
3.225	93.02	Tovar	Venezuela	1
		Lins	Brasil	0.5
3.227	92.97	Monrovia	Liberia	10
3.230	92.88	Johannesburg	South Africa	100/250
3.232	92.82	Brazzaville	Congo	25
		Tananarive	Malagasy Rep	30
3.235	92.74	Gauhati	Gauhati	10
		Marilia	Brasil	0.5
3.240	92.59	Islamabad	Pakistan	10/100
		Lima	Peru	
3.242	92.54	Baghdad	Iraq	50
3.245	92.45	Caracas	Venezuela	1
3.250	92.31	Bloemendal	South Africa	20

MHz	*Metres*	*Station*	*Country*	*kW*
		Tovar	Venezuela	
		El Tigre	Venezuela	1
3.255	92.17	Monrovia	Liberia	10
3.260	92.02	Niamey	Niger	4
		Kweiyang	China Rep	
3.265	91.88	Georgetown	Guyana	2
		L Marques	Mozambique	25/100
			Brasil	1
3.270	91.74	Peking	China Rep	
3.273	91.66	Quetta	Pakistan	
3.277	91.55		India	
3.285	91.32	Manila	Philippines	2.5
		Meyerton	South Africa	
		Pernambuco	Brasil	1
		Puerto Cabello	Venezuela	1
3.288	91.24	Tananarive	Malagasy Rep	100
3.290	91.19	Tristan da Cunha	Tristan da Cunha	0.04
		Peking	China Rep	
3.295	91.05	Delhi	India	20
		Trujillo	Venezuela	
		Lusaka	Zambia	10
		Accra	Ghana	10
3.300	90.91	Libreville	Gabon Rep	4/20
		Bujumbara	Burundi	25
		Guatemala	Guatemala	10
		Belmopan	Honduras Br	1
3.305	90.77	Daru	Papua	
3.307	90.72	Gwelo	Rhodesia	10/100
3.315	90.50	Bhopal	India	10
		Fort de France	Martinique	4
3.316	90.47	Freetown	Sierra Leone	10
3.320	90.36	Bloemfontein	South Africa	
		Pyongyang	Korea (N)	100
3.325	90.23	Maceio	Brasil	25
3.330	90.09	Dzaudazi	Comoro Is	15
		Peshawar	Pakistan	10
		Kigali	Rwanda	5
3.335	89.96	Wewak	Papua New Guinea	10
3.336	89.93	Ziguinchor	Senegal	4
3.338	89.87	Maputo	Mozambique	
3.339	89.85	Zanzibar	Tanzania	10
3.340	89.82	Kampala	Uganda	7.5
3.343	89.74	Nampula	Mozambique	
3.345	89.69	Manila	Philippines	40
		Uberlandia	Brasil	5
		Luanda	Angola	
3.346	89.66	Lusaka	Zambia	120
3.350	89.55	Tema	Ghana	20

MHz	*Metres*	*Station*	*Country*	*kW*
		Franceville	Gabon Rep	20
3.355	89.42	Noumea	New Caledonia	20
		Gaberone	Botswana	10
		Valencia	Venezuela	1
		Kurseong	India	20
		Luanda	Angola	
3.360	89.29	Milne Bay	Papua New Guinea	10
		Peking	China Rep	
3.365	89.15	Delhi	India	10
		Araraquara	Brasil	1
3.366	89.13	Tema	Ghana	10
3.370	89.02	Tananarive	Malagasy Rep	4
		Beira	Mozambique	10
3.375	88.89	Luanda	Angola	10
		Gauhati	India	10
3.380	88.76	Blantyre	Malawi	100
		Esmeraldas	Ecuador	10
3.385	88.63	Cayenne	Guyana Fr	4
		Colombo	Ceylon	10
		Rabaul	Papua New Guinea	10
		Barcelona	Venezuela	1
3.388	88.55	Bloemendal	South Africa	
3.390	88.50	Kabul	Afghanistan	100
		Peking	China Rep	20/120
		S Domingo	Ecuador	5
3.395	88.37	Colombo	Ceylon	
		Merida	Venezuela	1
3.396	88.34	Kaduna	Niger	10
		Gwelo	Rhodesia	100
3.397	88.31		Pakistan	
3.400	88.24	Fukien Front Stn	China Rep	
3.417	87.80	Met Station	USSR	
3.421	87.69	Medan	Indonesia	1
3.425	87.59	Khumaltar	Nepal	100
3.440	87.21	Met Station*	USSR	
3.450	86.96	Peking	China Rep	
3.460	86.71		USSR	
3.470	86.46	Met Station*	USSR	
3.500	85.71	Peking	China Rep	20/240
3.535	84.87	Fukien Front Stn	China Rep	
3.550	84.51	R Dili	Timor	
3.560	84.27	Pyongyang	Korea (N)	120
3.640	82.42	Fukien Front Stn	China Rep	
3.660	81.97	Peking	China Rep	
3.695	81.19	Pyongyang	Korea (N)	
3.700	81.08	Peking	China Rep	
3.830	78.33	Peking	China Rep	120

MHz	Metres	Station	Country	kW
3.850	77.92	Dili	Timor	
3.860	77.72	Karachi	Pakistan	
3.885	77.22	C Verde Is	Cape Verde Is	
3.890	77.12	Pyongyang	Korea (N)	
		Karachi	Pakistan	10
3.900	76.92	Hailar	Mongolian Rep	
		Ventiane	Laos	
3.905	76.82	Delhi	India	20
		Palang	Indonesia	
3.910	76.73	Tokio	Japan	10
		Surakarta	Indonesia	
3.915	76.63	Islamabad	Pakistan	10
		Tebrau	Malaysia	100/250
		Kashmir Radio		
3.920	76.53	Peking	China Rep	
3.925	76.43	Delhi	India	10
		Port Moresby	Papua New Guinea	2
		Tokio	Japan	50
		Jakarta	China Rep	
3.930	76.34	Huhetot	Mongolian Rep	
		Barlavento	Cape Verde Is	10
		Met Station*	USSR	
		Tokio	Japan	
			Korea (S)	
3.935	76.24	Semarang	Indonesia	10
3.940	76.14	Wunan	China Rep	
			Indonesia	
			USSR	
3.945	76.05	Denpassar	Indonesia	10
		Hokkaido	Japan	
3.950	75.95	Peking	China Rep	
		Baghdad	Iraq	100
		Jermate	Indonesia	
3.952	75.91	London	UK	100
		Peking	China Rep	
3.955	75.85	Bloemendal	South Africa	20
		Padang	Indonesia	
		Warszawa	Poland	
3.960	75.76	Peking	China Rep	120
		Padang	Indonesia	10
		Holzkirchen	Germany (W)	10
		Biblis	Germany (W)	100
		Lampertheim	Germany (W)	
		Vladivostock	USSR	
3.965	75.66	Bloemendal	South Africa	20
		Pontianak	Indonesia	10
		Allouis	France	4
		Karachi	Pakistan	

MHz	*Metres*	*Station*	*Country*	*kW*
3.970	75.57	London	UK	
		Buea	Cameroon	8
		Huhetot	Mongolian Rep	
		Holzkirchen	Germany (W)	10
		Riyadh	Saudi Arabia	
3.975	75.47	London	UK	
3.980	75.38	Ismaning	Germany (W)	
		Bloemendal	South Africa	100
		Surabaya	Indonesia	
3.985	75.28	Schwarzenburg	Switzerland	250
		Riobamba	Ecuador	1
		Peking	China Rep	
		Lampertheim	Germany (W)	20
		Biblis	Germany (W)	
3.990	75.19	Limassol	Cyprus	
		Monrovia	Liberia	250
		Biblis	Germany (W)	100
		Lampertheim	Germany (W)	20
		Holzkirchen	Germany (W)	10
			USSR	
		Urumchi	China Rep	
3.995	75.09	Vladivostock	USSR	
		Roma	Italy	5
		Julich	Germany (W)	100
		Bloemendal	South Africa	20
		Pontianak	Indonesia	
3.999	75.02	Godthaab	Greenland	1
4.000	75.00	Kabul	Afghanistan	100
		Hanoi	Vietnam	
		Phnom Penh	Cambodia	
4.005	74.91	Pontianak	Indonesia	
4.010	74.81	Frunze	USSR	15
4.020	74.63	Peking	China Rep	120
		Islamabad	Pakistan	10/100
4.030	74.44	Magadan	USSR	
		Kanggye	Korea (N)	
4.035	74.35	Lhasa	Tibet	
4.040	74.26	Erevan	USSR	
		Vladivostok	USSR	
4.045	74.17	Fukien Front Stn	China Rep	
4.050	74.07	Frunze	USSR	
		Yuzhno Sakhalinsk	USSR	
4.055	73.98	Petropavlovsk	USSR	50
4.060	73.89	Rawalpindi	Pakistan	
4.068	73.75	Huhetot	Mongolian Rep	
			China Rep	
4.070	73.71	Jakarta	Indonesia	
4.080	73.53	Ulan Bator	Mongolian Rep	50
4.110	72.99	Urumchi	China Rep	

MHz	Metres	Station	Country	kW
4.115	72.90	V Rev Pty for Reunification		
4.130	72.64	Peking	China Rep	
4.180	71.77	Peking	China Rep	
4.190	71.60	Urumchi	China Rep	
4.200	71.43	Peking	China Rep	120
4.220	71.09	Urumchi	China Rep	
4.250	70.59	Peking	China Rep	120
4.284	70.03	Vinh Phu	Vietnam	
4.320	69.44	Vientiane	Laos	
4.330	69.28	Fukien Front Stn	China Rep	
4.380	68.49	Fukien Front Stn	China Rep	
4.395	68.26	Yakutsk	USSR	50
4.410	68.03		China Rep	
4.425	67.80		USSR	
4.460	67.26	Peking	China Rep	
4.465	67.19	Vladivostock	USSR	
4.467	67.16	London	UK	
4.475	67.04	RFE/R Liberty*		
4.485	66.89	Petropavlovsk	USSR	
		Ufa	USSR	
4.500	66.67	Urumchi	China Rep	
4.505	66.59	RFE/R Liberty*		
4.520	66.37	Palana	USSR	
		Khanty-Mansiysk	USSR	
4.525	66.30	Ikechao/Tungsheng	Mongolian Rep	
		Standard Time Signal*	Germany (W)	
4.545	66.01	Alma Ata	USSR	50
4.557	65.83	V Rev Pty for Reunification		
4.565	65.72	RFE/R Liberty*		
4.600	65.22	Thu Dau Mot City	Vietnam	
4.603	65.17	Xieng Khouang	Laos	
4.620	64.94	Peking	China Rep	
4.630	64.79	Peking	China Rep	
4.635	64.72	Duchanbe	USSR	50
4.645	64.59	Vientiane	Laos	
			USSR	
		Met Station*	USSR	
4.647	64.56	Ha Bac	Vietnam	
4.650	64.52	Houa Phan	Laos	
4.654	64.46		USSR	
4.656	64.43	Met Station	USSR	
4.665	64.31	Houa Phan	Laos	
		Met Station	USSR	
4.675	64.17	V of NUFK		
4.678	64.13	Met Station*	USSR	
4.679	64.12	Espejo	Ecuador	

MHz	Metres	Station	Country	kW
4.680	64.10	Hue	Vietnam	
4.685	64.03	Met Station*	USSR	
4.692	63.94	Quang Ninh	Vietnam	
4.695	63.90	Met Station	USSR	
		RFE/R Liberty*		
4.700	63.83	Luang Prabang	Laos	
		Surabaya	Indonesia	
4.710	63.69		USSR	
4.720	63.56		USSR	
		Jakarta	Indonesia	
		Karachi	Pakistan	
		Bassacongo	Angola	0.5
		Sao Vicente	Cape Verde Is	1.5
4.725	63.49	Rangoon	Burma	50
4.730	63.42		USSR	
4.735	63.36	Karachi	Pakistan	
4.747	63.20		USSR	
4.750	63.16	Lubumbashi	Zaire	10
		Bertoua	Cameroon	20
4.755	63.09	Bogota	Colombia	11
		Jakarta	Indonesia	
4.760	63.03	Peking	China Rep	
		Mbabane	Swaziland	
		Dzhambul	USSR	
4.762	63.00	Gia Lai-Cong Tum	Vietnam	
4.763	62.99	Ulan Bator	Mongolian Rep	50
4.764	62.97	Medan	Indonesia	
4.765	62.96	Guayaquil	Ecuador	5
		Brazzaville	Congo	50
			USSR	
4.770	62.89	Peking	China Rep	
		Monrovia	Liberia	10
		Son La	Vietnam	
		Bolivar	Venezuela	1
		Pyongyang	Korea (N)	
		Jakarta	Indonesia	
4.775	62.83	Gauhati	India	10
		Kabul	Afghanistan	100
		Sibolga	India	50
		Libreville	Gabon Rep	100
		Cuiba	Brasil	
4.780	62.76	Djibuti	Afars & Issas	4
		Petrozavodsk	USSR	50
		Valencia	Venezuela	
		Luambo	Angola	
4.783	62.72	Bamako	Malawi	18
4.785	62.70	Baku	USSR	50
		Ibaque	Colombia	
		Dar-es-Salaam	Tanzania	50

MHz	Metres	Station	Country	kW
		Sao Luiz	Brasil	
		Yunnan	China Rep	
		Cao Lang	Vietnam	
4.790	62.63	Iquitos	Peru	
		Caracas	Ecuador	
		Ulan Bator	Mongolian Rep	
			Swaziland	
			Angola	
4.795	62.57	Ulan Ude	USSR	50
		Aquidauana	Brasil	
		R Neuva America	Bolivia	1
		Brazzaville	Congo	4
4.800	62.50	Peking	China Rep	
		Hyderabad	India	10
		Barquesimeto	Venezuela	10
		Maseru	Lesotho	10
4.804	62.45	Nairobi	Kenya	5
4.805	62.43	Djakarta	Indonesia	100
		Manaus	Brasil	10
		Yacuiba	Bolivia	
4.807	62.41	Sao Thome	St Thomas Is	10
		Santiago	Dominican Rep	
4.810	62.37	Erevan	USSR	
		Maracaibo	Venezuela	2
		Bloemfontein	South Africa	
4.815	68.31	Peking	China Rep	
		Londrina	Brasil	
		Ouagadougou	Upper Volta	20
		Vallendupar	Colombia	
			USSR	
4.820	62.24	Luanda	Angola	100
			USSR	50
		Barquesimeto	Venezuela	1
		Ha Tuyen	Vietnam	
		Tegucigalpa	Honduras Rep	1
		Calcutta	India	10
4.825	62.18	Moskva	USSR	100
		Achkhabad	USSR	5
		Dar-es-Salaam	Tanzania	
		Bamako	Malawi	18
		Braganca	Brasil	5
4.828	62.14	Gwelo	Rhodesia	100
4.830	62.11	Franceville	Gabon Rep	20
			USSR	
		Ulan Bator	Mongolian Rep	
4.832	62.09	San Jose	Costa Rica	1
		Bangkok	Thailand	10
4.835	62.05	Kuching	Malaysia	10
		Meyerton	South Africa	

MHz	Metres	Station	Country	kW
		Boa Vista	Brasil	
			USSR	
4.840	61.98	Bombay	India	10
		Valera	Venezuela	1
		Heeilunkiang	China Rep	
			USSR	
4.843	61.95	Point Noire	Congo	4
4.845	61.92	Gaberones	Botswana	10
		La Paz	Bolivia	
		Bucramanga	Colombia	
		Nouakchott	Mauretania	100
		Kuala Lumpur	Malaysia	50
4.850	61.86	Moskva	USSR	
		S Domingo	Dominican Rep.	3
		Tashkent	USSR	
		P Louis	Mauritius	10
		Yaounde	Cameroon	
		Paramaribo	Surinam	
		Peking	China Rep	
			India	
		Ulan Bator	Mongolian Rep	
4.853	61.82	Sanaa	Yemen	
4.855	61.79	Palembang	Indonesia	10
		Taubate	Brasil	
		L Marques	Mozambique	20
4.860	61.73	Delhi	India	10
		Moskva	USSR	
		Tchita	USSR	
		Saurimo	Angola	5
		Maracaibo	Venezuela	10
		Hanoi	Vietnam	
4.865	61.66	Belem	Brasil	
		Arauca	Colombia	1
		Lanchow	China Rep	
		Maputo	Mozambique	7.5
		Berakas	Brunei	10
4.870	61.60	Caracas	Venezuela	
		Cotonou	Benin	30/50
		Ekala	Ceylon	10
4.872	61.58	Peking	China Rep	
4.875	61.54	Bloemfontein	South Africa	100/250
		Rio de Janeiro	Brasil	
		Bamako	Malawi	
		S Crux del Sur	Bolivia	10
		Medellin	Colombia	
		Ibaque	Colombia	5
		Jakarta	Indonesia	
4.880	61.48	Barquisemeto	Venezuela	10
		Peking	China Rep	

MHz	Metres	Station	Country	kW
		Bloemendal	South Africa	
		V of Kawthulay		
		Dacca	Bangladesh	
4.881	61.46	Thanh Hoa	Vietnam	
4.882	61.45		Germany (W)*	
4.885	61.41	Pocas de Caldas	Brasil	
		Villavicencio	Colombia	1
		Peking	China Rep	
		Mombasa	Kenya	10
		Bukittinggi	Indonesia	
4.890	61.35	Caracas	Venezuela	
		Port Moresby	Papua New Guinea	10
		Dakar	Senegal	25
		Dacca	Bangladesh	
4.895	61.29	Tyumen	USSR	50
		Achkhabad	USSR	50
		Kuching	Malaysia	10
		Beira	Mozambique	100
		Kurseong	India	
		Huhetot	Mongolian Rep	
		Peking	China Rep	
4.896	61.27	Silva Porto	Angola	1
4.900	61.22	Barquisemeto	Venezuela	10
		Cordac	Burundi	2.5
		Ekala	Ceylon	
4.905	61.16	Peking	China Rep	
		Fort Lamy	Chad	100
		Rio de Janeiro	Brasil	
		Gedja	Ethiopia	100
4.908	61.12	Phnom Penh	Cambodia	
		V of Kampuchea People		
4.910	61.10	Conakry	Guinea	18
		Carora	Venezuela	
		Quito	Ecuador	10
		Lusaka	Zambia	10
4.915	61.04	Macapa	Brasil	
		Accra	Ghana	
		Langata	Kenya	100
		Nanning	China Rep	
		Gutapuri/ Valledupar	Colombia	
4.920	60.98	Brisbane	Australia	
		Caracas	Venezuela	
		Madras	India	
		Kiev	Ukraine	
		El Progreso	Honduras Rep	10
		Quito	Ecuador	5

MHz	*Metres*	*Station*	*Country*	*kW*
		Jakarta	Indonesia	
4.923	60.94	Quito	Ecuador	
4.925	60.91	Maputo	Mozambique	7.5
		Bata	Guinea	5
		Yaounde	Cameroon	
			Brasil	
		Aranca	Colombia	
4.927	60.89	Djambi	Indonesia	7.5
4.930	60.85		USSR	
		San Cristobal	Venezuela	
		Quito	Ecuador	
		Jakarta	Indonesia	
4.932	60.83	Jakarta	Indonesia	10
		Benin	Niger	10
		Hanoi	Vietnam	
4.934	60.80	Nairobi	Kenya	5
4.935	60.79	Tarapoto	Peru	
		Rawalpindi	Pakistan	10
4.940	60.73	Abidjan	Ivory Coast	25
		Kiev	Ukraine	50
		San Filipe	Venezuela	10
		Colombo	Ceylon	10
		Peking	China Rep	
		Hanoi	Vietnam	
4.945	60.67	Neiva	Colombia	2.5
		Pocas de Caldas	Brasil	5
		Hanoi	Vietnam	
		Ekala	Ceylon	
4.950	60.61	Nairobi	Kenya	10
4.952	60.58	Silinhot	Mongolian Rep	50
4.955	60.54	Bogota	Colombia	50
			USSR	
		Kuching	Malawi	
4.957	60.52	Baku	USSR	50
4.960	60.48	Cumana	Venezuela	1
		Jakarta	Indonesia	
		Peking	China Rep	
		Sucua	Ecuador	
		Delhi	India	
			Venezuela	
4.965	60.42	Bogota	Colombia	5
		Potosi	Bolivia	
		Uberaba	Brasil	
		Lusaka	Zambia	2.5
4.970	60.36	Villa de Cura	Venezuela	10
		Urumchi	China Rep	
		Jurong	Senegal	10
4.972	60.34	Yaounde	Cameroon	30
		Koya Kinabalu	Malaysia	10

MHz	*Metres*	*Station*	*Country*	*kW*
		Cayenne	Guyana Fr.	1
4.975	60.30	Blagoveshchensk	USSR	50
		Foochow	China Rep	
		Lima	Peru	
		Sabah	Malaysia	
		Sanaa	Yemen	
4.976	69.29	Kampala	Uganda	7.5
		Sao Luis	Brasil	
4.980	60.24	San Cristobal	Venezuela	10
		Tema	Ghana	
			Burma	
		Peking	China Rep	
4.985	60.18	Luanda	Angola	0.5
		Tananarive	Malagasy Rep	4
		Goiana	Brasil	5
		Penang	Malaysia	10
4.990	60.12	Alma Ata	USSR	50
		Erevan	USSR	
		Barquesimeto	Venezuela	15
		Lagos	Niger	50
		Choraya	Bolivia	
		Meyerton	South Africa	250
		Changsha	China Rep	
4.995	60.06	Hanoi	Vietnam	
		Goiana	Brasil	
		Andina	Peru	
		Ulan Bator	Mongolian Rep	
4.999	60.01	Bangui	Central African Rep	100
5.000	60.00	Rugby Std Freq	UK	
		Boulder " "	USA	
		Honolulu " "	Hawaii	
		Turin	Italy	
		Rome	Italy	
5.005	59.94	La Paz	Bolivia	
		Lalitpur	Nepal	100
5.010	59.88	Santo Domingo	Dominican Rep	
		Garoura	Cameroon	30
		Jurong	Senegal	10
		Islamabad	Pakistan	10/100
		Tananarive	Malagasy Rep	4
5.015	59.82	Arkhangelsk	USSR	15
		Vladivostock	USSR	50
		Rio de Janeiro	Brasil	
5.016	59.81	Gwelo	Rhodesia	100
5.020	59.76	Niamey	Niger	20
		Caracas	Venezuela	
5.025	59.70	Aquidauana	Brasil	
		Andahuaylas	Peru	

MHz	*Metres*	*Station*	*Country*	*kW*
5.026	59.69	Kampala	Uganda	7.5
5.030	59.64	Kuching	Malaysia	10
		Caracas	Venezuela	10
		Urumchi	China Rep	
5.035	59.58	Sao Paulo	Brasil	
		Tachkent	USSR	
		Florencia	Colombia	
5.038	59.55	Khartoum	Sudan	20
5.040	59.52	Villa Vicencio	Colombia	3
		Tbilisi	USSR	50
		Bissau	Guinea	
5.045	59.46	Rarotonga	Cook Is	1
		Prudente	Brasil	
5.047	59.44	Lome	Togo	100
5.050	59.41	Dar-es-Salaam	Tanzania	20
		Caracas	Venezuela	10
		Jakarta	Indonesia	
5.052	59.38	Jurong	Senegal	
5.054	59.36	San Jose	Costa Rica	
5.055	59.35	Ulan Bator	Mongolian Rep	50
		Cuiba	Brasil	
5.057	59.32	Tirane	Albania	
5.060	59.29	Aden	Yemen	7.5
		Urumchi	China Rep	
5.061	59.28	Islamabad	Pakistan	
		Quito	Ecuador	
5.065	59.23	Petrozavodsk	USSR	
5.075	59.11	Bogota	Colombia	25
		Peking	China Rep	
5.090	58.94	Peking	China Rep	120
5.095	58.88	Sutatenza	Colombia	50
5.110	58.71	V of People of Burma		
5.112	58.69		Pakistan	
5.125	58.54	RFE/R Liberty*		
		Peking	China Rep	
5.135	58.42	Peking	China Rep	
5.139	58.38	Phu Khanh	Vietnam	
5.145	58.31	Peking	China Rep	
5.160	58.14	Vientiane	Laos	
5.163	58.10	Peking	China Rep	
5.189	57.81	Rangoon	Burma	
5.195	57.75	Julich*	Germany (West)	
5.220	57.47	Peking	China Rep	
5.230	57.36		USA*	
5.240	57.25	Fukien Front Stn	China Rep	
5.250	57.14	Peking*	China Rep	
5.255	57.09		USSR*	
5.260	57.03	Alma Ata	USSR	

MHz	*Metres*	*Station*	*Country*	*kW*
		Riyadh*	Saudi Arabia	
5.265	56.98	Fukien Front Stn	China Rep	
5.290	56.71	Krasnoiarsk	USSR	
			USSR*	
5.295	56.66	Peking	China Rep	
		RFE/R Liberty*		
5.320	56.39	Peking	China Rep	
			USSR	
5.339	56.19	Sao Tome	St Thomas Is	
5.345	56.13	R Freedom from S Yemen		
5.390	55.66	Riyadh*	Saudi Arabia	
5.420	55.35	Peking	China Rep	
5.440	55.15	Urumchi	China Rep	
5.455	55.00		USSR*	
5.460	54.95	Tangier*	Morocco	
5.470	54.84		USSR*	
5.504	53.63	Hoang Lien Son	Vietnam	
5.610	53.48	Luanda	Angola	
5.700	52.63		USSR	
5.703	52.60		USA*	
5.745	52.22	Greenville*	USA	
5.770	51.99	Moscow	USSR	
5.790	51.81	RFE/R Liberty*		
5.794	51.78		USSR	
5.815	51.59		USSR*	
5.830	51.46		USSR*	
			Germany (W)*	
5.840	51.37	Peking	China Rep	
5.845	51.33	RFE/R Liberty*		
5.850	51.28	Peking	China Rep	
5.860	51.19	Peking	China Rep	
5.870	51.11	Pyongyang	Korea (N)	
5.872	51.09	London*	UK	
5.876	51.06	Riyadh	Saudi Arabia	
5.880	51.02	Peking	China Rep	
		Jakarta	Indonesia	
5.882	51.00	Jerusalem	Israel	
		Buenos Aires	Argentina	
5.885	50.98	Jakarta	Indonesia	
			USSR	
5.900	50.85	Fukien Front Stn	China Rep	
		Jerusalem	Israel	
		Moskva	USSR	
5.905	50.80	Moskva	USSR	
		Tula	USSR	
5.910	50.71	Moskva*	USSR	50
		Vologda	USSR	
5.914	50.73	Bizam Radio		

MHz	Metres	Station	Country	kW	M	M	S	N
5.915	50.72	Armavir	USSR					
		Kiev	Ukraine					
		Jerusalem	Israel					
		Peking	China Rep					
		Sofia	Bulgaria					
		Alma Ata	USSR					
5.920	50.68		USSR	50				
		Sverdlovsk	USSR	100				
5.925	50.63	Tashkent	USSR	50				
		Wien	Austria					
		Urumchi	China Rep					
5.927	50.61	Lai Chau	Vietnam					
5.930	50.59	Murmansk	USSR	15				
		Tbilisi	USSR					
		Prague	Czechoslovakia	100				
5.935	50.55	Riga	USSR	50				
			USSR					
		Peking	China Rep					
		Lhasa	Tibet					
5.940	50.51	Magadan	USSR					
		Moskva	USSR					
		Petropavlovsk	USSR					
		Sofia	Bulgaria					
5.945	50.46	Minsk	USSR					
		Tashkent	USSR					
		Monte Carlo	Monaco					
		RFE/R Liberty*						
		Wien	Austria					
		Tirane	Albania					
5.950	50.42	Peking	China Rep					
			USSR					
		Leningrad	USSR					
		Managua	Nicaragua					
		V of Iraqi Kurdistan						
5.955	50.38	London	UK	100		_		
		Dixon	USA	100/250	∠	_	_	_
		Bluefields	Nicaragua	0.5	∠	_	_	_
		Llallagua	Bolivia	1	∠	_	_	_
		Pakanbaru	Indonesia	10	∠	_	_	_
		Allouis	France	100	∠	_	_	_
		Pt Limon	Costa Rica	1	∠	_	_	_
		S Paulo	Brasil	7.5	∠	_	_	_
		Lopik	Netherlands	100	∠	_	_	_
		Tinang	Philippines	250	_	_	_	
		Kavalla	Greece	250	∠	_	_	_
		Athinai	Greece	100	/			
		Nauen	Germany (E)	50/500	∠			_
		Manzini	Swaziland	25	_			

MHz	*Metres*	*Station*	*Country*	*kW*	*M*	*M*	*S*	*N*
		Komsomolskamur	USSR	100				
		Diosd	Hungary	100	—			
		Santiago	Chile	1	/		—	—
		Ismaning	Germany (W)	100	/	—	—	
		Biblis	Germany (W)	100		—		—
		Holzkirchen	Germany (W)	10	/	—		
		Lampertheim	Germany (W)	100	∠	—	—	—
		Villavicencio	Colombia	5	∠	—	—	—
		Shepparton	Australia	100			—	—
		Franceville	Gabon	500			—	
5.960	50.34	Antigua	Br W Indies	250	—	—		
		Jammu	India	1	∠	—	—	—
		Delhi	India	100	—	—	—	—
		Godthaab	Greenland	1/10	∠	—	—	—
		Sisoguichi	Mexico	0.3	∠	—	—	—
		Alma Ata	USSR	100	∠	—	—	—
		Armavir	USSR	100		—	—	
		Sverdlovsk	USSR	100	∠			—
		Blagoevechtchen	USSR	100		—	—	
		Vladivostock	USSR	100	∠	—	—	—
		Sackville	Canada	250	∠	—	—	—
		Tirane	Albania					
		S Rosa Copan	Honduras Rep	1	/		—	—
		Warszawa	Poland	1	—			
		Wien	Austria	100	—	—	—	—
		Jaszbereny	Hungary	250		—	—	
		Bogota	Colombia	1	∠	—	—	—
		Julich	Germany (W)	100	/	—	—	—
		Wertachtal	Germany (W)	500	/	—	—	—
		Luanda	Angola	100	/		—	—
		Cyclops	Malta	250	/			
		Peking	China Rep					
5.965	50.29	London	UK	100/500	∠	—	—	—
		Huanuni	Bolivia	10	∠	—	—	—
		P Alegre	Brasil	7.5	∠	—	—	—
		Kajang	Malaysia	100	∠	—	—	—
		Ismaning	Germany (W)	100	∠	—	—	—
		Mt Carlo	Monaco	100	∠	—	—	—
		Tanger	Morocco	35	∠	—	—	—
		Malolos	Philippines	7.5	—	—	—	—
		Moskva	USSR	50	∠	—	—	—
		Armavir	USSR	100		—	—	
		Rhodos	Greece	50	∠	—	—	—
		Diosd	Hungary	100	∠	—	—	—
		S Pedro Sula	Honduras Rep	1	/		—	—
		Benin	Niger	10	—	—		
		Jos	Niger	10	/		—	—
		Rangoon	Burma	50	∠	—	—	—
		Wavre	Belgium	100	/	—	—	—

MHz	Metres	Station	Country	kW	M	M	S	N
		Domingo	Dominican Rep					
5.970	50.25	Bandjarmasin	Indonesia	10	∠	_	_	_
		Bogota	Colombia	1	∠	_	_	_
		Tula	USSR	240	_			
		Alma Ata	USSR	100	∠	_	_	_
		Moskva	USSR	100	∠			_
		Komsomolskamur	USSR	50		_	_	
		Nikolaevskamur	USSR	50	∠			_
		Tchita	USSR	240				_
		Gauhati	India	10	∠	_	_	_
		Aden	Aden	100				
		Arganda	Spain	100	_	_	_	_
		Holzkirchen	Germany (W)	10	_			
		Biblis	Germany (W)	100	∠	_	_	_
		Lisbonne	Portugal	100				_
		Diosd	Hungary	100	_			
		Shepparton	Australia	10		_	_	_
		Schwarzenburg	Switzerland	150			_	_
5.975	50.21	London	UK	100/500	∠	_	_	_
		Florianapolis	Brasil	1	∠	_	_	_
		Kyung San	Korea (S)	10	∠	_	_	_
		Cochabamba	Bolivia	1	∠	_	_	_
		Villarrica	Paraguay	3	∠	_	_	_
		Gwelo	Rhodesia	10/100	_	_		
		Komsomolskamur	USSR	50		_		
		Tachkent	USSR	100	∠			_
		Minsk	Bielorussia	100	∠			_
		Peking	China Rep					
5.980	50.17	Beyrouth	Lebanon	100	_	_	_	_
		Meyerton	South Africa	250/500	∠	_	_	_
		Godthaab	Greenland	1	∠	_	_	_
		Riazan	USSR	240	∠	_	_	_
		Tbilisi	USSR	240	∠		_	_
		Irkutsk	USSR	100	∠			_
		Quetta	Pakistan	10	∠	_	_	_
		Goderich	Sierra Leone	10	∠	_	_	_
		Waterloo	Sierra Leone	250	∠	_	_	_
		Redwood City	USA	30/250	_		_	_
		Linares	Mexico	0.5	∠	_	_	_
		Taipei	China Nat					
		Diosd	Hungary	100	∠	_	_	_
		Jaszbereny	Hungary	250	∠	_	_	_
		Szekesfehervar	Hungary	20				
		Medellin	Colombia	10	∠	_	_	_
		Cyclops	Malta	250	/			
		RFE						
		Sparendum	Guyana	2	/			
		Jakarta	Indonesia					

MHz	*Metres*	*Station*	*Country*	*kW*	*M*	*M*	*S*	*N*
5.985	50.13	Dar es Salaam	Tanzania	100				
		Lomas Mirador	Argentina	1	∠	_	_	_
		Lisbonne	Portugal	100/250	_			_
		Holzkirchen	Germany (W)	10			_	
		Lampertheim	Germany (W)	100	∠		_	
		Biblis	Germany (W)	100	∠	_	_	_
		Mexico	Mexico	10	∠	_	_	_
		Khabarovsk	USSR	120	_			_
		Scituate	USA	50/100	∠	_	_	_
		Rabaul	Papua New Guinea	10	∠	_	_	_
		Kavalla	Greece	250			_	_
		Rangoon	Burma	50	∠	_	_	_
		Allouis	France	100	∠	_	_	_
		Tunja	Colombia	10	∠	_	_	_
5.990	50.08	London	UK	100/250	∠	_	_	_
		Limassol	Cyprus	10	∠			_
		Bhopal	India	10	∠	_	_	_
		Ejura	Ghana	10	∠	_	_	_
		Serpukhov	USSR	100	∠			_
		Nikolaevskamur	USSR	50	∠	_	_	_
		Roma	Italy	60/100	∠	_	_	_
		Rio de Janeiro	Brasil	7.5/10	∠	_	_	_
		Menado	Indonesia	10	∠	_	_	_
		Bucuresti	Roumania	120	∠	_	_	_
		Tokyo Yamata	Japan	100	_			
		Sackville	Canada	50/250	_	_	_	_
		Allouis	France	100	_	_	_	_
		Cyclops	Malta	250		_		
		Arganda	Spain	100	_	_	_	_
		Gedja	Ethiopia	100	/	_	_	_
		Peking	China Rep					
		V. of Kurdistan						
5.995	50.04	London	UK	100	∠	_	_	_
		Bamako	Malawi	50		_	_	
		Greenville	USA	250/500	∠	_	_	_
		Mbandaka	Zaire	10	∠	_	_	_
		Panama	Panama	4	∠	_	_	_
		Ft de France	Martinique	4	∠	_	_	_
		Warszawa	Poland	8/10	_	_	_	_
		Limbe	Malawi	20	∠	_	_	_
		Sucre	Bolivia	1	_	_	_	_
		Lyndhurst	Australia	10	∠	_	_	_
		Julich	Germany (W)	100	∠	_	_	_
		Poro	Philippines	35	_			
		S M Galeria	Vatican	100	∠	_	_	_
		P J Caballero	Paraguay	2	∠	_	_	_
		S Pedro Sula	Honduras Rep	1	/		_	_
		Tula	USSR	50		_	_	

MHz	Metres	Station	Country	kW	M	M	S	N
		Allouis	France	100	∠	_	_	_
		Pereira	Colombia	1	∠	_	_	_
6.000	50.00	Montserrat	Br W Indies	15	_	_	_	_
		Singapore	Singapore	10/50	∠	_	_	_
		Belo Horizonte	Brasil	1/25	∠	_	_	_
		Montevideo	Uruguay	5	∠	_	_	_
		Innsbruck	Austria	10	_	_	_	_
		Tchita	USSR	100	∠	_	_	_
		Moskva	USSR	240	∠	_	_	_
		Kabul	Afghanistan	50/100	∠	_	_	_
		Diriyya	Saudi Arabia	50	∠	_	_	_
		Jaszbereny	Hungary	250	∠	_	_	_
		Szekesfehervar	Hungary	20	_	_	_	
		Cyclops	Malta	250	∠	_	_	_
		Arganda	Spain	100	_	_	_	_
6.005	49.96	London	UK	250	/			
		Ascension	Ascension	126/250	∠	_	_	_
		Buea	Cameroon	4	∠	_	_	_
		Ekala	Ceylon	10	∠	_	_	_
		La Paz	Bolivia	10	∠	_	_	_
		Marhubi	Zanzibar	3.5	_	_	_	_
		Montreal	Canada	0.5	∠	_	_	_
		Ismaning	Germany (W)	20	∠	_	_	_
		Berlin (RIAS)	Germany (W)	20	_	_	_	_
		S Jose	Costa Rica	1	∠	_	_	_
		Voronej	USSR	100	_	_	_	_
		Matsuyama	Japan	0.6	∠	_	_	_
		Carnarvon	Australia	250	∠	_	_	_
		Fredrikstad (Relay of Yemen Arab Rep)	Norway	100				_
6.010	49.92	London	UK	100/250	∠	_	_	_
		Kranji	Singapore	250	_	_	_	
		Limassol	Cyprus	100	∠	_	_	_
		Wavre	Belgium	50/100	∠	_	_	_
		Calcutta	India	10	_	_	_	_
		Mexico	Mexico	5	∠	_	_	_
		Montevideo	Uruguay	10	∠	_	_	_
		Moskva	USSR	240/500	∠	_	_	_
		Krasnoiarsk	USSR	240	∠	_	_	_
		Delano	USA	100				_
		Allouis	France	100/500	∠	_	_	_
		Managua	Nicaragua	0.1	∠	_	_	_
		Sines	Portugal	250				_
		Roma	Italy	60/100			_	_
		K Wusterhausen	Germany (E)	100	∠			_
		Nauen	Germany (E)	100		_	_	

MHz	Metres	Station	Country	kW	M	M	S	N
		Islamabad	Pakistan	100	/		_	_
		Holzkirchen	Germany (W)	10	/			
		Biblis	Germany (W)					
		Tinang	Philippines	250	_	_	_	
		Meyerton	South Africa	100	_			
		Pereira	Colombia	10	∠	_	_	_
		Quito	Ecuador	100	_			
6.015	49.88	London	UK	250	/			_
		Abidjan	Ivory Coast	100	∠	_	_	_
		Asuncion	Paraguay	1	∠	_	_	_
		Animas	Bolivia	5	∠	_	_	_
		Recife	Brasil	10	∠	_	_	_
		Rhodos	Greece	50	∠	_	_	_
		Dixon	USA	250	_		_	_
		Orcha	Bielorussia	100				_
		Tanger	Morocco	50/100	∠	_		_
		Velkekostolany	Czechoslovakia	120	∠	_	_	_
		Sines	Portugal	250	∠		_	_
		Delhi	India	100	∠	_	_	_
		Gedja	Ethiopia	100	_	_	_	
		Wien	Austria	100	_	_	_	_
		Fredrikstad	Norway	100/250	∠	_	_	_
		Tumaco	Colombia	1	∠	_	_	_
		S M Galeria	Vatican	100	∠	_	_	_
		Wavre	Belgium	250	∠		_	_
6.020	49.85	Limassol	Cyprus	100			_	
		Lopik	Netherlands	10/100	∠	_	_	_
		Khabarovsk	USSR	50	∠	_	_	_
		Kiev	Ukraine	50	∠	_	_	_
		Simla	India	2.5	∠	_	_	_
		Delhi	India	100	∠	_	_	_
		Vera Cruz	Mexico	5	∠	_	_	_
		Greenville	USA	250/500	∠	_	_	_
		Bonaire Noord	Neth Antilles	300	∠	_	_	_
		Talata Volon	Malagasy Rep	300	∠	_	_	_
		Gwelo	Rhodesia	20/100	_	_		
		Tegucigalpa	Honduras Rep	0.5	/		_	_
		Bogota	Colombia	10	∠	_	_	_
6.025	49.79	Asuncion	Paraguay	10/100	∠	_	_	_
		Kajang	Malaysia	50	∠	_	_	_
		S Gabriel	Portugal	100	∠	_	_	_
		S Pedromacoris	Dominican Rep	0.1	∠	_	_	_
		Enugu	Niger	10	∠	_	_	_
		Jaszbereny	Hungary	250	∠	_	_	_
		Diosd	Hungary	100	/			_
		La Paz	Bolivia	10	∠	_	_	_
		Tachkent	USSR	50	_			_
		Cyclops	Malta	250	∠	_	_	_
		Beira	Mozambique	10	∠			

MHz	*Metres*	*Station*	*Country*	*kW*	*M*	*M*	*S*	*N*
		National Voice of Iran						
6.030	49.75	London	UK	100/250	∠	_	_	_
		Masirah	Oman	200	/			
		Greenville	USA	50	∠	_	_	_
		Cincinnati	USA	175	_	_	_	_
		Tokyo Yamata	Japan	20/50	∠	_	_	_
		Bogota	Colombia	25	∠	_	_	_
		Calgary	Canada	0.1	∠	_	_	_
		Moskva	USSR	100	∠	_	_	_
		Komsomolskamur	USSR	240				_
		Simferopol	Ukraine	240	∠			_
		Muehlacker	Germany (W)	20	∠	_	_	_
		Bocaue	Philippines	25	∠	_	_	_
		Kampala	Uganda	250				_
		Kimjae	Korea (S)	100			_	_
		Islamabad	Pakistan	10/100	_	_	_	_
		Franceville	Gabon Rep	500			_	
6.035	49.71	Warszawa	Poland	60	∠	_	_	_
		Bombay	India	100	_	_	_	_
		La Paz	Bolivia	10	∠	_	_	_
		Careysburg	Liberia	250	∠	_	_	_
		Mt Carlo	Monaco	100	∠	_	_	_
		Montevideo	Uruguay	1	∠	_	_	_
		Vladivostock	USSR	100	_	_	_	_
		Rio de Janeiro	Brasil	10	∠	_	_	_
		Carnarvon	Australia	250	∠	_	_	_
		Tegucigalpa	Honduras Rep	0.5	/		_	_
6.040	49.67	London	UK	250	∠	_	_	_
		Antigua	Br W Indies	250	∠	_	_	_
		Dubai	United Arab Emirates	10	∠	_	_	_
		Ibague	Colombia	10	∠	_	_	_
		S Jose	Costa Rica	1	∠	_	_	_
		Kazan	USSR	100	_			
		Simferopol	Ukraine	100	∠	_	_	_
		Alotau	Papua New Guinea	10	∠	_	_	_
		Nauen	Germany (E)	500	∠	_	_	_
		Cincinnati	USA	175	_		_	_
		Allouis	France	100	∠	_	_	_
		Jaszbereny	Hungary	250	∠	_	_	_
		Szekesfehervar	Hungary	20	∠	_	_	_
		Ismaning	Germany (W)	100	/	_	_	
		Julich	Germany (W)	100		_	_	_
6.045	49.63	Athinai	Greece	5	∠	_	_	_
		Curityba	Brasil	7.5	∠	_	_	_
		David	Panama	1	∠	_	_	_
		Careysburg	Liberia	50/250	∠	_		_

MHz	*Metres*	*Station*	*Country*	*kW*	*M*	*M*	*S*	*N*
		Jakarta	Indonesia	100	∠	_	_	_
		Montevideo	Uruguay	2.5	∠	_	_	_
		Moskva	USSR	240	∠	_	_	_
		Novosibirsk	USSR	100	_			_
		Tchita	USSR	240/500	_			_
		S Luis Potosi	Mexico	0.3	∠	_	_	_
		Lopik	Netherlands	10/100	∠	_	_	_
		Lyndhurst	Australia	10	∠	_	_	_
		Arganda	Spain	100	∠	_	_	_
		Las Mesas	Canary Is	50	_	_	_	
		Schwarzenburg	Switzerland	150	∠		_	_
		Tungsheng	Mongolian Rep					
		Tambacounda	Senegal	4	/			
		Sackville	Canada	50	/			
6.050	49.59	London	UK	100/250	∠	_	_	_
		Limassol	Cyprus	100	∠	_	_	
		Kranji	Singapore	250	∠	_	_	_
		Sibu	Malaysia	10			_	
		Ibadan	Niger	10	∠	_	_	_
		Delhi	India	20	∠	_	_	_
		Irkutsk	USSR	50	∠	_	_	_
		Maputo	Mozambique	10/20	∠	_	_	_
		Quito	Ecuador	100	∠	_	_	_
		Roma	Italy	100	∠			_
		S Jose	Costa Rica	1	∠	_	_	_
		Noblejas	Spain	350	_	_	_	_
		Tegucigalpa	Honduras Rep	0.5	/		_	_
		Sanaa	Yemen	50				
6.055	49.55	Ascension	Ascension	125/250		_		
		Antigua	Br W Indies	250	_		_	
		Greenville	USA	50	∠	_	_	_
		Dixon	USA	250	/	_		_
		Kigali	Rwanda	50	∠	_	_	_
		Sulaibiyah	Kuwait	250	∠	_	_	_
		Melo	Uruguay	5	∠	_	_	_
		Erevan	USSR	100	∠			_
		Tchita	USSR	50				_
		Tallin	USSR	240			_	
		Starobelsk	Ukraine	100	∠	_	_	_
		La Paz	Bolivia	100	∠	_	_	_
		Velkekostolany	Czechoslovakia	120	∠	_	_	_
		Praha	Czechoslovakia	120/400	∠	_	_	_
		S Paulo	Brasil	7.5	∠	_	_	_
		Tokyo Nagara	Japan	50	∠	_	_	_
		Cali	Colombia	5	∠	_	_	_
6.060	49.50	London	UK	250	_	_	_	_
		Gral Pacheo	Argentina	50	∠	_	_	_

MHz	Metres	Station	Country	kW	M	M	S	N
		Caltanissetta	Italy	25	∠	_	_	_
		Roma	Italy	50	∠	_	_	_
		Alma Ata	USSR	100		_	_	
		Tbilisi	USSR	100	∠			_
		Komsomolskamur	USSR	240	∠	_	_	_
		Krasnoiarsk	USSR	100	_			
		Habana	Cuba	10/50	∠	_	_	_
		Lusaka	Zambia	20	∠	_	_	_
		Santiago	Dominican Rep	0.5	∠	_	_	_
		Kavalla	Greece	250	∠	_	_	_
		Mira	Malaysia	10			_	
		Ismaning	Germany (W)	100	∠	_	_	_
		Tanger	Morocco	35/100	_	_	_	_
		Tegucigalpa	Honduras Rep	1	/		_	_
		Jaszbereny	Hungary	250		_	_	
		Shepparton	Australia	100	/			_
6.065	49.46	London	UK	250	_		_	
		Kranji	Singapore	250	/		_	_
		Rio de Janeiro	Brasil	7.5	∠	_	_	_
		Brasilia	Brasil	10	_	_	_	
		Hoerby	Sweden	350	∠	_	_	_
		Karlsborg	Sweden	350	∠	_	_	_
		Kazan	USSR	100	∠			_
		Armavir	USSR	100	∠	_	_	_
		Serpukhov	USSR	100		_		
		Texmelucan	Mexico	1	∠	_	_	_
		Kohima	India	2	_	_	_	_
		Bogota	Colombia	2	∠	_	_	_
		Wertachtal	Germany (W)	500	∠	_	_	_
		Julich	Germany (W)	100	∠	_	_	_
		Arganda	Spain	100	_	_	_	_
		Noblejas	Spain	100	/			
		Sackville	Canada	50/250	∠	_	_	_
		Wavre	Belgium	100	_			
6.070	49.42	London	UK	250	/			_
		Limassol	Cyprus	100	∠		_	_
		Tema	Ghana	100	∠	_	_	_
		Bangkok	Thailand	10	∠	_	_	_
		Sofia	Bulgaria	50/100	∠	_	_	_
		Jajapura	Indonesia	0.5/20	∠	_	_	_
		Toronto	Canada	1	∠	_	_	_
		Oruro	Bolivia	5	∠	_	_	_
		Khabarovsk	USSR	100/150		_	_	_
		Tula	USSR	50		_		_
		Leipzig	Germany (E)	10	∠			_
		K Wusterhausen	Germany (E)	100		_	_	
		Nauen	Germany (E)	500	_			_
		Rhodos	Greece	50	∠	_	_	_
		Manzini	Swaziland	25	_	_		_

MHz	*Metres*	*Station*	*Country*	*kW*	*M*	*M*	*S*	*N*
		Quito	Ecuador	100	/		_	_
		Biblis	Germany (W)	100	_		_	
		Lisbonne	Portugal	100				_
		Arganda	Spain	100	/		_	_
6.075	49.38	Ekala	Ceylon	10	∠	_	_	_
		Julich	Germany (W)	100	∠	_	_	_
		Wertachtal	Germany (W)	500	∠	_	_	_
		Roma	Italy	60	∠			_
		Montevideo	Uruguay	2.5	∠	_	_	_
		Santiago	Dominican Rep	0.5	∠	_	_	_
		Bogota	Colombia	10/50	∠	_	_	_
		Sverdlovsk	USSR	50				_
		Tula	USSR	50			_	
		Volgograd	USSR	100		_	_	
		Quito	Ecuador	100	_	_	_	_
		Noblejas	Spain	350	_			
		Delhi	India	20	_	_	_	_
		Monrovia	Liberia	50	∠	_	_	_
		S Barbara	Honduras Rep	1	/		_	_
6.080	49.34	London	UK	250	∠	_	_	_
		Kranji	Singapore	250	∠	_	_	_
		K Wusterhausen	Germany (E)	100	∠			_
		Nauen	Germany (E)	50	∠	_	_	_
		Frunze	USSR	240	/			_
		Moskva	USSR	100	∠			_
		Komsomolskamur	USSR	50	∠	_	_	_
		Vancouver	Canada	0.1	∠	_	_	_
		Tanger	Morocco	35/100				_
		Riyadh	Saudi Arabia	350	∠	_	_	_
		Bangkok	Thailand	2	∠	_	_	_
		Daru	Papua New Guinea	10	∠	_	_	_
		Wavre	Belgium	100	/	_	_	_
		Greenville	USA	50/500	∠	_	_	_
		Alger	Algeria	100	∠	_	_	_
		Tambacounda	Senegal	4	_	_	_	_
		Peshawar	Pakistan	10	∠	_	_	_
		Hailar	Mongolian Rep					
		Diosd	Hungary	100	_			
		Jaszbereny	Hungary	250	_	_	_	
		Catavi	Bolivia					
		Shepparton	Australia	10				_
6.085	49.30	London	UK	100	_			_
		Limassol	Cyprus	100	_		_	_
		Antigua	Br W Indies	250	∠	_	_	_
		Lopik	Netherlands	100	∠	_	_	_
		Kisangani	Zaire	10	∠	_	_	_
		Delhi	India	100	_	_	_	_
		Madras	India	100	_	_	_	_

MHz	Metres	Station	Country	kW	M	M	S	N
		Ismaning	Germany (E)	10	∠	_	_	_
		Wertachtal	Germany (E)	500	∠	_	_	_
		Recife	Brasil	15	∠	_	_	_
		Tbilisi	USSR	100	∠			_
		Tallin	USSR	50	∠	_	_	_
		Kiev	Ukraine	50		_	_	
		Sackville	Canada	125/250	∠	_	_	_
		Noblejas	Spain	350	∠	_	_	_
		Riyadh	Saudi Arabia	350	∠	_	_	_
		Tegucigalpa	Honduras Rep	1	/		_	_
		Sofia	Bulgaria	50/100	∠	_	_	_
		Fredrikstad	Norway	100			_	
6.090	49.26	Hurlingham	Argentina	30	∠	_	_	_
		Cd Mante	Mexico	1	∠	_	_	_
		Simferopol	Ukraine	240		_	_	
		Orenburg	USSR	50	∠	_	_	_
		Nikolaevskamur	USSR	100	∠			_
		Irkutsk	USSR	50	∠	_	_	_
		Junglinster	Luxembourg	500	∠	_	_	_
		S Domingo	Dominican Rep	7.5	∠	_	_	_
		Jaji	Niger	10	∠	_	_	_
		Sydney	Australia	2	∠	_	_	_
		Beira	Mozambique	25	∠	_	_	_
		Kavalla	Greece	250	∠	_	_	_
		Las Mesas	Canary Is	50	_	_	_	_
		V of Kampuchea People						
6.095	49.22	Ismaning	Germany (E)	100	_		_	_
		Warszawa	Poland	30	∠	_	_	_
		Salman Pack	Iraq	50	_	_	_	_
		S Paulo	Brasil	25	∠	_	_	_
		Mogadiscio	Somalia	50	∠	_	_	_
		Espinal	Colombia	10	∠	_	_	_
		Kavalla	Greece	250			_	_
		Dixon	USA	100	_	_	_	_
		Quito	Ecuador	100	∠	_	_	_
		Tanger	Morocco	35/50	∠	_	_	_
		Serpukhov	USSR	100	∠			_
		Tegucigalpa	Honduras Rep	1	/		_	_
		Phnom Penh	Cambodia					
6.100	49.18	Belgrade	Yugoslavia	100	∠	_	_	_
		Wertachtal	Germany (W)	500	∠	_	_	_
		Kajang	Malaysia	100	∠	_	_	_
		Nueva Segovia	Nicaragua	0.4	_			
		Ocotal	Nicaragua	0.4	/	_	_	_
		Kursk	USSR	50	_	_	_	
		Kaunas	USSR	50	∠	_	_	_
		Petropavlo Kam	USSR	240	∠	_	_	_
		Vinnitsa	Ukraine	100				_

MHz	*Metres*	*Station*	*Country*	*kW*	*M*	*M*	*S*	*N*
		Maiduguri	Niger	10	∠	_	_	_
		Kurseong	India	20	_	_	_	_
		Hoerby	Sweden	350	_			
		Cyclops	Malta	250	∠	_	_	_
		Arganda	Spain	100	∠	_	_	_
		Tanger	Morocco	50	_	_		
		Rabat	Morocco	50	/			
		Tripoli	Libya					
		Kathmandu	Nepal	100	∠		_	_
		R of the Patriots						
		Caracas	Venezuela					
6.105	49.14	Fortaleza	Brasil	5	∠	_	_	_
		Medellin	Colombia	10	∠	_	_	_
		Merida	Mexico	1	∠	_	_	_
		Armavir	USSR	100	_			_
		Kalinin	USSR	120	∠			_
		Duchanbe	USSR	100	∠			_
		Delhi	India	100	_	_	_	_
		Dar es Salaam	Tanzania	50	_	_	_	_
		Cite Vatican	Vatican	80	∠	_	_	_
		Wellington	New Zealand	7.5	∠	_	_	_
		Diosd	Hungary	100	/			_
		Jaszbereny	Hungary	100	_	_	_	_
		Szekesfehervar	Hungary	20	/			_
		Biblis	Germany (W)	100	∠	_	_	_
		Holzkirchen	Germany (W)	10	_	_	_	
		Lampertheim	Germany (W)	100	∠		_	_
		Lisbonne	Portugal	250	_		_	
		Sackville	Canada	50/250			_	
6.110	49.10	London	UK	250	∠	_	_	_
		Ascuncion	Paraguay	3	∠	_	_	_
		Baku	USSR	50	∠	_	_	_
		Kazan	USSR	50				_
		S Domingo	Dominican Rep	1	∠	_	_	_
		Szekesfehervar	Hungary	20	∠			_
		Diosd	Hungary	100	/			_
		Jaszbereny	Hungary	250		_	_	
		Srinagar	India	7.5	_	_	_	_
		Tinang	Philippines	50/250	∠	_	_	_
		Ismaning	Germany (W)	100	∠	_	_	_
		Comayaguela	Honduras Rep	0.5	/		_	_
		Tanger	Morocco					
6.115	49.06	London	UK	100	_			_
		K Wusterhausen	Germany (E)	5	∠	_	_	_
		Hermosillo	Mexico	1	∠	_	_	_
		Khabarovsk	USSR	50/240	∠	_	_	_
		Lvov	Ukraine	240	∠			_
		Montevideo	Uruguay	5	∠	_	_	_
		Maputo	Mozambique	25	∠	_	_	_

MHz	*Metres*	*Station*	*Country*	*kW*	*M*	*M*	*S*	*N*
		Biblis	Germany (W)	100	∠	_	_	_
		Holzkirchen	Germany (W)	10	∠	_	_	_
		Lisbonne	Portugal	250			_	_
		Sines	Portugal	250	∠	_	_	_
		Brazzaville	Congo	50	∠	_	_	_
		Rio de Janeiro	Brasil	10	∠	_	_	_
		Karachi	Pakistan	50	∠	_	_	_
		Tokyo Nagara	Japan	10	∠	_	_	_
		Villavicencio	Colombia	1	∠	_	_	_
		Szekesfehervar	Hungary	20				
6.120	49.02	London	UK	100/250	_	_	_	_
		Limassol	Cyprus	7.5	∠	_	_	_
		Hyderabad	India	10	_	_	_	_
		Delhi	India	10/100	∠	_	_	_
		S Fernando	Argentina	10	∠	_	_	_
		Julich	Germany (W)	100	∠	_	_	_
		Wertachtal	Germany (W)	500	∠	_	_	_
		Bluefields	Nicaragua	1	∠	_	_	_
		Bocaue	Philippines	50	∠	_	_	_
		Pori	Finland	15/250	∠	_	_	_
		Tapachula	Mexico	0.5	∠	_	_	_
		Armavir	USSR	240	∠	_	_	_
		Moskva	USSR	100	_			_
		Novosibirsk	USSR	100	_			_
		Duchanbe	USSR	120	∠			_
		Santiago	Dominican Rep	0.5	∠	_	_	_
		Noblejas	Spain	350	∠	_	_	_
		Surabaja	Indonesia	10	∠	_	_	_
		Sackville	Canada	50/250	∠	_	_	_
		Bata	Guinea	50	∠	_	_	_
		Shepparton	Australia	50	/			
6.125	48.98	London	UK	100/500	∠	_	_	_
		Limassol	Cyprus					
		Greenville	USA	250/500	∠	_	_	_
		Dixon	USA	250	_		_	_
		Cincinnati	USA	250	∠	_	_	_
		Kananga	Zaire	10	∠	_	_	_
		Montevideo	Uruguay	10	∠	_	_	_
		La Paz	Bolivia	1	∠	_	_	_
		S Paulo	Brasil	10	∠	_	_	_
		Achkhabad	USSR	100	∠	_	_	_
		Lvov	Ukraine	240				_
		Krasnoiarsk	USSR	100	∠	_	_	_
		Bogota	Colombia	1	∠	_	_	_
		Jaszbereny	Hungary	250		_	_	
		S Pedro Sula	Honduras Rep	1	/			_
6.130	48.94	London	UK	100	∠	_	_	_
		Limassol	Cyprus	100		_	_	
		Ekala	Ceylon	10	∠	_	_	_

MHz	Metres	Station	Country	kW	M	M	S	N
		Julich	Germany (W)	100	∠	_	_	_
		Wertachtal	Germany (W)	500	∠	_	_	_
		Halifax	Canada	0.5	∠	_	_	_
		Armavir	USSR	100	/			
		Moskva	USSR	240	_	_	_	_
		Novosibirsk	USSR	100	∠	_	_	_
		Simferopol	Ukraine	100	∠			_
		Greenville	USA	250	∠	_	_	_
		S Domingo	Dominican Rep	0.3	∠	_	_	_
		Tema	Ghana	100	∠	_	_	_
		Kumamoto	Japan	1	∠	_	_	_
		Vientiane	Laos	10	∠	_	_	_
		Karlsborg	Sweden	350	_			
		Quito	Ecuador	100	∠	_	_	_
		Rawalpindi	Pakistan	10	/	_	_	_
		Gauhati	India	10	_	_	_	_
		Tanger	Morocco	100				
6.135	48.90	Montserrat	Br W Indies	15	_	_	_	_
		Baku	USSR	100	∠	_	_	_
		Alma Ata	USSR	50	∠	_	_	
		Khabarovsk	USSR	100		_	_	
		Biblis	Germany (W)	100	/			
		Holzkirchen	Germany (W)	10/50	_		_	_
		Lisbonne	Portugal	250	_			_
		Warszawa	Poland	40/100	∠	_	_	_
		Papeete	Tahiti	4/20	∠	_	_	_
		P Alegre	Brasil	7.5	∠	_	_	_
		Suwon	Korea (S)	10	∠	_	_	_
		Samarinda	Indonesia	7.5	∠	_	_	_
		S Cruz	Bolivia	1	∠	_	_	_
		Tananarive	Malagasy Rep	30	/	_	_	_
		Schwarzenburg	Switzerland	150	∠	_	_	_
		Santiago	Chile	100				_
		Concepcion	Chile	10	∠	_	_	
		La Ceiba	Honduras Rep	0.5	/		_	_
		Greenville	USA	50	∠	_	_	_
6.140	48.86	London	UK	100/500	∠	_	_	_
		Masirah	Oman	100/200				_
		Ascension	Ascension	250	_	_		
		Limassol	Cyprus	100	_	_	_	_
		Chihuahua	Mexico	0.3	∠	_	_	_
		Voronej	USSR	100	∠	_	_	_
		Kiev	Ukraine	100	∠	_	_	_
		Arganda	Spain	100	∠	_	_	_
		Montevideo	Uruguay	10	∠	_	_	_
		Perth	Australia	10	∠	_	_	_
		Bujumbura	Burundi	10/25	∠	_	_	_
		Wewak	Papua New Guinea	10	∠	_	_	_

MHz	Metres	Station	Country	kW	M	M	S	N
		Kavalla	Greece	250	∠	_	_	_
		Athinai	Greece	100	∠	_	_	_
		Ranchi	India	2	∠	_	_	_
		Delhi	India	100	_	_	_	_
		Sackville	Canada	50/250	∠	_	_	_
		Bogota	Colombia	5	∠	_	_	_
6.145	48.82	Montserrat	Br W Indies	15/50	_	_	_	_
		Delhi	India	20/100	∠	_	_	_
		Julich	Germany (W)	100	∠	_	_	_
		Wertachtal	Germany (W)	500	∠	_	_	_
		Moskva	USSR	100	∠	_	_	_
		Khabrovsk	USSR	100	∠			_
		Starobelsk	Ukraine	240	∠			_
		Tlaxiaco	Mexico	0.3	∠	_	_	_
		Delano	USA	250	_		_	_
		Allouis	France	500	∠	_	_	_
		Calabar	Niger	10	∠	_	_	_
		Alger	Algeria	50	∠	_	_	_
		Tarija	Bolivia	1	∠	_	_	_
		Juticalpa	Hungary	0.5	/		_	_
		Dacca	Bangladesh	7.5	_	_	_	
		Popayan	Colombia	1	∠	_	_	_
6.150	48.78	London	UK	100/500	∠	_	_	_
		Limassol	Cyprus	100	_		_	
		Belgrade	Yugoslavia	10/50	∠	_	_	_
		Bucuresti	Roumania	120	∠	_	_	_
		Ismaning	Germany (W)	100	∠	_	_	_
		Lyndhurst	Australia	10	∠	_	_	_
		S Jose	Costa Rica	1	∠	_	_	_
		Omdurman	Sudan	20	_	_		_
		Kavalla	Greece	250	∠	_	_	_
		Ekala	Ceylon	10	∠	_	_	_
		Male	Maldives	2.7	_	_	_	_
		Serpukhov	Ukraine	50	/			_
		Komsomolskamur	USSR	240	_			_
		Kazan	USSR	100		_		
		Delhi	India	20	_	_	_	_
		Santiago	Chile	5/100	∠			_
		Sackville	Canada	50		_		
		Neiva	Colombia	1	∠	_	_	_
6.155	48.74	Limassol	Cyprus	100	/	_		_
		Salman Pack	Iraq	100	_	_	_	_
		La Paz	Bolivia	1	∠	_	_	_
		S Gabriel	Portugal	50	∠	_	_	_
		Krasnoiarsk	USSR	50	∠			_
		Nikolaevskamur	USSR		∠	_	_	_
		Montevideo	Uruguay	10	∠	_	_	_
		Singapore	Singapore	50	∠	_	_	_
		Tokyo	Japan	10	∠	_	_	_

MHz	*Metres*	*Station*	*Country*	*kW*	*M*	*M*	*S*	*N*
		Wien	Austria	100	L	_	_	_
		Warszawa	Poland	100	/	_	_	_
		Bucuresti	Roumania	18/120	L	_	_	_
		Conakry	Guinea	18/100	L	_	_	_
		Scituate	USA	50/100	_	_		_
		Cincinnati	USA	175	_	_	_	
		Salvador	Brasil	10	L	_	_	_
		Dacca	Bangladesh	7.5	/			_
6.160	48.70	London	UK	250/500	L	_	_	_
		Delhi	India	100	L	_	_	_
		Sofia	Bulgaria	50/120	L	_	_	_
		Bogota	Colombia	10	L	_	_	_
		St Johns	Canada	0.3	L	_	_	_
		Vancouver	Canada	0.5	L	_	_	_
		Malargue	Argentina	0.3/3	L	_	_	_
		Wavre	Belgium	50/100	_			
		Moskva	USSR	100	L	_	_	_
		Alger	Algeria	50	L	_	_	_
		Kigali	Rwanda	250	L	_	_	_
		Jaszbereny	Hungary	250	_	_	_	
		Arganda	Spain	100	/		_	_
6.165	48.66	Lenk	Switzerland	250	L	_	_	_
		Vladivostock	USSR	100	/	_	_	_
		Kiev	Ukraine	100	L			_
		Lusaka	Zambia	20	L	_	_	_
		Mexico	Mexico	10	L	_	_	_
		S Paulo	Brasil	7.5	L	_	_	_
		Bonaire Noord	Neth Antilles	300	L	_	_	_
		Tegucigalpa	Honduras Rep	0.5/1	/		_	_
		Ho Chi Minh City	Vietnam					
		Jaszbereny	Hungary	250	_	_	_	
		N Djamena	Chad	100				_
6.170	48.62	London	UK	100	_			_
		Holzkirchen	Germany (W)	10	_			_
		Biblis	Germany (W)	100	/		_	_
		Lampertheim	Germany (W)	20/ 100		_	_	_
		Cayenne	Guyana Fr	4	L	_	_	_
		Lucknow	India	10	L	_	_	_
		Marulas	Philippines	10	L	_	_	_
		Montevideo	Uruguay	1	L	_	_	_
		Jigulevsk	USSR	240	L			_
		Armavir	USSR	100	L			_
		Starobelsk	Ukraine	240	L			_
		Tanger	Morocco	50/100	_	_		_
		Tananarive	Malagasy Rep	30	/	_	_	_
		Kavalla	Greece	250	_	_	_	_
		Abu Zaabal	Egypt	60/100			_	
		Caracas	Venezuela	10	L	_	_	_

MHz	Metres	Station	Country	kW	M	M	S	N
		Florencia	Colombia	2.5	/_	_	_	_
		Kimjae	Korea (S)	100	/_	_	_	_
6.175	48.58	Antigua	Br W Indies	125/ 250	/_	_	_	_
		Belo Horizonte	Brasil	10	/	_	_	_
		Kajang	Malaysia	100	/_	_	_	_
		Khabarovsk	USSR	100		_	_	
		Kazan	USSR	120	/_	_		
		Krasnoiarsk	USSR	100				_
		Moskva	USSR	100		_	_	_
		Vinnitsa	Ukraine	240/500	/_			_
		Allouis	France	100	/_	_	_	_
		Hiroshima	Japan	1/5	/_	_	_	_
		S Cruz	Bolivia	5	/_	_	_	_
		Kaduna	Niger	10/20	/_	_	_	_
		Sackville	Canada	50/250	_	_	_	
		Seeb	Oman	50	/_	_	_	_
			China Rep					
		Luanda	Angola	100	/		_	_
		Koebenhaven	Denmark	50			_	
		Faro Caribe	Costa Rica	2.5	/			_
6.180	48.54	London	UK	250	/_	_	_	_
		Limassol	Cyprus	100	/_	_	_	_
		Mendoza	Argentina	10	/_	_	_	_
		Alma Ata	USSR	100	/	_	_	_
		Tula	USSR	100	/_	_	_	_
		Careysburg	Liberia	50	/_	_	_	_
		Bogota	Colombia	25	/	_	_	_
		Fredrikstad	Norway	100	/_		_	
		Bucuresti	Roumania	18/250	/_	_	_	_
		Ziguinchor	Senegal	4	_	_	_	_
		Tambacounda	Senegal	4	/			
		Tanger	Morocco	100	_			
		Guatemala City	Guatemala					
		Islamabad	Pakistan	100	/	_	_	_
		Dacca	Bangladesh	250	/			_
		Tirana	Albania					
6.185	48.50	London	UK	250	_			
		Gedja	Ethiopia	100	_			
		Ekala	Ceylon	10	/_	_	_	_
		Julich	Germany (W)	100	/_	_	_	_
		Wertachtal	Germany (W)	500	/_	_	_	_
		La Paz	Bolivia	1	/_	_	_	_
		Tripoli	Libya	100	/_	_	_	
		Manokwari	Indonesia	1/10	/_	_	_	_
		Mexico	Mexico	1	/_	_	_	_
		S Paulo	Brasil	10	/_	_	_	_
		Delano	USA	200/250	_		_	_
		Dixon	USA	200				_

MHz	Metres	Station	Country	kW	M	M	S	N
		Riazan	USSR	240	∟	_	_	_
		Novosibirsk	USSR	240/500	_			_
		Petropavlo Kam	USSR	240	∟			_
		Poro	Philippines	50	_	_		
		Tirana	Albania					
		S Pedro Sula	Honduras Rep	0.5	/		_	_
		Mahe	Seychelles	100	∟	_	_	_
		Ankara	Turkey	250	/			_
		Athinai	Greece					
6.190	48.47	S M Galeria	Vatican	100	∟	_	_	_
		Cite Vatican	Vatican	50/80	∟	_	_	_
		Bremen	Germany (W)	10	∟	_	_	_
		Bucuresti	Roumania	250	∟	_	_	_
		Greenville	USA	250/500	∟	_	_	_
		Delhi	India	10	∟	_	_	_
		Omsk	USSR	100	∟			_
		Nikolaevskamur	USSR	50	∟	_	_	_
		Pt Plata	Dominican Rep	0.1	∟	_	_	_
		Sebaa Aioun	Morocco	10/50	∟	_	_	_
		Osaka	Japan	0.5/6	∟	_	_	_
		Padang	Indonesia	10	∟	_	_	_
		Noblejas	Spain	350	∟	_	_	_
		Tirana	Albania					
		Bonaire Zuid	Neth Antilles	10/ 100	_			
		Santiago	Chile	25/100	∟	_	_	_
		Kavalla	Greece	250				_
6.195	48.43	London	UK	100/250	∟	_	_	_
		Masirah	Oman	100/200				_
		Kranji	Senegal	50/250	/	_	_	_
		Antigua	Br W Indies	250	∟	_	_	_
		Limassol	Cyprus	20/100	_	_	_	
		Tebrau	Malaysia	100/250	_	_	_	_
		Baku	USSR	50	∟	_	_	_
		Cali	Colombia	1	∟	_	_	_
		Rio de Janeiro	Brasil	7.5	∟	_	_	_
		La Paz	Bolivia	5	∟	_	_	_
		Sokoto	Niger	10	∟	_	_	_
		Leipzig	Germany (E)	100	_			_
		Nauen	Germany (E)	100	_			_
		Ismaning	Germany (W)	100	∟	_	_	_
		La Ceiba	Honduras Rep	0.5	/		_	_
		Sackville	Canada	250	∟	_	_	_
		Warszawa	Poland	1	/	_	_	_
		Pt au Prince	Haiti	50				_
6.200	48.39	Tirane	Albania	240				
		Moskva	USSR					
		Leningrad	USSR					
		Padang	Indonesia					

MHz	Metres	Station	Country	kW
		V Communist Party of Turkey		
		Houa Phan	Laos	
		Bizam Radio		
		Tripoli	Libya	
		Jurong	Singapore	
6.205	48.35		USSR	
		Peking	China Rep	
6.206	48.34	Mebo II		
6.208	48.32	Milan	Italy	
6.210	48.31	Peking	China Rep	
		Vientiane	Laos	
		Vatican	Vatican	
		Paris	France	
		Tirane	Albania	
6.215	48.27		USSR	
		Monte Carlo	Monaco	
		Andorra	Andorra	
6.221	48.22	Wien	Austria	
		Vatican	Vatican	
6.225	48.19	Peking	China Rep	
6.230	48.15	Cairo	Egypt	50
			USSR	
		Vatican	Vatican	
		Kabul	Afghanistan	
6.235	48.12	Karachi	Pakistan	
6.240	48.08	Irkutsk	USSR	
		Seoul	Korea (S)	
6.243	48.05		USSR	
6.250	48.00	Pyongyang	Korea (N)	50
		Malabo	Guinea	10
		V of Peace		
6.260	47.92	Peking	China Rep	
		Cao Lang	Vietnam	
6.270	47.85	Peking	China Rep	
6.280	47.77	Peking	China Rep	
		Nghe Tinh	Vietnam	
6.285	47.73		USSR	
		R Bayrak	Cyprus	
6.290	47.69	Pyongyang	Korea (N)	
		Peking	China Rep	
6.304	47.59	V of People of Burma		
6.320	47.47	Peking	China Rep	
6.330	47.39	Peking	China Rep	
6.331	47.39	Son La	Vietnam	
6.338	47.33	Pyongyang	Korea (N)	
6.340	47.32		Turkey*	
6.345	47.28	Peking	China Rep	

MHz	Metres	Station	Country	kW
6.348	47.26	R Echo of Hope		
6.385	46.99	Ulan Bator	Mongolian Rep	
6.390	46.95		USSR	
6.400	46.88	Pyongyang	Korea (N)	
			USSR*	
6.405	46.84	Taipei	China Nat	
6.410	46.80	Peking	China Rep	
6.426	46.69	Hanoi	Vietnam	
6.430	46.66	Peking	China Rep	
6.435	46.62	Hanoi	Vietnam	
6.453	46.49	German Figure Groups*		
6.495	46.19	Peking	China Rep	
6.520	46.01	Peking	China Rep	
6.540	45.87	Peking	China Rep	
6.548	45.82		USSR	
6.550	45.80	Peking	China Rep	
		Voice of Lebanon		
6.554	45.77		USSR	
6.555	45.77	Peking	China Rep	
6.560	45.73	Met Station*	USSR	
6.568	45.68		USSR	
6.576	45.62	Pyongyang	Korea (N)	
6.585	45.56	Peking	China Rep	
6.590	45.52	Peking	China Rep	
6.596	45.48	Met Station*	USSR	
6.600	45.45	Pyongyang	Korea (N)	
		Pakse	Laos	
6.615	45.35	Met Station*	USSR	
6.617	45.34	Met Station*	USSR	
6.645	45.15	Peking	China Rep	
6.665	45.01	Peking	China Rep	
6.675	44.94	Xieng Khouang	Laos	
6.730	44.58	Met Station*	USSR	
6.745	44.48	Paris*	France	
6.750	44.44	Peking	China Rep	
6.765	44.35	Fukien Front Stn	China Rep	
6.770	44.31		USSR*	
		Pyongyang	Korea (N)	
6.790	44.18	Peking	China Rep	
6.807	44.07	Riyadh*	Saudi Arabia	
6.808	44.07		USSR*	
6.810	44.05	Peking	China Rep	
6.822	43.98		USSR*	
6.825	43.96	Peking	China Rep	
			USSR*	
		V of Democratic Kampuchea		
6.838	43.87	London*	UK	
6.840	43.86	Huhetot	Mongolian Rep	

MHz	Metres	Station	Country	kW
6.852	43.78		USSR*	
		V of Lebanon		
6.860	43.73	Peking	China Rep	
6.865	43.70	Shepparton*	Australia	
6.870	43.67	Alma Ata	USSR	
6.873	43.65	Greenville*	USA	
6.876	43.63	Paris*	France	
6.880	43.60	Peking	China Rep	
6.884	43.,59	Bac Thai	Vietnam	
6.890	43.54	Peking	China Rep	
			USSR*	
6.900	43.48	Ankara	Turkey	
6.905	43.45		USSR	
6.910	43.42	Udomsai	Laos	
6.920	43.35		USSR*	
6.935	43.26	Peking	China Rep	
6.937	43.25		China Rep	
6.940	43.23	V of Iraqi Kurdistan		
6.941	43.22	Rhodes	Greece	
6.945	43.20	Moscow	USSR	
6.955	43.13	Peking	China Rep	
6.970	43.04	RFE/R Liberty*		
6.975	43.01	Julich*	Germany (W)	
		RFE/R Liberty*		
		Luang Prabang	Laos	
6.980	42.98		USSR*	
6.987	42.94		USSR*	
6.995	42.89	Peking	China Rep	
		RFE/R Liberty*		
7.010	42.80	Peking	China Rep	
			USSR	
7.015	42.77	V of NUFK		
7.020	42.74	Novosibirsk	USSR	
7.025	42.70	Peking	China Rep	
7.030	42.67		USSR	
		Peking	China Rep	
7.035	42.64	Peking	China Rep	
7.040	42.61	Thimpu	Bhutan	0.3/10
			USSR	
		Peking	China Rep	
		Vatican	Vatican	
7.045	42.58		USSR	
		Peking	China Rep	
7.050	42.55	Cairo	Egypt	100
		Urumchi	China Rep	
7.055	42.52	Peking	China Rep	
		Algiers	Algeria	
7.060	42.49	Peking	China Rep	
7.065	42.46	Tirane	Albania	240

MHz	*Metres*	*Station*	*Country*	*kW*	*M*	*M*	*S*	*N*
		Peking	China Rep					
7.075	42.40	Cairo	Egypt	100				
		Tirane	Albania	120				
7.080	42.37	Bac Thai	Vietnam					
		Peking	China Rep	120				
			USSR					
		Tirane	Albania	120				
7.085	42.34	Karachi	Pakistan					
7.090	42.31	Tirane	Albania					
7.095	42.28	Peking	China Rep					
		Karachi	Pakistan					
		Laos	Laos					
7.100	42.25		USSR					
7.103	42.24	V of the People						
7.105	42.22	London	UK	100/250	∠	_	_	_
		Ascension	Ascension	250	∠	_	_	_
		Orcha	Bielorussia	50/240	∠			_
			USSR					
		Simferopol	Ukraine	100/500	/			_
		Brazzaville	Congo	4	∠	_	_	_
		Delhi	India	10/100	_	_	_	_
		Kohima	India	2	/			
		Yogyakarta	Indonesia	20/100	∠	_	_	_
		Arganda	Spain	100	/			
		Noblejas	Spain	350	_	_	_	_
		Tananarive	Malagasy Rep	4/30	/	_	_	_
		Ekala	Ceylon	35		_	_	
		Mt Carlo	Monaco	100	∠	_	_	_
		Kathmandu	Nepal	100	∠		_	_
		Julich	Germany (W)	100	∠	_	_	_
		Wertachtal	Germany (W)	500	_	_		
		Quetta	Pakistan	10				_
7.110	42.19	London	UK	100	∠	_	_	_
		Limassol	Cyprus	100	/			_
		Khabarovsk	USSR	100		_	_	
		Omsk	USSR	50/100	∠	_	_	_
		Rhodos	Greece	50	∠	_	_	_
		Warszawa	Poland	20/100	∠	_	_	_
		Bamako	Malawi	50		_	_	
		Ekala	Ceylon	10/35	∠	_	_	_
		Jakarta	Indonesia	50/100	∠	_	_	_
		Riyadh	Saudi Arabia	350	∠		_	
		Maputo	Mozambique	10	_		_	_
		Gedja	Ethiopia	100		_	_	_
		Arganda	Spain	100	/		_	_
		Delhi	India	10	/			
7.115	42.16	Bangkok	Thailand	5	∠	_	_	_
		Bandundu	Zaire	10	∠	_	_	_
		Biblis	Germany (W)	100	∠	_	_	_

MHz	Metres	Station	Country	kW	M	M	S	N
		Lisbonne	Portugal	100	∠		_	_
		Tbilisi	USSR	240				_
		Duchanbe	USSR	100				_
		Tchita	USSR	100/240	_			
		Ivanofrankovsk	Ukraine	240	∠			_
		Sofia	Bulgaria	500				_
		Gedja	Ethiopia	100		_	_	
		Kimjae	Korea (S)	250			_	
7.120	42.13	London	UK	100/250	∠	_	_	_
		Denpassar	Indonesia	10	∠	_	_	_
		N Djamena	Chad	100	∠			_
		Tirane	Albania					
		Bucuresti	Roumania	18	_			
		Mogadiscio	Somalia	50	∠	_	_	_
		Novosibirsk	USSR	100	∠	_		
		Tula	USSR	100		_		
		Lvov	Ukraine	240/500	∠			_
		Sulaibiyah	Kuwait	250	∠	_	_	_
		Delhi	India	100	∠	_	_	_
		Cyclops	Malta	250	/			
7.125	42.11	Conakry	Guinea	100	∠	_	_	_
		Delhi	India	50/100	∠	_	_	_
		Aligarh	India	250	_	_	_	_
		Nairobi	Kenya	5	∠	_	_	_
		Warszawa	Poland	40/100	∠	_	_	_
		Kazan	USSR	100		_	_	
		Kenga	USSR	100	∠			_
		Khabarovsk	USSR	240	∠	_	_	
		Jerusalem	Israel	50/300	_			
		Mt Carlo	Monaco	100			_	_
		Athinai	Greece	100	/			
		Kavalla	Greece	250	/			_
7.130	42.08	London	UK	100/250	∠	_	_	_
		Limassol	Cyprus	100	_			
		Stapok	Malaysia	10			_	
		Wertachtal	Germany (W)	500	∠	_	_	_
		Julich	Germany (W)	100	∠	_	_	_
		Minsk	Bielorussia	100		_	_	
		Erevan	USSR	100	∠			_
		Vinnitsa	Ukraine	240	_			_
		Vladivostock	USSR	100				_
		Sines	Portugal	250	_			_
		Limbe	Malawi	100	∠		_	_
		Noblejas	Spain	350	∠	_	_	_
		Taipei	China Nat					
		Kavalla	Greece	250	/		_	
			Pakistan					
7.135	42.05	Careysburg	Liberia	250	∠	_	_	_
		Mt Carlo	Monaco	30	∠	_	_	_

MHz	Metres	Station	Country	kW	M	M	S	N
		Islamabad	Pakistan	250	/_		_	_
		Karachi	Pakistan	10	_	_		
		Duchanbe	USSR	240	_	_	_	_
		Moskva	USSR	240				_
		Komsomolskamur	USSR	240	/_			_
		Simferopol	Ukraine	240/500	/_			_
		Tinang	Philippines	250	/_	_	_	_
		Allouis	France	100/500	/_	_	_	_
		Kimjae	Korea (S)	100	/_			
		Delhi	India	100	/			
		Cyclops	Malta	250	/			
		Franceville	Gabon Rep	500			_	
		Thessaloniki	Greece	35	/			
7.140	42.02	London	UK	100/250	_			_
		Masirah	Oman	100	/		_	_
		Limassol	Cyprus	20/100	/_	_	_	_
		Amboina	Indonesia	10	/_	_	_	_
		Hyderabad	India	10	/_	_	_	_
		Nairobi	Kenya	100	/_	_	_	_
		Kazan	USSR	100		_	_	
		Riga	USSR	240	/_			_
		Alma Ata	USSR	240		_	_	
		Tokyo Yamata	Japan	20/100	/_	_	_	_
		Nampula	Mozambique	5			_	
7.145	41.99	Stapok	Malaysia	10			_	
		Warszawa	Poland	40/100	/_	_	_	_
		Tachkent	USSR	100		_	_	
		Novosibirsk	USSR	100	/	_		_
		Tula	USSR	100		_	_	
		Quelimane	Mozambique	0.3	/_	_	_	_
		Alger	Algeria	50	/_	_	_	_
		Karlsborg	Sweden	350	_			
		Kavalla	Greece	250	_	_		
		Delhi	India	100	_	_	_	_
		Lisbonne	Portugal	100/250	_		_	_
		Biblis	Germany (W)	100				
		Lampertheim	Germany (W)	100		_		
		Mt Carlo	Monaco	100	_	_		
7.150	41.96	London	UK	100/250	/_	_	_	_
		Ascension	Ascension	125		_		
		Krasnoiarsk	USSR	100		_	_	
		Tchita	USSR	100/500	/_			_
		Serpukhov	USSR	240		_	_	
		Lvov	Ukraine	240/500	/_		_	_
		Nairobi	Kenya	10	/_	_	_	_
		Julich	Germany (W)	100	/	_	_	
		Wertachtal	Germany (W)	500	/_	_	_	_
		Ismaning	Germany (W)	100				_
		Pemba	Mozambique	0.3	/_	_	_	_

MHz	Metres	Station	Country	kW	M	M	S	N
		Kavalla	Greece	250	_			
		Quetta	Pakistan	10	/			
		Dacca	Bangladesh	10/100	_	_	_	
		Suwon	Korea (S)	50	_	_	_	_
		Bata	Guinea	50	∠	_	_	_
		Sofia	Bulgaria	100/500	∠			_
		Gauhati	India	10	/	_	_	_
7.155	41.93	London	UK	100/250	∠	_	_	_
		Amman	Jordan	100	∠	_	_	_
		Tananarive	Mongolian Rep	5/30	/	_	_	_
		S M Galeria	Vatican	100	∠	_	_	_
		Armavir	USSR	100	∠			_
		Poro	Philippines	100	∠	_	_	_
		Szekesfehervar	Hungary	20	_	_	_	
		Jaszbereny	Hungary	250	∠	_	_	_
		Arganda	Spain	100	∠	_	_	_
		Playa de Pals	Spain	250/500				_
		Biblis	Germany (W)	100			_	
		Holzkirchen	Germany (W)	10			_	
		Lampertheim	Germany (W)	100	∠	_	_	
		Ismaning	Germany (W)	100			_	
7.160	41.90	Stapok	Malaysia	10			_	
		Madras	India	10	_	_	_	_
		Kazan	USSR	240	/	_	_	_
		Omsk	USSR	100	_	_	_	_
		Tula	USSR	240	/	_	_	_
		Petropavlo Kam	USSR	100				_
		Lvov	Ukraine	240	_			
		Allouis	France	100/500	∠	_	_	_
		Hargeisa	Somalia	10	∠	_	_	_
		Tinang	Philippines	250	_	_		_
		Kavalla	Greece	250	∠	_	_	_
		Julich	Germany (W)	100	∠	_	_	_
		Cyclops	Malta	250	∠	_	_	_
		S M Galeria	Vatican	100	∠	_	_	_
		Huambo	Angola	10	/		_	_
7.165	41.87	Tripoli	Libya	100	∠	_	_	
		Dar es Salaam	Tanzania	20	_	_	_	_
		Delhi	India	20/100	∠	_	_	_
		Jayapura	Indonesia	1/5	∠	_	_	_
		Lampertheim	Germany (W)	100	/			
		Biblis	Germany (W)	100	/	_	_	
		Holzkirchen	Germany (W)	10	∠	_	_	_
		Playa de Pals	Spain	250			_	
		Lisbonne	Portugal	100/250	∠	_	_	_
		Kiev	Ukraine	240	∠	_	_	_
		Serpukhov	USSR	240				_
		Gedja	Ethiopia	100	/	_	_	_
		Islamabad	Pakistan	100	_	_	_	_

MHz	Metres	Station	Country	kW	M	M	S	N
		Karachi	Pakistan	50	_	_		_
		Poro	Philippines	50	_	_	_	
		Kathmandu	Nepal	100	/_		_	_
7.170	41.84	London	UK	250	_		_	
		Kavalla	Greece	250	/_	_	_	_
		Dakar	Senegal	30	/_	_	_	_
		Noumea	New Caledonia	20	/_	_	_	_
		Singapore	Singapore	10	/_	_	_	_
		Armavir	USSR	240	_		_	_
		Kazan	USSR	100		_	_	
		Novosibirsk	USSR	100	/_			_
		Tachkent	USSR			_	_	
		Meyerton	South Africa	100	_	_	_	_
		Tanger	Morocco	100				_
		Wien	Austria	100	/_	_	_	_
		Ankara	Turkey	250	/	_	_	_
		Ranchi	India	2	/			
7.175	41.81	Brazzaville	Congo	4/25	/_	_	_	_
		Caltanissetta	Italy	5	/_	_	_	_
		Wertachtal	Germany (W)	500	/_	_	_	_
		Starobelsk	Ukraine	240/500	/_	_		_
		Gorkii	USSR	240	_	_	_	_
		Khabarovsk	USSR	100	/_			_
		Kinghisepp	USSR	240	/			
		Careysburg	Liberia	250	/_	_	_	_
		Bucuresti	Roumania	18/120	/_	_	_	_
		Gwelo	Rhodesia	10/100	/		_	
7.180	41.78	London	UK	100/250	_		_	
		Kranji	Singapore	250	/_	_	_	_
		Warszawa	Poland	100	/_	_	_	_
		Abu Ghraib	Iraq	100	_	_	_	_
		Babel	Iraq	500	_	_		
		Bhopal	India	10	/_	_	_	_
		Kazan	USSR	100	_			_
		Tanger	Mongolian Rep	100		_		_
		Gedja	Ethiopia	100	/_			
		Biblis	Germany (W)	100	_			_
		Lampertheim	Germany (W)	100	/		_	
		Kavalla	Greece	250				_
		V of the Namibian People						
7.185	41.75	London	UK	100/250	/_	_	_	_
		K Wusterhausen	Germany (E)	5	/_	_	_	_
		Merauke	Indonesia	1	/_	_	_	_
		Gorkii	USSR	100	/			_
		Tbilisi	USSR	100	_		_	_
		Alma Ata	USSR	100				_
		Novosibirsk	USSR	240/500	/_	_	_	_
		Rangoon	Burma	50	/_	_	_	_

MHz	Metres	Station	Country	kW	M	M	S	N
		Noblejas	Spain	350	_			
7.190	41.72	Limassol	Cyprus	100		_		
		Ekala	Ceylon	10	∠	_	_	_
		Jayapura	Indonesia	10	∠	_	_	_
		Armavir	USSR	100	_			_
		Moskva	USSR	240		_	_	
		Peking	USSR					
		Tanger	Morocco	100	∠	_	_	_
		Aden	Aden	100				
		Bata	Guinea	50	∠	_	_	_
		Lisbonne	Portugal	100/250	∠	_	_	_
		Holzkirchen	Germany (W)	10			_	
		Lampertheim	Germany (W)	100				_
		Biblis	Germany (W)	100	∠	_	_	
		Playa de Pals	Spain	500				_
		Poro	Philippines	50	/	_		_
7.195	41.70	Bucuresti	Roumania	18/250	∠	_	_	_
		Delhi	India	100	∠	_	_	_
		Careysburg	Liberia	50/250	∠	_	_	_
		Tula	USSR	100	∠		_	_
		Simferopol	Ukraine	500				_
		Alger	Algeria	50	∠	_	_	_
		Noblejas	Spain	350	∠	_	_	_
		Tokyo Yamata	Japan	20	/	_	_	_
		Islamabad	Pakistan	100	_	_		
		Karachi	Pakistan	10	/		_	_
7.200	41.67	London	UK	250	∠	_	_	_
		Penang	Malaysia	10	∠	_	_	_
		Irkutsk	USSR	50	∠	_	_	_
		Kazan	USSR	240				_
		Jigulevsk	USSR	100		_	_	
		Vladivostock	USSR	50	∠	_	_	_
		Belgrade	Yugoslavia	10	∠	_	_	_
		Omdurman	Sudan	120	_	_		_
		Kabul	Afghanistan	10	∠	_	_	_
		Holzkirchen	Germany (W)	10				_
		Lampertheim	Germany (E)	100	_			
		Lisbonne	Portugal	250				_
		Diosd	Hungary	100	∠	_	_	_
		Arganda	Spain	100	_	_	_	_
		Silinhot	Mongolian Rep					
		Tripoli	Libya					
		Jerusalem	Israel					
		V of Blackman's Resistance	Philippines					
7.205	41.64	London	UK	250	_	_	_	_
		Armavir	USSR	100/500	∠	_	_	_
		Frunze	USSR	100				_

MHz	*Metres*	*Station*	*Country*	*kW*	*M*	*M*	*S*	*N*
		Moskva	USSR	240	∠		_	_
		Athinai	Greece	100	∠	_	_	_
		Kavalla	Greece	250	/			_
		Rhodos	Greece	50	∠	_	_	_
		Warszawa	Greece	1/100	/	_	_	_
		Yaounde	Cameroon	30	∠	_	_	_
		Lubumbashi	Zaire	10	∠	_	_	_
		Tokyo Yamata	Japan	20	∠	_	_	_
		Ismaning	Germany (W)	100	/		_	
		Kamalabad	Iran	100		_		
7.210	41.61	London	UK	100/250	∠	_	_	_
		Limassol	Cyprus	100	/	_	_	_
		Beromunster	Switzerland	150	∠	_	_	
		Schwarzenburg	Switzerland	150				_
		Calcutta	India	10	∠	_	_	_
		Moskva	USSR	100	∠	_	_	
		Kazan	USSR	240			_	_
		Khabarovsk	USSR	50	∠	_	_	_
		Sverdlovsk	USSR	100			_	
		Nairobi	Kenya	10	∠	_	_	_
		Biak	Indonesia	1	∠	_	_	_
		Lopik	Netherlands	100	∠	_	_	_
		Wertachtal	Germany (W)	500	_			
		Julich	Germany (W)	100	∠	_	_	_
		Dakar	Senegal	100	∠	_	_	_
		Ziguinchor	Senegal	30	/			
		Tanger	Morocco	100	_		_	_
		Karachi	Pakistan	50	_			
7.215	41.58	Abidjan	Ivory Coast	10	∠	_	_	_
		Delhi	India	100/250	_	_	_	_
		Biblis	Germany (W)	100	/			
		Lisbonne	Portugal	50/100	_			_
		Jaszbereny	Hungary	250	∠	_	_	_
		Vinnitsa	Ukraine	240/500	_			_
		Armavir	USSR	240		_	_	
		Komsomolskamur	USSR	100			_	
		Athinai	Greece	100	_	_	_	_
		Kavalla	Greece	100/250	_	_	_	_
		Kamalabad	Iran	100	_	_	_	_
		Berakas	Brunei	10	_	_	_	
		Sofia	Bulgaria	50	_			
		Manzini	Swaziland	25	_	_	_	_
		Mt Carlo	Monaco					
7.220	41.55	London	UK	250	/	_		
		Lusaka	Zambia	20	∠	_	_	_
		Diriyya	Saudi Arabia	50	∠	_	_	_
		Bangui	Central African Rep	4/100		_	_	_

MHz	Metres	Station	Country	kW	M	M	S	N
		Tanger	Morocco	50/100	_		_	_
		Tachkent	USSR	100				_
		Tchita	USSR	240/500	∠	_	_	_
		Vladivostock	USSR	100		_		
		Jakarta	Indonesia	1	∠	_	_	_
		Biblis	Germany (W)	100		_	_	_
		Lampertheim	Germany (W)	100	∠	_	_	_
		Holzkirchen	Germany (W)	10	/		_	
		Playa de Pals	Spain	250/500	∠			_
		Salman Pak	Iraq	100	_			
7.225	41.52	Bucuresti	Roumania	120/250	∠	_	_	_
		Delhi	India	100	_	_	_	_
		Aligarh	India	250	∠	_	_	_
		Bocaue	Philippines	25/50	∠	_	_	_
		Sebaa Aioun	Morocco	10	∠	_	_	_
		Kigali	Rwanda	250	∠	_	_	_
		Abu Zabaal	Egypt	100	_			
		Male	Maldives	2.7	_	_	_	_
		Arganda	Spain	100	_	_	_	_
		Sfax	Tunisia					
7.230	41.49	London	UK	100/250	∠	_	_	_
		Masirah	Oman	100				
		Limassol	Cyprus	20/100	∠	_	_	_
		Kazan	USSR	100	∠		_	_
		Frunze	USSR	100	_			
		Krasnoiarsk	USSR	50	∠	_	_	_
		Nikolaevskamur	USSR	50/100	_	_	_	
		Kiev	Ukraine	50	_		_	_
		Lvov	Ukraine	240		_		
		Mt Carlo	Monaco	100	∠	_	_	_
		Tananarive	Malagasy Rep	10	/			_
		Ouagadougou	Upper Volta	20	_			
		Tanger	Morocco	50/100	∠	_	_	_
		Dar es Salaam	Tanzania	50	_	_	_	_
		Kamalabad	Iran	100	_		_	_
		Kurseong	India	20	/			
7.235	41.47	London	UK	100	∠	_	_	_
		Julich	Germany (W)	100	∠	_	_	_
		Wertachtal	Germany (W)	500	∠	_	_	_
		Delhi	India	50	_	_	_	_
		Roma	Italy	100	∠	_	_	_
		Enugu	Niger	10	∠	_	_	_
		S M Galeria	Vatican	100	∠	_	_	_
		Lusaka	Zambia	50	∠	_	_	_
		Riga	USSR	100		_	_	
		Sverdlovsk	USSR	100	∠			_
		Poro	Philippines	35/50	/	_		_
		Luanda	Angola	100	/		_	_
		R Kulmis						

MHz	*Metres*	*Station*	*Country*	*kW*	*M*	*M*	*S*	*N*
7.240	41.44	London	UK	150	/		_	_
		Garoua	Cameroon	4	∠	_	_	_
		Maputo	Mozambique	25/100	∠	_	_	_
		Belgrade	Yugoslavia	10	∠	_	_	_
		Bombay	India	10	∠	_	_	_
		Medan	Indonesia	7.5	∠	_	_	_
		Tula	USSR	240/500	∠	_	_	_
		Nairobi	Kenya	10	∠	_	_	_
		Kavalla	Greece	250	∠		_	
		Carnarvon	Australia	100	_	_	_	_
		Shepparton	Australia	10	_	_	_	_
		Lyndhurst	Australia					
		Tanger	Morocco	100				_
		Poro	Philippines	50	_	_	_	_
		Lopik	Netherlands	100	∠	_	_	_
		Kimjae	Korea (S)	250	∠	_	_	_
		V of Eritrea Revln						
		Salman Pack	Iraq	100		_	_	_
7.245	41.41	London	UK	100	_	_		
		Nouakchott	Mauretania	4/100				_
		Arganda	Spain	100	/		_	_
		Playa de Pals	Spain	250	_			
		Holzkirchen	Germany (W)	10	∠	_	_	_
		Biblis	Germany (W)	100	/			
		Lampertheim	Germany (W)	100	∠			_
		Lisbonne	Portugal	50			_	
		Luanda	Angola	100			_	_
		S Denis	Reunion	4	∠	_	_	_
		Vinnitsa	Ukraine	240	_			
		Khabarovsk	USSR	240		_	_	
		Krasnoiarsk	USSR	50				_
		Mt Carlo	Monaco	100	∠	_		_
		Alger	Algeria	50	∠	_	_	_
		Praha	Czechoslovakia	120	∠			_
		Kimjae	Korea (S)	100		_		
		Dacca	Bangladesh	10	/			_
7.250	41.38	London	UK	100	_	_	_	_
		Limassol	Cyprus	20/100	/			_
		Masirah	Oman	100	∠	_	_	_
		Singapore	Singapore	50	∠	_	_	_
		Sverdlovsk	USSR	100				_
		Tula	USSR	240/500		_	_	
		S M Galeria	Vatican	100	∠	_	_	_
		Lusaka	Zambia	120	∠	_	_	_
		Tokyo Yamata	Japan	100				_
		Kimjae	Korea (S)	100	∠	_	_	
		Lucknow	India	10	∠	_	_	_
7.255	41.35	London	UK	100	∠	_	_	_
		Nampula	Mozambique	0.3	∠	_	_	_

MHz	Metres	Station	Country	kW	M	M	S	N
		Ikorodu	Niger	100	/		_	_
		Sogunle	Niger	10	_	_		
		Sofia	Bulgaria	50	∠	_	_	_
		Biblis	Germany (W)	100	∠	_	_	_
		Lampertheim	Germany (W)	100				_
		Holzkirchen	Germany (W)	10		_	_	
		Lisbonne	Portugal	100	∠		_	_
		Kenga	USSR	50			_	
		Tbilisi	USSR	50	∠			_
		Manzini	Swaziland	25			_	
		Kinshasa	Zaire	10	∠	_	_	_
		Poro	Philippines	50/100		_	_	
7.260	41.32	London	UK	100/250	∠	_	_	_
		Limassol	Cyprus	20/100	∠	_	_	_
		Mt Carlo	Monaco	100	∠	_	_	_
		Port Vila	New Hebrides	2	_	_	_	
		Bombay	India	100	/			
		Delhi	India	10/50	_	_	_	_
		Madras	India	100	_	_	_	_
		Iujnsakhalinsk	USSR	100	_			_
		Minsk	Bielorussia	100	∠	_	_	_
		Ulan Bator	Mongolian Rep	25				
		Nauen	Germany (E)	500	∠	_	_	_
7.265	41.29	London	UK	250				_
		Togblekope	Togo	100	∠		_	_
		Riazan	USSR	240	_			_
		Iakutsk	USSR	100	∠	_	_	_
		Komsomolskamur	USSR	240		_	_	
		Armavir	USSR	240	_	_	_	
		Volgograd	USSR	100			_	
		Rohrdorf	Germany (W)	20	∠	_	_	_
		Karachi	Pakistan	10/50	_	_	_	_
		Islamabad	Pakistan	100/240	/			
		Tanger	Morocco	50			_	_
		Cyclops	Malta	250	∠	_	_	_
		Sanaa	Yemen	50			_	
		Luanda	Angola		/			
7.270	41.27	Ascension	Ascension	250	_	_	_	
		Limassol	Cyprus	100			_	
		Stapok	Malaysia	10			_	
		Meyerton	South Africa	20/500	∠	_	_	_
		Jakarta	Indonesia	50/100	∠	_	_	_
		Erevan	USSR	100	_			_
		Kenga	USSR	100/500	∠			_
		Moskva	USSR	50		_	_	
		Kavalla	Greece	250	∠	_	_	_
		Warszawa	Poland	100	∠	_	_	_
		Sofia	Bulgaria	50/100				_
		Ankara	Turkey	250		_	_	_

MHz	Metres	Station	Country	kW	M	M	S	N
		Kamalabad	Iran	100		_		
		Delhi	India	100			_	_
		Franceville	Gabon	500			_	
		Ismaning	Germany (W)					
7.275	41.24	Masirah	Oman	200	/	_	_	_
		Limassol	Cyprus	100	_	_		_
		Ikorodu	Niger	100	_	_		
		Tinang	Philippines					_
		Poro	Philippines	50	_	_	_	
		Roma	Italy	60/100	/_	_	_	_
		Kenga	USSR	100	_	_		
		Komsomolskamur	USSR	240	/			
		Krasnoiarsk	USSR	100				_
		Duchanbe	USSR	50	/_	_	_	_
		Ismaning	Germany (W)	100				_
		Wertachtal	Germany (W)	500	/_	_	_	_
		Julich	Germany (W)	100	/_	_	_	_
		Tirane	Albania					
		Cyclops	Malta	250	/_	_	_	_
		Mt Carlo	Monaco	100				_
		Sfax	Tunisia	100			_	_
		Jaszbereny	Hungary	250	/_			_
		Arganda	Spain	100	/_	_	_	_
		Tanger	Morocco	100	_	_		
		Kavalla	Greece	250	/_	_	_	_
		Suwon	Korea (S)	50	/			
7.280	41.21	London	UK	250	_			
		Moskva	USSR	200	/_	_	_	_
		Komsomolskamur	USSR	240	_	_		_
		Dar es Salaam	Tanzania	10	_	_	_	_
		Careysburg	Liberia	250	/_	_	_	_
		Tirane	Albania					
		Thessaloniki	Greece	35	_	_	_	_
		Manzini	Swaziland	25	_			_
		Gauhati	India	10	/			
		Delhi	India	20	/_	_	_	_
		Aligarh	India	250	/_	_	_	_
		Allouis	France	100/500	/_	_	_	_
		Dacca	Bangladesh	100	/			
7.285	41.18	London	UK	100	/_	_	_	_
		Warszawa	Poland	15	/_	_	_	_
		Tula	USSR	100			_	
		Kazan	USSR	100	_			
		Krasnoiarsk	USSR	100	/			_
		Moskva	USSR			_		
		Allouis	France	100	_	_	_	_
		Ibadan	Niger	1/10	/_	_	_	_
		Wertachtal	Germany (W)	500	/_	_	_	_
		Julich	Germany (W)	100	_	_	_	

MHz	Metres	Station	Country	kW	M	M	S	N
		Sines	Portugal	250	∠	_	_	_
		Talata Volon	Malagasy Rep	600	∠	_	_	_
		Bamako	Malawi	18		_	_	
		Gwelo	Rhodesia	20/100	/		_	
		Poro	Philippines	100				_
7.290	41.15	Moskva	USSR	240/500	∠	_	_	_
		Petropavlo Kam	USSR	240	∠			_
		Delhi	India	60/100	∠	_	_	_
		Roma	Italy	60/100	∠	_	_	_
		Tirane	Albania					
		Ismaning	Germany (W)	100/ 250	∠	_		
		Islamabad	Pakistan	100/250	∠	_	_	_
		Kimjae	Korea (S)	100	∠	_	_	_
		Dacca	Bangladesh	100	/			/
7.295	41.12	London	UK	100/250	∠	_	_	_
		Accra	Ghana	10	∠	_	_	_
		K Wusterhausen	Germany (E)	100	∠	_	_	_
		Kajang	Malaysia	100	∠	_	_	_
		Moskva	USSR	100/240	∠	_	_	_
		Blagovechtchen	USSR	100		_	_	
		Menado	Indonesia	0.5	∠	_	_	_
		Mbujimayi	Zaire	10	∠	_	_	_
		Nairobi	Kenya	5	∠	_	_	_
		Sines	Portugal	250		_		
		Tanger	Morocco	35	/	_		_
		Biblis	Germany (W)	100	∠		_	_
		Holzkirchen	Germany (W)	10	∠			
		Lampertheim	Germany (W)	50/ 100	_		_	_
		Playa de Pals	Spain	250				_
		Lisbonne	Portugal	100		_		
7.300	41.10	Tirane	Albania	240				
		Berlin	Germany (E)					
		Kiev	USSR					
		Moskva	USSR					
7.305	41.07		USSR					
		Leningrad	USSR					
		V of Malayan Revln						
7.310	41.04	Kalinin	USSR					
			USSR					
7.315	41.01	Peking	China Rep					
		Chita	USSR					
7.320	40.98	London	UK					
		Magadan	USSR					
		Vladivostock	USSR					
7.325	40.96	London	UK					
		Moskva	USSR					
		Lanchow	China Rep					

MHz	Metres	Station	Country	kW
7.330	40.93	Minsk/Duchanbe	USSR	50
		Peking	China Rep	
7.335	40.90		USSR	
		Ottawa*	Canada	
7.340	40.87	Moskva	USSR	
7.345	40.84	Praha	Czechoslovakia	100
			USSR	
7.350	40.82	Moskva	USSR	
		V of Kampuchean People		
7.355	40.79	Moskva	USSR	
7.360	40.76	Peking	China Rep	
		Moskva/Kiev	USSR	
7.365	40.73	Rabat*	Morocco	
7.370	40.71	Minsk	USSR	100
			USSR	
7.375	40.68	Hanoi	Vietnam	
		Karachi	Pakistan	
		Peking	China Rep	
7.380	40.65	Moskva	USSR	100
		Peking	China Rep	
7.385	40.62	Peking	China Rep	
		Savannakhet	Laos	
		Hanoi	Vietnam	
			USSR	
7.390	40.60	Kiev	Ukraine	240
7.395	40.57	Moskva	USSR	
		Jerusalem	Israel	
7.400	40.54	Moskva	USSR	
7.410	40.49	Moskva*	USSR	
7.412	40.48	Jerusalem	Israel	
		Delhi	India	100
7.415	40.46	Hanoi	Vietnam	
7.420	40.43	Minsk	USSR	
		Moskva	USSR	100
7.430	40.38	Moskva	USSR	
7.437	40.34		USSR*	
7.440	40.32	Peking	China Rep	
		Moskva	USSR	100
		R.F.E.		
7.442	40.31	Monrovia	Liberia	
7.443	40.31	Geneva (UNO)	Switzerland	25
7.465	40.19	Jerusalem	Israel	
7.470	40.16	Peking	China Rep	
		Hanoi	Vietnam	
7.480	40.11	Peking	China	
		Vientiane	Laos	
7.490	40.05	Vladivostock*	USSR	
7.500	40.00	Lyndhurst Time Sig	Australia	

MHz	Metres	Station	Country	kW
7.504	39.98	Peking	China Rep	
7.512	39.94	Hanoi	Vietnam	
7.530	39.84		USSR*	
7.540	39.79		USSR*	
7.550	39.74		USSR*	
		Seoul	Korea (S)	
		Peking	China Rep	
7.588	39.54	R Sandino		
7.590	39.53	Peking	China Rep	120
7.605	39.45		USSR	
7.620	39.37	Peking	China Rep	
7.649	39.22	Hakkari	Turkey	1
7.651	39.21	Greenville*	USA	
7.660	39.16	Peking	China Rep	
7.670	39.11	Sofia	Bulgaria	
			USSR	
7.675	39.09	Julich	Germany (W)	
7.700	38.96	Peking	China Rep	
7.727	38.82	Ismaning*	Germany (W)	
7.740	38.76		USSR	
7.767	38.62	Julich*	Germany (W)	
7.768	38.62	Greenville*	USA	
7.770	38.61	Peking	China Rep	
		Greenville*	USA	
7.775	38.59	Peking	China Rep	
7.780	38.56	Peking	China Rep	
7.800	38.46	Peking	China Rep	
7.815	38.39	Peking	China Rep	
7.820	39.36	Peking	China Rep	
7.827	38.33	Peking	China Rep	
7.844	38.25	London*	UK	
7.848	38.23	London*	UK	
7.850	38.22	Fukien Front Stn	China Rep	
7.855	38.19	Peking	China Rep	
7.893	38.01	London*	UK	
7.925	37.85		USSR*	
7.935	37.81	Peking	China Rep	
7.948	37.75		USSR*	
7.973	37.63	London*	UK	
7.976	37.61	London*	UK	
7.991	37.54	London*	UK	
8.007	37.47	Peking	China Rep	
8.063	37.21	Algiers	Algeria	
8.125	36.92		USSR*	
8.240	36.41	Peking	China Rep	
8.260	36.32	Peking	China Rep	
8.300	36.14	Peking	China Rep	
8.320	36.06	Peking	China Rep	
8.345	35.95	Peking	China Rep	

MHz	*Metres*	*Station*	*Country*	*kW*
8.360	35.89	V of Arabian Peninsula People		
8.395	35.74	Luang Prabang	Laos	
8.425	35.61	Peking	China Rep	
8.450	35.50	Peking*	China Rep	
8.490	35.34	Peking	China Rep	
8.565	35.03	Peking	China Rep	
8.600	34.88	Peking	China Rep	
8.660	34.64	Peking*	China Rep	
			Laos	
8 903	33.70		USSR	
8.910	33.67		USSR	
8.917	33.64		USSR	
8.965	33.46			
8.970	33.44		USSR	
9.009	33.30	Jerusalem	Israel	50/100
9.020	33.26	Peking	China Rep	
9.022	33.25	Teheran	Iran	350
9.030	33.22	Peking	China Rep	
9.064	33.10	Peking	China Rep	
9.080	33.04	Peking	China Rep	
9.090	33.00	RFE*		
9.091	33.00	RFE*		
9.130	32.86	Moskva*	USSR	
9.150	32.79	Moskva*	USSR	
9.170	32.72	Peking	China Rep	
		RFE/R Liberty*		
9.200	32.61		USSR*	
9.210	32.57		USSR*	
9.240	32.47		USSR*	
9.250	32.43	RFE/R Liberty*		
9.290	32.29	Peking	China Rep	
9.317	32.20	London*	UK	
9.323	32.17	London*	UK	
9.330	32.15			
9.336	32.13	Peking	China Rep	
9.340	32.12	Peking	China Rep	
			USSR	
9.345	32.10		USSR	
9.355	32.07	Jerusalem	Israel	
9.360	32.05	Madrid	Spain	50
9.365	32.03	Peking	China Rep	
9.375	32.00	Tirane	Albania	
9.380	31.98	Peking	China Rep	
9.390	31.95	Peking	China Rep	
			USSR	
9.400	31.91	Peking	China Rep	
9.410	31.88	London	UK	
9.417	31.86	Peking	China Rep	

MHz	*Metres*	*Station*	*Country*	*kW*	*M*	*M*	*S*	*N*
9.420	31.85	Pyongyang	Korea (N)					
		V of People of Thailand						
9.425	31.83	Jerusalem	Israel					
9.430	31.81	Tirane	Albania					
9.435	31.80	Jerusalem	Israel					
9.440	31.78	Peking	China Rep					
			USSR					
9.445	31.76	Jerusalem	Israel					
9.450	31.75	Moskva	USSR	100				
9.455	31.73	Cairo	Egypt	10				
		Peking	China Rep					
9.460	31.71	Peking	China Rep					
			USSR					
		Karachi	Pakistan					
9.465	31.70	Karachi	Pakistan					
9.470	31.68	Peking	China Rep					
		V of Democratic Kampuchea						
9.475	31.66	Cairo	Egypt					
9.480	31.65	Peking	China Rep	120				
		Moskva	USSR	120				
		Tirane	Albania					
9.485	31.63	Tirane	Albania					
9.490	31.61	Lhasa	Tibet					
		Moskva	USSR					
		Peking	China Rep					
9.495	31.60	Cairo	Egypt	100				
		Jerusalem	Israel					
9.500	31.58	Berlin	Germany (E)	100				
		Moskva	USSR	100				
		Bizam Radio		15				
		Tirane	Albania	240				
		Budapest	Hungary	100/250		_		_
		Tripoli	Libya					
		Dacca	Bangladesh					
		R of the Patriots						
9.505	31.56	London	UK	100		_		
		La Paz	Bolivia	5/100	∠	_	_	_
		Bocaue	Philippines	50	∠	_	_	_
		Belgrade	Yugoslavia	10	∠	_	_	_
		K Wusterhausen	Germany (E)	100	∠	_	_	_
		Omdurman	Sudan	50	_	_		_
		Velkekostolany	Czechoslovakia	120	∠	_	_	_
		S Domingo	Dominican Rep	20	∠	_	_	_
		Tokyo Yamata	Japan	100	∠	_	_	_
		S Paulo	Brasil	7.5	∠	_	_	_
		Alma Ata	USSR	100	∠	_	_	_
		Komsomolskamur	USSR	240	∠			_

MHz	Metres	Station	Country	kW	M	M	S	N
		Kenga	USSR	500	/			
		Allouis	France	100/500		_	_	
		Noumea	New Caledonia	4	∠	_	_	_
		Arganda	Spain	100	∠	_	_	_
		Noblejas	Spain	350	_	_	_	
		Redwood City	USA	250		_		
		Shepparton	Australia	100	∠	_	_	_
		Lisbonne	Portugal	50/250	∠		_	_
		Biblis	Germany (W)	100	∠	_	_	_
		Holzkirchen	Germany (W)	10	∠	_		
		Lampertheim	Germany (W)	100				_
		Sulaibiyah	Kuwait	250	∠	_	_	_
		Islamabad	Pakistan	100	/			
9.510	31.55	London	UK	100/250	∠	_	_	_
		Antigua	Br W Indies	250	∠	_	_	_
		Sackville	Canada	50/250	∠		_	
		Alger	Algeria	50	∠	_	_	_
		Bucuresti	Roumania	18/250	∠	_	_	_
		Madras	India	100	∠	_	_	_
		Serpukhov	USSR	100	/			
		Vladivostock	USSR	100		_	_	
		Irkutsk	USSR	240/500	_			_
		Kazan	USSR	100	∠	_	_	_
		Greenville	USA	50/250	_	_	_	_
		Tanger	Morocco	100	_			
		K Wusterhausen	Germany (E)	100	_			_
		Taipei	China Nat					
		Noblejas	Spain	350	_		_	_
		Santiago	Chile	100	∠	_		_
		Allouis	France	100/500	∠	_	_	_
		V of Arab Syria						
		V of the One Lebanon						
		Kimjae	Korea (S)	100				_
		Mt Carlo	Monaco	100	/		_	_
		Wertachtal	Germany	100		_	_	
		Padang	Indonesia					
9.515	31.53	London	UK	250				_
		Kajang	Malaysia	50	∠	_	_	_
		Ankara	Turkey	100/250	/	_	_	_
		Caltanissetta	Italy	25	∠	_	_	_
		Erevan	USSR	100		_	_	
		Simferopol	Ukraine	100	_	_		
		Mexico	Mexico	20	∠	_	_	_
		Montevideo	Uruguay	10	∠	_	_	_
		Rio de Janeiro	Brasil	1/7.5	∠	_	_	_
		Kampala	Uganda	250				_
		Las Mesas	Canary Is	50	_	_	_	_
		Malolos	Philippines	100	_	_	_	_
		Tirana	Albania					

MHz	*Metres*	*Station*	*Country*	*kW*	*M*	*M*	*S*	*N*
		Peking	China Rep					
		Meyerton	South Africa	100				_
		Athinai	Greece	100	/			_
		Madras	India	100	/			
9.520	31.51	Pt Moresby	Papua New Guinea	10/50	∠	_	_	_
		Bonanza	Niçaragua	0.1	∠	_	_	_
		Manzini	Swaziland	25		_		
		Abu Zaabal	Egypt	100			_	
		Blagovechtchen	USSR	100		_	_	
		Kazan	USSR	100				_
		Armavir	USSR	150/500	∠	_	_	_
		Tachkent	USSR	100	_			_
		Holzkirchen	Germany (W)	10			_	
		Lampertheim	Germany (W)	50/100	/	_	_	_
		Biblis	Germany (W)	100	_	_	_	_
		Playa de Pals	Spain	250/500	_	_	_	_
		Noblejas	Spain	350	∠	_	_	_
		Santiago	Chile	100	/			_
9.525	31.50	Scituate	USA	50/100	_			
		Okeechobee	USA	100				_
		Cincinnati	USA	250	∠	_	_	_
		Habana	Cuba	50	∠	_	_	_
		Tokyo Yamata	Japan	100	∠	_	_	_
		Warszawa	Poland	100	∠	_	_	_
		Aligarh	India	250	_	_	_	_
		Bombay	India	100	∠	_	_	_
		Delhi	India	100	/			
		Mt Carlo	Monaco	100	_			_
		Allouis	France	100/500	_	_	_	_
		Las Mesas	Canary Is	50	_	_	_	_
		Kimjae	Korea (S)	180/250	∠	_	_	_
		Peking	China Rep					
9.530	31.48	London	UK	100/250				_
		Limassol	Cyprus	100				_
		Amman	Jordan	100	∠	_	_	_
		Greenville	USA	50/500	∠	_	_	_
		Redwood City	USA	250		_		
		Dar es Salaam	Tanzania	20	_	_	_	_
		Vinnitsa	Ukraine	50		_		
		Moskva	USSR	240/500	∠	_	_	_
		Vladivostock	USSR	100	∠			_
		Komsomolskamur	USSR	100			_	
		Frunze	USSR	100/500				_
		Poro	Philippines	50	_	_	_	_
		Tokyo Yamata	Japan	100	∠	_	_	_
		Tanger	Morocco	35/100			_	
		Bucuresti	Roumania	18/120	∠	_	_	_

MHz	Metres	Station	Country	kW	M	M	S	N
		Santiago	Chile	10/100	_			
		Athinai	Greece	100	/_	_	_	_
		Kavalla	Greece	250				_
		Peking	China Rep					
		Sackville	Canada	50/250	_	_		_
		Sofia	Bulgaria	500	_	_	_	_
		Noblejas	Spain	350	/_	_	_	_
		Singapore	Singapore	50	/_	_	_	_
		Cyclops	Malta	250		_		
		Dacca	Bangladesh	100	/			_
		Franceville	Gabon Rep	500			_	
		Ismaning	Germany (W)	100				_
		Calcutta	India	10	/			
9.535	31.46	Stapok	Malaysia	10			_	
		Schwarzenburg	Switzerland	150	_			
		Sarnen	Switzerland	250	/_	_	_	_
		Aligarh	India	250	/			_
		Delhi	India	50/100	_	_	_	_
		Luanda	Angola	100	/		_	_
		Nagoya	Japan	0.6	/_	_	_	_
		Simferopol	Ukraine	100	/_			_
		Sackville	Canada	50/250	/_	_		_
		Bonaire Zuid	Neth Antilles	50/ 250		_		_
		K Wusterhausen	Germany (E)	100				
		Allouis	France	100/500	_	_	_	
		Kimjae	Korea (S)	100	/_	_	_	_
		Kamalabad	Iran	100	_			
		Malolos	Philippines	100	/_	_	_	_
		Dacca	Bangladesh	100	/			
		Peking	China Rep					
9.540	31.45	London	UK	250				_
		Limassol	Cyprus	100	_	_	_	_
		Lyndhurst	Australia	10	/_	_	_	_
		Minsk	Bielorussia	50	/	_	_	
		Moskva	USSR	50/240	/_	_	_	_
		Tachkent	USSR	100	/_	_	_	_
		Petropavlo Kam	USSR	200	/_	_	_	_
		Warszawa	Poland	8/100	/_	_	_	_
		Velkekostolany	Czechoslovakia	120	/_	_	_	_
		Tanger	Morocco	25/100				_
		Kavalla	Greece	250	/_	_	_	
		Allouis	France	100/500	_	_	_	_
		Ismaning	Germany (W)	100		_	_	
		Lampertheim	Germany (W)	100				_
		Malolos	Philippines	100	/_	_	_	_
		Greenville	USA	50	/_	_	_	_
		Bucuresti	Roumania	120/250	/_			_
		Kimjae	Korea (S)	100		_		

MHz	Metres	Station	Country	kW	M	M	S	N
		Jerusalem	Israel	20/300				_
		Peking	China Rep					
		Careysburg	Liberia	250		_		
		Dacca	Bangladesh	100	/			_
9.545	31.43	Antigua	Br W Indies	250	∠	_	_	_
		Montserrat	Br W Indies	15/50	∠	_	_	_
		Khabarovsk	USSR	100		_	_	
		Tula	USSR	50	/			_
		Tema	Ghana	100	∠	_	_	_
		Curityba	Brasil	7.5	∠	_	_	_
		Julich	Germany (W)	100	∠	_	_	_
		Wertachtal	Germany (W)	500	∠	_	_	_
		Delano	USA	250	∠	_	_	_
		Dixon	USA	200				_
		Vera Cruz	Mexico	0.5	∠	_	_	_
		Tinang	Philippines	250	∠	_	_	_
		Islamabad	Pakistan	10/250	∠	_	_	_
		Honiara	Solomon Is	5	_	_	_	
		Mt Carlo	Monaco	100	_		_	
		Mahe	Seychelles	25	_	_		
		Kavalla	Greece	250	∠	_	_	
		Beyrouth	Lebanon	100	_	_	_	_
9.550	31.41	Dar es Salaam	Tanzania	100	_	_	_	_
		Habana	Cuba	10/50	∠	_	_	_
		Makassar	Indonesia	7.5	∠	_	_	_
		Pori	Finland	15/250	_			_
		Fredrikstad	Norway	100	∠		_	_
		Moskva	USSR	240/500	∠	_	_	_
		Tachkent	USSR	100/500	∠			_
		Peking	China Rep					
		S M Galeria	Vatican	100				_
		Dacca	Bangladesh	7.5	∠			_
		Monrovia	Liberia	50	∠	_	_	_
		Male	Maldives	15	_	_	_	_
		Noblejas	Spain	350	/	_	_	_
		Bucuresti	Roumania	18/250	∠	_	_	_
		Allouis	France	100/500	∠	_	_	_
		Tokyo Kawagu	Japan	1	∠	_	_	_
		Shepparton	Australia	100	_	_	_	
		Bonaire Zuid	Neth Antilles	50		_	_	
		Santiago	Chile	100	/	_	_	_
		Bombay	India	10	/			
9.555	31.40	London	UK	100/250	∠	_	_	_
		Tula	USSR	100		_		
		Irkutsk	USSR	100	∠	_	_	_
		Kenga	USSR	240	/	_		
		Mexico	Mexico	0.5/1	∠	_	_	_
		Tinang	Philippines	250	∠	_	_	_
		La Paz	Bolivia	10	∠	_	_	_

MHz	Metres	Station	Country	kW	M	M	S	N
		Okeechobee	USA	100				_
		Redwood City	USA	250	∠	_	_	
		Salman Pack	Iraq	50	_	_	_	_
		Kimjae	Korea (S)	250	∠	_	_	_
		Bata	Guinea	50	∠	_	_	_
		Lisbonne	Portugal	50			_	
		Biblis	Germany (W)	100		_	_	_
		Lampertheim	Germany (W)	20/100	∠	_	_	_
		Kavalla	Greece	250	∠	_	_	
9.560	31.38	Amman	Jordan	100	∠	_	_	_
		Serpukhov	USSR	240	∠	_	_	_
		Krasnoiarsk	USSR	100	_			
		Allouis	France	100	_	_	_	_
		Sofia	Bulgaria	50/500	∠	_	_	_
		Meyerton	Australia	100	∠	_	_	_
		Carnarvon	Australia	100/250	∠	_		
		Shepparton	Australia	100	_	_	_	_
		Schwarzenburg	Switzerland	150	∠	_	_	_
		Quito	Ecuador	100	_	_	_	_
		Tinang	Philippines	250	_	_	_	
		Noblejas	Spain	350	/			_
9.565	31.36	London	UK	250	∠	_	_	
		Greenville	USA	50/250	∠	_	_	_
		Delano	USA	200	∠	_	_	_
		Tripoli	Libya	100	∠	_	_	
		Kigali	Rwanda	250	∠	_	_	_
		Simferopol	Ukraine	100			_	
		Irkutsk	USSR	240		_	_	
		Frunze	USSR	50/500	∠	_		_
		Recife	Brasil	10/15	∠	_	_	_
		Julich	Germany (W)	100	∠	_	_	_
		Cyclops	Malta	250	∠	_	_	_
		Holzkirchen	Germany (W)	10	∠	_	_	_
		Lisbonne	Portugal	50/250	∠	_	_	_
		Sines	Portugal	250	_		_	
		Kimjae	Korea (S)	250		_		
		Aligarh	India	250	/			_
		Delhi	India	20	/			
		Pori	Finland	100/250			_	_
9.570	31.35	London	UK	250	∠	_	_	_
		Kranji	Singapore	125	/			_
		Bucuresti	Roumania	18/250	∠	_	_	_
		Doha	Qatar	100	∠	_	_	_
		Shepparton	Australia	50	∠	_	_	_
		Arganda	Spain	100	/	_	_	_
		Noblejas	Spain	350	∠	_		
		Jaji	Niger	10	∠	_	_	_
		Warszawa	Poland	30/100	∠	_	_	_
		Bonaire Zuid	Neth Antilles	50	∠	_	_	_

MHz	Metres	Station	Country	kW	M	M	S	N
		Malolos	Philippines	100	_			
		Santiago	Chile	10	∠	_	_	_
			USSR					
9.575	31.33	Bombay	India	100	∠	_	_	_
		Delhi	India	20/100	_	_	_	_
		Taipei	China Nat	10				
		Roma	Italy	60/100	∠	_	_	_
		Godthaab	Greenland	10	∠	_	_	_
		Armavir	USSR	240	∠			_
		Tchita	USSR	100				_
		Irkutsk	USSR	100	∠	_	_	
		Vinnitsa	Ukraine	240	_			
		Pt Moresby	Papua New Guinea	10	∠	_	_	_
		Mt Carlo	Monaco	100	/		_	_
		Kimjae	Korea (S)	100			_	
		Kamalabad	Iran	350	_			
		Pori	Finland	100/250	/		_	_
		Sackville	Canada	250				_
			Philippines					
		Sulaibiyah	Kuwait					
9.580	31.32	London	UK	100/250	∠	_	_	_
		Ascension	Ascension	125/250	∠	_	_	_
		Kranji	Singapore	250	/		_	_
		Limassol	Cyprus	100	_	_	_	
		Kazan	USSR	200	∠	_	_	_
		Blagovechtchen	USSR	100	∠			_
		Starobelsk	Ukraine	240/500	∠			
		Kiev	Ukraine					
		Malolos	Philippines	50	_	_	_	_
		Shepparton	Australia	50/100	∠	_	_	_
		Lusaka	Zambia	50	∠	_	_	_
		Sackville	Canada	50	_	_	_	_
		Greenville	USA	50	∠	_	_	_
		Arganda	Spain	100	/			
		Noblejas	Spain	350	_	_	_	_
		Suwon	Korea (S)	50	∠	_	_	_
		Tanger	Morocco					
		Roma	Italy	60/100	/	_	_	_
9.585	31.30	S Paulo	Brasil	7.5/50	∠	_	_	_
		Mogadiscio	Somalia	5	∠	_	_	_
		Bizam Radio						
		Meyerton	South Africa	250/500	∠	_	_	_
		Jaszbereny	Hungary	250		_	_	
		Tinang	Philippines	250				_
		Diosd	Hungary	100	∠	_	_	_
		Tokyo Yamata	Japan	100	∠		_	_
		Quito	Ecuador	100/500	_	_	_	

MHz	Metres	Station	Country	kW	M	M	S	N
		V of Communist Party of Turkey						
		Bata	Guinea	50	∠	_	_	_
		Kavalla	Greece	250			_	
		Wien	Austria	100	/	_	_	_
		Sines	Portugal	250	_	_	_	
		Dacca	Bangladesh	100			_	
		Franceville	Gabon Rep	500			_	
		Pori	Finland	100				_
9.590	31.28	Montserrat	Br W Indies	15/50	_	_	_	_
		Masirah	Oman	100	/		_	_
		Kranji	Singapore	250				_
		Limassol	Cyprus	7.5/100	∠	_	_	_
		Bucuresti	Roumania	120/250	∠	_	_	_
		Bonaire Noord	Neth Antilles	300	∠	_	_	_
		Aligarh	India	250	_	_	_	_
		Delhi	India	100	_	_	_	_
		Madras	India	10	/			
		Kazan	USSR	240	_	_		_
		Irkutsk	USSR	240		_		
		Leningrad	USSR	100	∠			_
		Omsk	USSR	240	∠	_	_	_
		Starobelsk	Ukraine	240	/	_	_	
		Wertachtal	Germany (W)	500	∠	_	_	
		Julich	Germany (W)	100	/	_		
		Fredrikstad	Norway	100/250	/	_	_	_
		S Domingo	Dominican Rep	0.3	∠	_	_	_
		Manzini	Swaziland	25/250		_	_	_
		Mt Carlo	Monaco	100	/		_	
		Kathmandu	Nepal	100	∠		_	_
		Hoerby	Sweden	350	_		_	_
		Malolos	Philippines	50/100	/		_	_
		Sackville	Canada	50/250	/	_	_	_
		Islamabad	Pakistan	10				
		V of Front for Redemption of Somalia						
9.595	31.27	Montevideo	Uruguay	20	∠	_	_	_
		Biblis	Germany (W)	100	∠	_	_	_
		Lisbonne	Portugal	50/100	∠	_	_	_
		Tokyo Nagara	Japan	50	∠	_	_	_
		Salvador	Brasil	10	∠	_	_	_
		Bucuresti	Roumania	18/120	/	_	_	_
		Noblejas	Spain	350	_	_	_	_
		Dar es Salaam	Tanzania	50	_	_	_	_
		Kimjae	Korea (S)	100		_		
		R Kulmis	USSR					
9.600	31.25	London	UK	100/250	∠	_	_	_
		Ascension	Ascension	250	∠	_	_	_

MHz	*Metres*	*Station*	*Country*	*kW*	*M*	*M*	*S*	*N*
		Limassol	Cyprus	100		_		_
		Praha	Czechoslovakia	250	/_	_	_	_
		Mexico	Mexico	1	/_	_	_	_
		K Wusterhausen	Germany (E)	100	/_			_
		Sorong	Indonesia	5	_			
		Moskva	USSR	240	/_	_	_	_
		Tachkent	USSR	50	/_	_	_	_
		Kenga	USSR	240		_	_	
		Okhotsk	USSR	50	/_	_	_	_
		Redwood City	USA	250	/			
		Greenville	USA	500				_
		Shepparton	Australia	100	/_	_	_	_
		Noblejas	Spain	350	/_	_	_	_
		Taipei	China Nat					
9.605	31.23	Limassol	Cyprus	100	_		_	_
		Allouis	France	100/500	_	_	_	_
		Stapok	Malaysia	10			_	
		Malolos	Philippines	100	/_	_	_	_
		Tinang	Philippines	50	_	_	_	_
		Brasilia	Brasil	10	/			_
		Serpukhov	USSR	50	/_	_	_	_
		Duchanbe	USSR	500				_
		S Cruz	Bolivia	5	/_	_	_	_
		Praha	Czechoslovakia	120	/_	_	_	_
		Wertachtal	Germany (W)	500	/_	_	_	_
		Julich	Germany (W)	100	/_	_	_	_
		Sackville	Canada	250	/_	_	_	_
		Tokyo Yamata	Japan	100	/_	_	_	_
		Mahe	Seychelles	30/100	_			
		S M Galeria	Vatican	100	/_	_	_	_
		Hoerby	Sweden	350	/_	_	_	_
		Karlsborg	Sweden	350			_	_
		Dacca	Bangladesh	100	_	_		
		Quito	Ecuador	100				
		Fredrikstad	Norway	120/250	_	_	_	_
		Riyadh	Saudi Arabia	350	/_	_	_	_
		Cyclops	Malta	250	/	_	_	
		Sines	Portugal	250	/_	_	_	
		Wien	Austria	100	_	_		
9.610	31.22	Limassol	Cyprus	100	/			
		Gedja	Ethiopia	10/100	_			
		Fredrikstad	Norway	100	_	_		_
		Julich	Germany (W)	100	/_	_	_	_
		Wertachtal	Germany (W)	500	/_	_	_	_
		Alma Ata	USSR	240/500	/		_	_
		Armavir	USSR	240/500				_
		Tula	USSR	240	/_	_	_	
		Khabarovsk	USSR	240	/_	_	_	_
		Nouakchott	Mauretania	30				_

MHz	*Metres*	*Station*	*Country*	*kW*	*M*	*M*	*S*	*N*
		Perth	Australia	10/50	/_	_	_	_
		Rio de Janeiro	Brasil	10	/_	_	_	_
		Brazzaville	Congo	50	/_	_	_	_
		Alger	Algeria	40/100	/_	_	_	_
		Sines	Portugal	250	_	_	_	_
		Cyclops	Malta	250	/_	_	_	_
		Wavre	Belgium	250	_	_		
		Allouis	France	100				_
		Mt Carlo	Monaco	100	/_	_	_	_
		Kimjae	Korea (S)	250				_
		Athinai	Greece	100			_	_
		Meyerton	South Africa	500	/		_	_
		Bonaire Zuid	Neth Antilles	50				_
9.615	31.20	London	UK	100	/	_	_	_
		Limassol	Cyprus	20	/		_	_
		Aligarh	India	250	_	_	_	_
		Delhi	India	100	/_	_	_	_
		Redwood City	USA	50	/_	_	_	_
		S Jose	Costa Rica	3/50	/_	_	_	_
		Marulas	Philippines	2.5	/_	_	_	_
		Tinang	Philippines	250		_	_	_
		Tanger	Morocco	50	/			_
		Rabat	Morocco	50/100	/			
		S M Galeria	Vatican	100	/_	_	_	_
		S Gabriel	Portugal					
		Sines	Portugal	250	/_	_	_	_
		Julich	Germany (W)	100				_
		Kavalla	Greece	250	/_	_	_	_
		Careysburg	Liberia	250				_
		Dacca	Bangladesh	7.5	_	_	_	
		Wavre	Belgium	100	/		_	_
		Mahe	Seychelles	25		_		
		Gedja	Ethiopia	100	/	_	_	_
		Allouis	France	100				_
		Alger	Algeria					
9.620	31.19	London	UK	250	/		_	_
		Limassol	Cyprus	100	/			
		Belgrade	Yugoslavia	100	/_	_	_	_
		Armavir	USSR	240	/_			_
		Moskva	USSR	240	/_	_	_	_
		Vladivostock	USSR	100	/_	_	_	_
		Montevideo	Uruguay	20	/_	_	_	_
		Ho Chi Minh City	Vietnam	50				
		Maputo	Mozambique	10	/_	_	_	_
		Leipzig	Germany (E)	100		_	_	
		Abu Dhabi	United Arab Emirates	120/ 250	/_	_	_	_
		Greenville	USA	50/250	/_	_	_	_
		Ismaning	Germany (W)	100	/_	_	_	

MHz	Metres	Station	Country	kW	M	M	S	N
		Quito	Ecuador	100/500	L	_	_	_
		V of Malayan Revln						
		Abu Zaabal	Egypt	100			_	
		Wien	Austria	100	_	_	_	_
		Bonaire Zuid	Neth Antilles	50	_	_		
		Wellington	New Zealand	7.5	_	_	_	
		S Gabriel	Portugal	100	/	_	_	_
		Allouis	France	100				_
9.625	31.17	London	UK	100				_
		Sackville	Canada	50/250	L	_	_	_
		Aligarh	India	100	_	_	_	_
		Delhi	India	20/50	/			
		S M Galeria	Vatican	100	L	_	_	_
		Schwarzenburg	Switzerland	100	_			
		Bucuresti	Roumania	18/120	L	_	_	_
		Moskva	USSR	240				_
		Tbilisi	USSR	240		_	_	
		Tchita	USSR	240	L			_
		Tinang	Philippines	250		_	_	_
		Peking	China Rep					
		Biblis	Germany (W)	20/100	L	_		
		Lampertheim	Germany (W)	100			_	
		Playa de Pals	Spain	100	/			_
		Greenville	USA	50/250		_	_	_
		Redwood City	USA	250			_	
		Mt Carlo	Monaco	100	_			
		Ho Chi Minh City	Vietnam					
9.630	31.15	Noblejas	Spain	350/700	L	_	_	_
		Delhi	India	20/100	L	_	_	_
		Velkekostolany	Czechoslovakia	120	L	_	_	_
		Moskva	USSR	50		_		
		Serpukhov	USSR	200	L		_	_
		Roma	Italy	60/100				_
		Tinang	Philippines	250	L	_	_	_
		Karlsborg	Sweden	350	L	_	_	_
		Jerusalem	Israel	50/300	L	_	_	_
		Mt Carlo	Monaco	100	_	_		
		Santiago	Chile	10	L	_	_	_
		Tanger	Morocco	100	/	_	_	
		Bonaire Noord	Neth Antilles	300	L	_	_	_
		Careysburg	Liberia	250				_
		Okeechobee	USA	100				
		Julich	Germany (W)	100	L	_	_	_
		Lopik	Netherlands	100	/		_	_
		Riyadh	Saudi Arabia	350				_
9.635	31.14	London	UK	100/250	L	_	_	_
		Aparacida	Brasil	7.5	L	_	_	_
		Bamako	Malawi	18		_	_	

MHz	Metres	Station	Country	kW	M	M	S	N
		Greenville	USA	250/500	∠	_	_	_
		Singapore	Singapore	50	∠	_	_	_
		Ivanofrankovsk	Ukraine	240				_
		Kaunas	USSR	50		_	_	
		Vladivostock	USSR	50	∠	_	_	_
		Bogota	Colombia	25	∠	_	_	_
		Babel	Iraq	500	_	_		
		Beira	Mozambique	100	∠	_	_	_
		Hoerby	Sweden	350	_			
		Sines	Portugal	250	∠	_	_	_
		S Gabriel	Portugal	100	_	_	_	
		Sackville	Canada	250				
		Quito	Ecuador					
9.640	31.12	London	UK	100/250	∠	_	_	_
		Antigua	Br W Indies	250	∠	_	_	_
		Montserrat	Br W Indies	50	_			
		Limassol	Cyprus	7.5/100	∠	_	_	_
		Greenville	USA	250/500	∠			_
		Julich	Germany (W)	100	∠	_	_	_
		Wertachtal	Germany (W)	500	∠	_	_	_
		Montevideo	Uruguay	10	∠	_	_	_
		Moskva	USSR	240	∠	_	_	_
		Bucuresti	Roumania	120	/			_
		Suwon	Korea (S)	50	∠	_	_	_
		Mt Carlo	Monaco	100	_		_	
		Caracas	Venezuela	10	∠	_	_	_
		Carnarvon	Australia	100	∠	_	_	_
		Alger	Algeria	50	∠	_	_	_
		Malolos	Philippines	100	_			
		Kavalla	Greece	250	/			
		Athinai	Greece	100	/			
9.645	31.10	London	UK	100/250		_		
		Fredrikstad	Norway	100	∠	_	_	_
		Karachi	Pakistan	10/100	∠	_	_	_
		Islamabad	Pakistan	100/250	_		_	_
		S Paulo	Brasil	7.5	∠	_	_	_
		Pocas Caldas	Brasil	7.5	∠	_	_	_
		S Jose	Costa Rica	1	∠	_	_	_
		S M Galeria	Vatican	100	∠	_	_	_
		Novosibirsk	USSR			_	_	
		Khabarovsk	USSR	50	∠	_		_
		Nauen	Germany (E)	500		_	_	
		Delhi	India	50	_	_	_	_
		Malolos	Philippines	100	∠	_	_	_
		Tinang	Philippines	250		_	_	_
		Pori	Finland	100	_			
		Wavre	Belgium	100	/			
			Australia					
9.650	31.09	Greenville	USA	50/500	∠	_	_	_

MHz	*Metres*	*Station*	*Country*	*kW*	*M*	*M*	*S*	*N*
		Delano	USA	250	/		_	_
		Dixon	USA	250	_	_		
		Conakry	Guinea	100	/_	_	_	_
		Baku	USSR	100		_	_	
		Moskva	USSR	240	/_	_	_	_
		Montevideo	Uruguay	10	/_	_	_	_
		Poro	Philippines	50	/			_
		Tinang	Philippines	50/250	/_	_	_	_
		Ismaning	Germany (W)	100	/_		_	_
		Julich	Germany (W)	100	/_	_	_	_
		Sines	Portugal	250	/_	_	_	_
		Magwa	Kuwait	50/250	/_	_	_	_
		Cyclops	Malta	250	/_	_	_	_
		Wien	Austria	100	_	_	_	_
		Tanger	Morocco	100	/_		_	_
		Santiago	Chile	10	_			
		Mahe	Seychelles	25	_	_		
		Tripoli	Libya					
		Kimjae	Korea (S)	250			_	_
		Meyerton	South Africa	250		_		
		R of the Patriots						
		Mt Carlo	Monaco					
		Quito	Ecuador	100	/			
		Jerusalem	Israel					
9.655	31.07	Patumthani	Thailand	100	/_	_	_	_
		Tripoli	Libya	100	/_	_	_	_
		Sackville	Canada	50/250	/_	_	_	_
		Jaszbereny	Hungary	250	/	_	_	_
		Diosd	Hungary	100	/_			_
		Armavir	USSR	100	/_			_
		Frunze	USSR	240		_	_	
		Orcha	Bielorussia	100		_	_	
		Mt Carlo	Monaco	100	/			
		Habana	Cuba	10/50		_	_	
		Athinai	Greece	100	/_	_	_	_
		Kavalla	Greece	250	_	_	_	_
		Thessaloniki	Greece	35	/			
		Bogota	Colombia	10	/_	_	_	_
		Wavre	Belgium	100/250	_	_	_	_
		Lhasa	Tibet					
9.660	31.06	London	UK	250	/_	_	_	_
		Luanda	Angola	100	/			
		Brisbane	Australia	10	/_	_	_	_
		Tachkent	USSR	240	_			
		Kavalla	Greece	250	/_	_	_	_
		Dixon	USA	250	/_	_	_	_
		Okeechobee	USA	100			_	
		Delano	USA	250				_
		Tanger	Morocco	100		_		

MHz	*Metres*	*Station*	*Country*	*kW*	*M*	*M*	*S*	*N*
		Kinshasa	Zaire	50	∟	_	_	_
		Islamabad	Pakistan	100	∟	_	_	
		Karachi	Pakistan	10/250	_		_	_
		Lopik	Netherlands	100	_	_	_	_
		Karlsborg	Sweden	350	/		_	_
		Hoerby	Sweden	350	/	_	_	_
		Allouis	France	100/500	∟	_	_	_
		Lampertheim	Germany (W)	100		_	_	
		Biblis	Germany (W)	100	∟			_
		Playa de Pals	Spain	250			_	_
		Pori	Finland	100	_			
		Sorong	Indonesia	5	/	_	_	_
		Wien	Austria	100	_			
		Mt Carlo	Monaco	100		_	_	
		Sines	Portugal	250	_			
		Schwarzenburg	Switzerland	150/ 250	/			_
		Kimjae	Korea (S)	100				_
		Tripoli	Libya					
		Caracas	Venezuela					
9.665	31.04	Kajang	Malaysia	50	∟	_	_	_
		Brasilia	Brasil	7.5/10	/			_
		Nairobi	Kenya	100	∟	_	_	_
		Nauen	Germany (E)	500	/			_
		K Wusterhausen	Germany (E)	50	∟			_
		Leipzig	Germany (E)	100	_			_
		Voronej	USSR	240	_			_
		Kinghisepp	USSR	240		_		
		Tchita	USSR	100	∟			_
		Ivanofrankovsk	Ukraine	240	∟	_	_	_
		Vinnitsa	Ukraine	100		_		
		Karlsborg	Sweden	350	_	_		
		Hoerby	Sweden	350	_	_		
		Suwon	Korea (S)	50	_	_		
		Ankara	Turkey	250	/	_	_	_
		Allouis	France	100		_		_
		Quito	Ecuador	100	/	_		_
		Sines	Portugal	250	/		_	_
9.670	31.02	Greenville	USA	50/500	∟	_	_	_
		Montevideo	Uruguay	10	∟	_	_	_
		Jeddah	Saudi Arabia	50/ 100	∟	_	_	_
		Irkutsk	USSR	100	∟	_	_	_
		Kiev	Ukraine	100	∟	_	_	_
		Sines	Portugal	250	∟	_	_	_
		Peking	China Rep					
		Careysburg	Liberia	250	/	_	_	_
		Tinang	Philippines	250				_
		Cyclops	Malta	250	/	_	_	

MHz	Metres	Station	Country	kW	M	M	S	N
		Tokyo Yamata	Japan	200	/_	_	_	_
		Shepparton	Australia	10/100	/_	_	_	_
9.675	31.01	Florianapolis	Brasil	10	/_	_	_	_
		Tokyo Yamata	Japan	50/100	/_	_	_	_
		Armavir	USSR	100/240	/_	_	_	_
		Voronej	USSR	240	/_	_		_
		Novosibirsk	USSR	100			_	
		Simferopol	Ukraine	240	_			_
		Warszawa	Poland	100	/_	_	_	_
		Aligarh	India	250	/_	_	_	_
		Delhi	India	20/100	/_	_	_	_
		Abu Zaabal	Egypt	100				
		Mt Carlo	Monaco	100	/_	_	_	_
		Kimjae	Korea (S)	50	_	_		
		Athinai	Greece	100	/_	_	_	_
		Santiago	Chile	45	_			
		Arganda	Spain	100	/		_	_
		Sines	Portugal	250	/		_	_
9.680	30.99	London	UK	250	/_	_	_	_
		Lyndhurst	Australia	10	/_	_	_	_
		Meyerton	South Africa	100	/_	_	_	_
		Montevideo	Uruguay	10	/_	_	_	_
		Tchita	USSR	100		_	_	
		Kavalla	Greece	250	/_	_	_	_
		Jakarta	Indonesia	50	/_	_	_	_
		Quito	Ecuador	100	/_	_	_	_
		Delano	USA	250		_		
		Greenville	USA	250	/_	_	_	_
		Ismaning	Germany (W)					
		Lampertheim	Germany (W)	100		_	_	
		Holzkirchen	Germany (W)	10				_
		Biblis	Germany (W)	100	_		_	
		Playa de Pals	Spain	100	/_	_	_	_
		Wavre	Belgium	100	_			
		Alger	Algeria	100	/_	_	_	_
		Dacca	Bangladesh	100			_	
		Tanger	Morocco	100				_
9.685	30.98	Alger	Algeria	50/100	/_	_	_	_
		Erevan	USSR	100	/			
		Moskva	USSR	240/500	/_	_	_	_
		Irkutsk	USSR	240/500	/_			_
		Lvov	Ukraine	100		_	_	
		Panama	Panama	1	/_	_	_	_
		S Paulo	Brasil	7.5	/			_
		Habana	Cuba	50	/_		_	_
		Bucuresti	Roumania	120	/_			_
		Kampala	Uganda	250				_
		Schwarzenburg	Switzerland	150	_			
		Wavre	Belgium	100	/	_	_	_

MHz	Metres	Station	Country	kW	M	M	S	N
		Arganda	Spain	100	∠	_	_	_
		Taipei	China					
		Redwood City	USA	250			_	
		Cincinnati	Portugal	175	∠	_	_	_
		Sines	Portugal			_		
		Santiago	Chile	100	/			_
		Quito	Ecuador					
		Sofia	Bulgaria	500				
9.690	30.96	London	UK	250	∠	_	_	
		Antigua	Br W Indies	250	∠	_	_	_
		Limassol	Cyprus	20/100	∠	_	_	_
		Julich	Germany (W)	100	∠	_	_	_
		Tula	USSR	100	∠	_	_	_
		Gral Pacheo	Argentina	100	∠	_	_	_
		Islamabad	Pakistan	100/250	_		_	
		Karachi	Pakistan	50	∠	_	_	_
		Okeechobee	USA	100	∠		_	_
		Redwood City	USA	250	/			
		Scituate	USA	50	_			
		Bucuresti	Roumania	18/250	∠	_	_	_
		Santiago	Chile	10	_			
		Malolos	Philippines	50/100	_			
		Noblejas	Spain	350	_			
		Kavalla	Greece	250	∠	_	_	_
		Allouis	France	100/500	/	_	_	_
		Hoerby	Sweden	350		_		
		Karlsborg	Sweden	350			_	_
		Pt au Prince	Haiti	100				_
9.695	30.94	Limassol	Cyprus	100		_		
		Manaos	Brasil	7.5	∠	_	_	_
		Holzkirchen	Germany (W)	10			_	
		Biblis	Germany (W)	100	∠	_	_	_
		Lisbonne	Portugal	50/250	∠	_	_	_
		Kazan	USSR	240	_			_
		Frunze	USSR	100	_			_
		Petropavlo Kam	USSR	100	/			
		Tbilisi	USSR	240	_			_
		Karlsborg	Sweden	350	/		_	_
		Hoerby	Sweden	350	∠	_	_	_
		Allouis	France	100/500	_	_	_	_
		Noblejas	Spain	350	∠	_	_	_
		Rarotonga	Cook Is	0.5		_	_	
		Franceville	Gabon Rep	500			_	
		Phnom Penh	Cambodia					
9.700	30.93	London	UK	250				_
		Ascension	Ascension	125/250	_		_	
		Dixon	USA	100	∠	_	_	_
		Sofia	Bulgaria	50/145	∠	_	_	_
		Tchita	USSR	50	/	_	_	_

MHz	Metres	Station	Country	kW	M	M	S	N
		Tula	USSR	50	_			
		Kazan	USSR	100		_		
		Kenga	USSR	50	/			
		Carnarvon	Australia	250	_	_		_
		Julich	Germany (W)	100	/_	_	_	_
		Kigali	Rwanda	250	/_	_	_	_
		Tokyo Yamata	Japan	100/200	/_	_	_	_
		Kavalla	Greece	250	/_	_	_	_
		Malolos	Philippines	100	_			
		Careysburg	Liberia	250		_		
		Schwarzenburg	Switzerland	250	_			_
9.705	30.91	London	UK	100	/			
		Limassol	Cyprus	100			_	
		Gedja	Ethiopia	100		_	_	
		Aligarh	India	250	/_	_	_	_
		Delhi	India	20/100	/			_
		Holzkirchen	Germany (W)	10			_	
		Biblis	Germany (W)	100	/_	_		
		Lampertheim	Germany (W)	100				
		Playa de Pals	Spain	250			_	
		Lisbonne	Portugal	50	/_	_	_	_
		Tokyo Yamata	Japan	20	/_	_	_	_
		Rio de Janeiro	Brasil	7.5	/_	_	_	_
		Mexico	Mexico	10	/_	_	_	_
		Lvov	Ukraine	240	_			_
		Alger	Algeria	50	/_	_	_	_
		Sofia	Bulgaria					
		Allouis	France	100/500	_	_		_
		Abis	Egypt	250			_	
		Quito	Ecuador					
9.710	30.90	S Fernando	Argentina	6	/_	_	_	_
		Malherbes	Mauritius	10				_
		Mauritius	Mauritius	10	/	_	_	
		Kaunas	USSR	100	/	_	_	_
		Kiev	Ukraine	240	/_	_	_	
		Komsomolskamur	USSR	240		_		
		Frunze	USSR	240/500	/_	_		_
		Vladivostock	USSR	50				_
		Petropavlo Kam	USSR	100		_	_	
		Roma	Italy	60/100	/_	_	_	_
		Penang	Malaysia	10	/_	_	_	_
		Meyerton	South Africa	100	/_	_	_	_
		Thessalonika	Greece	35	_	_	_	_
		Koebenhaven	Denmark	50		_		
		Kimjae	Korea (S)	250	/		_	_
9.715	30.88	London	UK	100/250	_	_	_	_
		Limassol	Cyprus	20	_			
		Quito	Ecuador	100	/_	_	_	_
		Bocaue	Philippines	10/50	/_	_	_	_

MHz	Metres	Station	Country	kW	M	M	S	N
		Iba	Philippines	100	/			
		Lopik	Netherlands	100	∠	_	_	_
		Brazzaville	Congo	50	∠	_	_	_
		Kiev	Ukraine	100/240				_
		Orcha	Bielorussia	100	∠	_	_	_
		Sucre	Bolivia	2	∠	_	_	_
		Bonaire Noord	Neth Antilles	300	∠	_	_	_
		Tanger	Morocco	50	∠	_	_	_
		Greenville	USA	250				_
		Okeechobee	USA	100	_			
		Scituate	USA	50/100	∠	_	_	_
		Allouis	France	100/500	∠	_	_	_
		Sackville	Canada	50/250	_			
		Ismaning	Germany (W)	100	∠	_		
		Lampertheim	Germany (W)	100				_
		Playa de Pals	Spain	100				_
		Careysburg	Liberia	250	/	_	_	
		Santiago	Chile	100	/	_		_
		Madras	India	100	/			
		Sines	Portugal	250	/	_	_	_
9.720	30.86	Ekala	Ceylon	10/35	∠	_	_	_
		Riazan	USSR	240/500	∠	_	_	_
		Kiev	Ukraine	240	_	_		
		Vinnitsa	Ukraine	240/500	/		_	_
		Diriyya	Saudi Arabia	50	∠	_	_	_
		Kimjae	Korea (S)	100/250	∠	_	_	_
		Kamalabad	Iran	100	_			_
		Dacca	Bangladesh					
		Hyderabad	India	10	/			
9.725	30.85	Kranji	Singapore	250	∠	_	_	_
		Nikolaevskamur	USSR	50	_	_		_
		Greenville	USA	50	_			
		Schwarzenburg	Switzerland	150	∠	_	_	_
		Tinang	Philippines	250	∠	_	_	_
		Noblejas	Spain	350	∠	_	_	_
		Mt Carlo	Monaco	100	_	_		
		Wien	Austria	100	∠	_	_	_
		Lisbonne	Portugal	100	∠	_	_	_
		Holzkirchen	Germany (W)	10	/	_		
		Biblis	Germany (W)	100	/	_	_	_
		Lampertheim	Germany (W)	50				_
9.730	30.83	Dixon	USA	100	∠	_	_	_
		Nauen	Germany (E)	50	∠	_	_	_
		Leipzig	Germany (E)	50	∠	_	_	_
		Serpukhov	USSR	100		_	_	
		Tula	USSR	100	∠			_
		Riyadh	Saudi Arabia	350	_		_	
		Rangoon	Burma	50	∠	_	_	_
		Wavre	Belgium	100	_		_	_

MHz	Metres	Station	Country	kW	M	M	S	N
		Kampala	Uganda	250				
		S M Galeria	Vatican	100				_
		Delhi	India	100	_	_	_	_
		Mt Carlo	Monaco	100	_			_
		Sackville	Canada	50				_
		Manzini	Swaziland	25		_	_	_
		Quito	Ecuador	100			_	
9.735	30.82	London	UK	100/250	/_	_	_	_
		Antigua	Br W Indies	125/250	/_	_	_	_
		Asuncion	Paraguay	3/100	/_	_	_	_
		Wertachtal	Germany (W)	500	/_	_	_	_
		Kigali	Rwanda	250	/_	_	_	_
		Lvov	Ukraine	500				_
		Tula	USSR	240	_	_	_	
		Duchanbe	USSR	500				_
		Nikolaevskamur	USSR	240	/_	_	_	_
		Novosibirsk	USSR	240/500	/_			_
		Irkutsk	USSR	100/500	_	_	_	
		Mt Carlo	Monaco	100	/	_	_	_
		Cyclops	Malta	250	/_	_	_	_
		Tanger	Morocco	100				_
		Kavalla	Greece	250				_
		Quetta	Pakistan	10	/			
9.740	30.80	Kranji	Singapore	250	/		_	_
		Lomas Mirador	Argentina	10	/_	_	_	_
		S Gabriel	Portugal	100	/_	_	_	_
		Moskva	USSR	120/240		_	_	
		Simferopol	Ukraine	100		_	_	
		Careysburg	Liberia	250	/_	_	_	
		Abis	Egypt	250				
		Velkekostolany	Czechoslovakia	120/200	/_	_	_	_
		Malolos	Philippines	100	/_	_	_	_
		Jerusalem	Israel	20/300	_		_	_
9.745	30.79	London	UK	100		_		
		Moskva	USSR	240	/_	_	_	_
		Kenga	USSR	500	/			_
		Quito	Ecuador	30/100	/_	_	_	_
		S Paulo	Brasil	1/7.5	/_	_	_	_
		Jayapura	Indonesia	5	/_	_	_	_
		Babel	Iraq	100/500	_	_		_
		Salman Pack	Iraq	500			_	
		Hoerby	Sweden	350	_			_
		Greenville	USA	250	_	_		
		Sofia	Bulgaria	500			_	_
		Yaounde	Cameroon					
		Noblejas	Spain	350	/			_
9.750	30.77	London	UK	100/250	/_	_	_	_
		Limassol	Cyprus	20/100	/_	_	_	_

MHz	Metres	Station	Country	kW	M	M	S	N
		Careysburg	Liberia	250	∠	_	_	_
		Madras	India	100	∠	_	_	_
		Krasnoiarsk	USSR	100	_	_		_
		Sverdlovsk	USSR	100	/	_	_	
		Petropavlo Kam	USSR	100		_	_	
		Tachkent	USSR	240			_	
		Vladivostock	USSR	100				_
		Komsomolskamur	USSR	50	∠			
		Simferopol	Ukraine	240		_	_	
		Papeete	Tahiti	4	∠	_	_	_
		Dar es Salaam	Tanzania	50	_			_
		Santiago	Chile	10	∠		_	_
		Tirane	Albania					
		Kajang	Malaysia		/			
		Dacca	Bangladesh	100	_	_		
		Bucuresti	Roumania	18/120	∠	_	_	_
		Playa de Pals	Spain	100	/			_
		Biblis	Germany (W)	100	∠		_	
		Holzkirchen	Germany (W)	10		_		
		Lampertheim	Germany (W)	100	∠		_	_
		Lisbonne	Portugal	250	_			
		Allouis	France	100	_		_	_
		Redwood City	USA	250	_			
9.755	30.75	Mokattam	Egypt	100			_	
		Goiana	Brasil	7.5	∠	_	_	_
		Duchanbe	USSR	100		_	_	
		Irkutsk	USSR	240			_	
		Nikolaevskamur	USSR	240	∠	_	_	_
		K Wusterhausen	Germany (E)	100	_	_		
		Cincinnati	USA	175/250	_	_		
		Delhi	India	100	∠	_	_	_
		Warszawa	Poland	30/60	∠	_	_	_
		Wavre	Belgium	250	∠	_	_	_
		Bonaire Zuid	Neth Antilles	50		_		
		Iba	Philippines	100	∠	_	_	
		Bocaue	Philippines	50	_	_		
		Santiago	Chile	100		_		
		Pori	Finland	100			_	_
		Sackville	Canada	50/250			_	_
		Allouis	France	100	/			
		V of Kampuchean People						
9.760	30.74	London	UK	100/250	∠	_	_	_
		Ascension	Ascension	250			_	
		Tanger	Morocco	35/100	∠	_	_	_
		Ivanofrankovsk	Ukraine	240			_	
		Khabarovsk	USSR	240	/	_	_	_
		Kinghisepp	USSR	240	∠			_
		Tbilisi	USSR	240	∠	_		_

MHz	Metres	Station	Country	kW	M	M	S	N
		Tokyo Nagara	Japan	10	∠	_	_	_
		Hurlingham	Argentina	20	∠	_	_	_
		Tirane	Albania					
		Tinang	Philippines	250	∠	_	_	_
		Quito	Ecuador	100				
		Shepparton	Australia	100	∠	_	_	_
		Athinai	Greece	100	∠	_	_	_
		Kavalla	Greece	250	∠	_	_	_
		Alger	Algeria	100	∠	_	_	_
		Cairo	Egypt					
9.765	30.72	Ascension	Ascension	250	_	_	_	
		Antigua	Br W Indies	125/ 250	/			_
		Julich	Germany (W)	100	∠	_	_	_
		Wertachtal	Germany (W)	500	∠	_	_	_
		Armavir	USSR	100		_	_	
		Kazan	USSR	100			_	
		Leningrad	USSR	240	/	_		_
		Lvov	Ukraine	240/500	∠	_	_	_
		Greenville	USA	500	∠	_	_	_
		Quito	Ecuador	100	∠	_	_	_
		Kamalabad	Iran	100		_	_	
		Habana	Cuba	10				_
		Kimjae	Korea (S)	100	∠	_	_	_
		Bocaue	Philippines	50	_			
		Schwarzenburg	Switzerland	500	/			
		Sottens	Switzerland	500	_			_
		Sofia	Bulgaria	500	_	_		
		Taipei	China Nat					
		Wien	Austria	100	_			
		Santiago	Chile	100	/			_
9.770	30.71	London	UK	100	∠	_	_	_
		Limassol	Cyprus	100	/	_	_	
		Ascension	Ascension	250	_			
		Jakarta	Indonesia	20	∠	_	_	
		Montevideo	Uruguay	10	∠	_	_	_
		Rio de Janeiro	Brasil	7.5/10	∠	_	_	_
		Serpukhov	USSR	100	/			
		Vladivostock	USSR	100	_			_
		Kenga	USSR	100	/			
		Wien	Austria	100	∠	_	_	_
		Poro	Philippines	100	∠	_	_	_
		Kinshasa	Zaire	10	∠	_	_	_
		Bonaire Noord	Neth Antilles	300	∠	_	_	_
		Kavalla	Greece	250	∠	_	_	_
		Habana	Cuba	10	∠	_	_	_
		Nauen	Germany (E)	100/ 500	/	_	_	_
		Tanger	Morocco	100	_		_	_

MHz	Metres	Station	Country	kW	M	M	S	N
		Wellington	New Zealand	7.5	_	_	_	
		Abu Zaabal	Egypt	100			_	
		Kamalabad	Iran	100	_			_
		Sanaa	Yemen	50			_	
		Lopik	Netherlands	100	_	_		
		Lampertheim	Germany (W)	50/100				_
		Shepparton	Australia	50/100	/	_	_	
		Cyclops	Malta	250		_	_	
		Rawalpindi	Pakistan	100	/			
9.775	30.69	Moskva	USSR	200				
9.780	30.67	Moskva	USSR					
		Tirane	Albania	240				
		Sanaa	Yemen					
9.785	30.66	Moskva	USSR	50/240				
9.790	30.64	Tirane	Albania	240				
		Moskva	USSR					
		Vladivostock	USSR					
		Karachi	Pakistan					
		Madrid	Spain					
9.795	30.63	Kazan	USSR	100				
		Minsk	USSR					
9.800	30.61	Moskva	USSR	100				
9.805	30.60	Cairo	Egypt	250				
9.810	30.58	Moskva	USSR	100				
9.815	30.57	Jerusalem	Israel					
			USSR					
9.820	30.55	Peking	China Rep					
		Jerusalem	Israel					
			USSR					
9.825	30.53	London	UK	100/250				
9.830	30.52		USSR					
9.833	30.51	Budapest	Hungary					
		Jerusalem	Israel					
9.840	30.49	Hanoi	Vietnam	100				
		Kuwait	Kuwait					
9.850	30.46	Cairo	Egypt	100				
		Moskva	USSR					
		Peking	China Rep					
9.858	30.43	London*	UK					
9.860	30.43	Peking	China Rep					
		Alger	Algeria					
9.870	30.40	Seoul	Korea (S)					
9.875	30.38	Julich*	Germany (W)					
9.880	30.36	Peking	China Rep					
9.887	30.34	Hanoi	Vietnam					
		Riyadh	Saudi Arabia					
9.893	30.32	Peking	China Rep					
9.895	30.32	Hilversum	Netherlands					
			China Rep					

MHz	Metres	Station	Country	kW
9.900	30.30	Peking	China Rep	
9.912	30.27	Delhi	India	100
9.915	30.26	London	UK	
9.920	30.24	Peking	China Rep	
9.929	30.21	(AFRTS)*	USA	
9.940	30.18	Peking	China Rep	
9.945	30.17	Peking	China Rep	
9.950	30.15	Delhi	India	
9.953	30.14	R Freedom for South Yemen		
9.965	30.11	Peking	China Rep	
9.977	30.07	Pyongyang	Korea (N)	
9.985	30.05	V of NUFC		
9.995	30.02	Cairo	Egypt	
10.000	30 00	Rugby Freq Std	UK	
		Honolulu Freq Std	Hawaii	
		Boulder Freq Std	USA	
10.009	29.98		USSR	
10.010	29.97	Kathmandu	Nepal	
		Hanoi	Vietnam	
10.040	29.88	Hanoi	Vietnam	
10.055	29.84	V of Iraqi Kurdistan		
10.060	29.82	Hanoi	Vietnam	
10.080	29.76	V of NUFK		
10.120	29.64	V of NUFC		
		Moskva*	USSR	
10.125	29.63			
10.190	29.44	R Liberty*		
10.225	29.34	Hanoi	Vietnam	
10.235	29.31	Greenville*	USA	
10.245	29.28	Peking	China Rep	
10.260	29.24	Peking	China Rep	
10.315	29.08	RFE/R Liberty*		
10.335	29.03	Delhi	India	
10.338	29.02	Moskva*	USSR	
10.356	28.97	Allouis	France	
10.380	28.90	Greenville*	USA	
10.385	28.89		Saudi Arabia*	
10.420	28.79	RFE/R Liberty*		
10.454	28.70	Greenville*	USA	
10.537	28.47		USA*	
10.615	28.26			
10.620	28.25		USSR*	
10.660	28.14		USSR*	
10.690	28.06		USSR*	
10.695	28.05		USSR*	
10.740	27.93	Moskva*	USSR	
10.761	27.88		Germany (W)*	
10.855	27.64		USSR*	

MHz	*Metres*	*Station*	*Country*	*kW*
10.865	27.61	Peking	China Rep	
10.869	27.60	Bethany*	USA	
		Greenville*	USA	
10.880	27.57	Monrovia*	Liberia	
10.905	27.51	Lisbon*	Portugal	
10.922	27.47		Germany (W)*	
10.972	27.34	Tangier*	Morocco	
11.000	27.27	Peking	China Rep	
11.033	27.19	Paris*	France	
11.040	27.17	Peking	China Rep	
11.100	27.03	Peking	China Rep	
11.290	26.57	Peking	China Rep	
11.300	26.55	Peking	China Rep	
11.330	26.48	Peking	China Rep	
11.340	26.46	Moskva	USSR	
11.350	26.43	Pyongyang	Korea (N)	
11.375	26.37	Peking	China Rep	
11.445	26.21	Peking	China Rep	
11.455	26.19	Peking	China Rep	
11.495	26.10		USSR*	
11.500	26.09	Peking	China Rep	
		Riyadh	Saudi Arabia	
11.505	26.08	Peking	China Rep	
11.515	26.05	Peking	China Rep	240
11.530	26.02	Peking	China Rep	
11.533	26.01	Pyongyang	Korea (N)	
11.568	25.93	Pyongyang	Korea (N)	
11.575	25.92	RFE/R Liberty		
			USSR*	
11.585	25.90		USSR	
11.600	25.86	Peking	China Rep	120
		V of Democratic Kampuchea		
11.602	25.85	Jerusalem	Israel	
11.610	25.84	Peking	China Rep	
11.620	25.82	Delhi	India	
		Jerusalem	Israel	
11.625	25.81	Jerusalem	Israel	
11.630	25.80	Cairo	Egypt	100
		Moskva	USSR	240
		Peking	China Rep	
11.635	25.78	Peking	China Rep	
		Jerusalem	Israel	
11.640	25.77	Karachi	Pakistan	
		Jerusalem	Israel	
11.645	25.76	Hargeisa	Somalia	5
11.650	25.75	Peking	China Rep	
		Dacca	Bangladesh	
11.655	25.74	Jerusalem	Israel	300

MHz	Metres	Station	Country	kW	M	M	S	N
11.660	25.73	Peking	China Rep					
		V of Democratic Kampuchea						
11.665	25.72	Peking	China Rep					
11.670	25.71		USSR					
11.672	25.70	Karachi	Pakistan					
11.675	25.70	Peking	China Rep	240				
		RFE*						
11.680	25.68	London	UK					
		Karachi	Pakistan					
11.685	25.67	Peking	China Rep					
		Riyadh	Saudi Arabia					
11.690	25.66	Moskva	USSR					
		Peking	China Rep					
		Kiev	Ukraine					
11.695	25.65	Peking*	China Rep	120/240				
11.700	25.64		USSR	50				
		Vatican	Vatican					
		Berlin	Germany	250				
		S Domingo	Dominican Rep	50				
		Tripoli	Libya					
		R of the Patriots						
		Jerusalem	Israel					
11.705	25.63	Limassol	Cyprus	100	/			
		Montserrat	Br W Indies	50	/			_
		Okeechobee	USA	100			_	
		Simferopol	Ukraine	500	/			
		Khabarovsk	USSR	100		_	_	
		Serpukhov	USSR	240	∠	_	_	_
		Karlsborg	Sweden	350	∠		_	_
		Hoerby	Sweden	350	/	_	_	_
		Nauen	Germany (E)	50	∠	_	_	_
		K Wusterhausen	Germany (E)	100	_			_
		Tokyo Yamata	Japan	100	∠	_	_	_
		S M Galeria	Vatican	100	∠	_	_	_
		Wellington	New Zealand	7.5	_			
		Mt Carlo	Monaco	100	∠	_	_	_
		Allouis	France	100/500	∠	_	_	_
		Jerusalem	Israel	300	_		_	_
		Shepparton	Australia	50/100	∠	_	_	_
		Carnarvon	Australia	100	_	_	_	
		Sackville	Canada	250	_			
		Santiago	Chile	25/100	_			
		Islamabad	Pakistan	100		_	_	_
		Karachi	Pakistan	50	∠			
		Mahe	Seychelles	100	_	_	_	
		Careysburg	Liberia	250		_	_	
		Bucuresti	Roumania	250		_	_	
		Bonaire Zuid	Neth Antilles	50		_	_	

MHz	Metres	Station	Country	kW	M	M	S	N
		Peking	China Rep					
		Julich	Germany (W)	100				
11.710	25.62	London	UK	250	∠	_	_	_
		Limassol	Cyprus	100		_	_	_
		Kranji	Singapore	250	∠		_	_
		Gral Pacheco	Argentina	100	∠	_	_	_
		Armavir	USSR	240	∠			_
		Frunze	USSR	50		_	_	
		Kazan	USSR	100	/			
		Komsomolskamur	USSR	100	_			
		Tbilisi	USSR	240		_	_	
		Noumea	New Caledonia	4	∠	_	_	_
		Tanger	Morocco	35/100				_
		Bucuresti	Roumania	120	∠	_		_
		Bonaire Zuid	Neth Antilles	50	/	_	_	
		Manzini	Swaziland	25			_	
		Careysburg	Liberia	250				_
		Peking	China Rep					
		Mt Carlo	Monaco					
11.715	25.61	London	UK	100	∠	_	_	
		Schwarzenburg	Switzerland	150	∠	_	_	_
		Delhi	India	20/100	_	_	_	_
		Jakarta	Indonesia	100/120	∠	_	_	_
		Orcha	Bielorussia	240		_	_	
		Komsomolskamur	USSR	50	/			
		Duchanbe	USSR	240/500	∠	_	_	_
		Sverdlovsk	USSR	100	/			
		Poro	Philippines	50	_	_		_
		Tinang	Philippines	250	/	_	_	_
		Abis	Egypt	250			_	
		S M Galeria	Vatican	100	∠	_	_	_
		Greenville	USA	500	/			_
		Noblejas	Spain	350	∠	_	_	_
		Peking	China Rep					
		Bata	Guinea	50	∠	_	_	_
		Quito	Ecuador	100	∠	_	_	_
		Riyadh	Saudi Arabia	350	_	_		
		Jerusalem	Israel	300	_		_	_
		Tangier	Morocco	100	/			_
		Santiago	Chile	100		_	_	
		Careysburg	Liberia	250		_		
		Carnarvon	Australia	250		_		
11.720	25.60	London	UK	100	_	_	_	
		Limassol	Cyprus	7.5/20	∠	_	_	
		Kinshasa	Zaire	10	∠	_	_	_
		Lvov	Ukraine	240		_		
		Erevan	USSR	100	_	_	_	
		Tula	USSR	50/240			_	_
		Frunze	USSR	100	∠	_		

MHz	*Metres*	*Station*	*Country*	*kW*	*M*	*M*	*S*	*N*
		Khabarovsk	USSR	240	/	_	_	
		Sackville	Canada	250	∠	_	_	_
		Peking	China Rep					
		Sottens	Switzerland	500	∠	_	_	_
		Schwarzenburg	Switzerland	100/ 150	/			_
		Leipzig	Germany (E)	100		_	_	
		Okeechobee	USA	100	/		_	_
		Shepparton	Australia	10/100	_	_	_	
		Wien	Austria	100	/			_
		Sofia	Bulgaria	100/500	∠	_	_	_
		Wertachtal	Germany (W)	500	∠	_	_	_
		Lopik	Netherlands	100	/	_	_	_
		Santiago	Chile	100	/	_	_	_
		Riyadh	Saudi Arabia	350	_			
11.725	25.59	S M Galeria	Vatican	100	/	_	_	_
		Delhi	India	20/100	/			
		Aligarh	India	250	_	_	_	_
		Lampertheim	Germany (W)	100		_		
		Biblis	Germany (W)	100	_			
		Lisbonne	Portugal	25/250	∠	_	_	_
		Vologda	USSR	10		_		
		Habana	Cuba	10	∠	_	_	_
		Shepparton	Australia	50	∠	_	_	_
		Malolos	Philippines	100	∠	_	_	_
		Abu Gharib	Iraq	250	_	_	_	_
		Bucuresti	Roumania	250	∠	_	_	_
		Mahe	Seychelles	100	∠	_	_	_
		Tokyo Yamata	Japan	100		_	_	
		Peking	China Rep					
		Lopik	Netherlands	100				
		Voice of Democratic Kampuchea						
		Dacca	Bangladesh	100	/			
11.730	25.58	Lopik	Netherlands	100	∠	_	_	_
		Talata Volon	Malagasy Rep	300/ 600	∠	_	_	_
		Kenga	USSR	100		_	_	
		Khabarovsk	USSR	240	_			
		Tachkent	USSR	50		_	_	
		Vinnitsa	Ukraine	100/500	∠			_
		Greenville	USA	50/500	_	_	_	_
		Allouis	France	100/500	∠	_	_	_
		Noblejas	Spain	350	∠	_	_	_
		Kavalla	Greece	250	∠	_	_	_
		Poro	Philippines	50		_	_	_
		Tinang	Philippines	250	_	_		_
		Agana	Guam	100	/			_
		Quito	Ecuador	100	_	_	_	

MHz	Metres	Station	Country	kW	M	M	S	N
		Riyadh	Saudia Arabia	350	∠			
		Peking	China Rep					
11.735	25.56	Limassol	Cyprus	20	∠	_	_	_
		Peking	Peking					
		Vinnitsa	Ukraine	240		_	_	_
		Kazan	USSR	100	∠			_
		Belgrade	Yugoslavia	10/100	∠	_	_	_
		Goiana	Brasil	1/7.5	∠	_	_	_
		Montevideo	Uruguay	5	∠	_	_	_
		Tanger	Morocco	50/100	_	_		
		Delhi	India	100	∠	_	_	_
		Allouis	France	100/500	_	_	_	_
		Kamalabad	Iran	100	_	_	_	_
		Jerusalem	Israel	100/300	/		_	
		Bucuresti	Roumania	250	∠	_	_	_
		Sofia	Bulgaria	100	∠	_	_	_
		Wavre	Belgium	250	∠	_	_	_
		Pori	Finland	100	∠			
		Talata Volon	Malagasy Rep	300	∠	_	_	_
		Fredrikstad	Norway	100		_		
		Sackville	Canada	250				_
		Mt Carlo	Monaco	100	/			_
		Sottens	Switzerland	500				
11.740	25.55	London	UK	250		_		
		Limassol	Cyprus	7.5/20	_	_	_	_
		Masirah	Oman	100	/		_	_
		Aligarh	India	250	_	_	_	_
		Delhi	India	20	/			
		Mexico	Mexico	5	∠	_	_	_
		Frunze	USSR	100/500	∠	_	_	_
		Novosibirsk	USSR	50	∠	_	_	_
		Shepparton	Australia	50	∠	_	_	_
		S M Galeria	Vatican	100	∠	_	_	_
		Cincinnati	USA	250	∠		_	_
		Dixon	USA	250	∠	_	_	_
		Greenville	USA	250		_		
		Ulan Bator	Mongolian Rep	300	_			
		Careysburg	Liberia	250	/	_	_	
		Lopik	Netherlands	100	∠	_	_	_
		Kavalla	Greece	250	_	_	_	_
		Kimjae	Korea (S)	100	∠	_	_	
		Noblejas	Spain	350	_	_	_	_
		Alger	Algeria	100	∠	_	_	_
		Talata Volon	Malagasy Rep	300	∠	_	_	_
		Mt Carlo	Monaco	100		_	_	_
		Bucuresti	Roumania	250		_	_	
11.745	25.54	Minsk	Bielorussia	240	∠	_	_	_
		Alma Ata	USSR	500		_		
		Armavir	USSR	500	/			

MHz	*Metres*	*Station*	*Country*	*kW*	*M*	*M*	*S*	*N*
		Allouis	France	100/500	_	_	_	_
		Quito	Ecuador	100	∠	_	_	_
		Godthaab	Greenland	1	∠	_	_	_
		Ekala	Ceylon	35/100	_	_	_	
		Kamalabad	Iran	100				_
		Aligarh	India	250	/			
		Delhi	India	50	_	_	_	_
		Taipei	China					
			Egypt					
11.750	25.53	London	UK	100/250	∠	_	_	_
		Ascension	Ascension	250				_
		Limassol	Cyprus	100	_			_
		Kranji	Singapore		/			_
		Makassar	Indonesia	1.5	∠	_	_	_
		Tokyo	Japan	1/10	∠	_	_	_
		Kazan	USSR	100		_	_	
		Moskva	USSR	100	∠			_
		Lvov	Ukraine	240		_	_	
		Caracas	Venezuela	10	∠	_	_	_
		Karachi	Pakistan	50	/		_	
		Islamabad	Pakistan	250	∠	_	_	_
		Mt Carlo	Monaco	100	∠	_	_	_
		Sofia	Bulgaria	500		_	_	
		Agana	Guam	100	/		_	_
		Tinang	Philippines	50				_
		Greenville	USA	250	/			_
11.755	25.52	S Fernando	Argentina	7	∠	_	_	_
		Tripoli	Libya	100	∠	_	_	
		Tbilisi	USSR	240/500	∠		_	_
		Leningrad	USSR	240	∠	_	_	_
		Tchita	USSR	100	_			
		Vladivostock	USSR	100		_	_	
		Mahe	Seychelles	100	_	_	_	
		Pori	Finland	15/250	∠	_	_	_
		Allouis	France	100		_	_	
		Santiago	Chile	100		_	_	
		Ekala	Ceylon	35		_		
		Tinang	Philippines					
		Delhi	India		/			
11.760	25.51	London	UK	100		_		
		Limassol	Cyprus	100	∠	_	_	_
		Tinang	Philippines	250	∠	_	_	_
		Erevan	USSR	100		_		
		Krasnoiarsk	USSR	100	_			
		Serpukhov	USSR	100	/			
		Sverdlovsk	USSR	100				_
		Kharkov	Ukraine	100		_	_	
		Tanger	Morocco	35/100	∠	_	_	_
		Habana	Cuba	100	∠	_	_	_

MHz	Metres	Station	Country	kW	M	M	S	N
		Manzini	Swaziland	25	_	_	_	_
		Kavalla	Greece	250	/_	_	_	_
		Athinai	Greece	100	/_	_	_	_
		Carnarvon	Australia	250	/	_	_	_
		Shepparton	Australia	100	/	_		
		Jerusalem	Israel	100/300	_	_		_
		Santiago	Chile	100	/	_	_	_
		Rarotonga	Cook Is	0.5		=	_	
		Ekala	Ceylon	35	/			_
		Delhi	India	100	/			
11.765	25.50	Kazan	USSR	100	_			
		Leningrad	USSR	500				_
		Irkutsk	USSR	100	/_			_
		Bombay	India	100	/			
		Delhi	India	100	/_	_	_	_
		Julich	Germany (W)	100	/_	_	_	_
		Wertachtal	Germany (W)	500	/_	_	_	_
		S Paulo	Brasil	10/25	/_	_	_	_
		Sofia	Bulgaria	50	/_	_	_	_
		Cyclops	Malta	250	/_	_	_	_
		Bocaue	Philippines	50	/_	_	_	_
		Schwarzenburg	Switzerland	250	_	_	_	_
		S M Galeria	Vatican	100	_		_	
		Kigali	Rwanda	250	/	_		
		Santiago	Chile	100	/	_	_	_
		Dacca	Bangladesh					
11.770	25.49	Ascension	Ascension	250	_	_	_	
		Jakarta	Indonesia	20/100	/_	_	_	_
		Ikorodu	Niger	100	/		_	_
		Omsk	USSR	100		_	_	
		Kiev	Ukraine	100	/	_	_	_
		Lisbonne	Portugal	100	/_			
		Biblis	Germany (W)	100	/	_	_	
		Lampertheim	Germany (W)	100	_	_	_	
		Playa de Pals	Spain	100/250			_	_
		Tanger	Morocco	25/50		_	_	_
		Mexico	Mexico	10	/_	_	_	_
		Aligarh	India	250	_	_	_	_
		Delhi	India	100	/			
		Kamalabad	Iran	350	_	_		
		Allouis	France	100	/_	_	_	_
		Kabul	Afghanistan	100	/			
		Schwarzenburg	Switzerland	150	/	_	_	_
		Okeechobee	USA	100				_
		Alger	Algeria					
11.775	25.48	London	UK	100	/			_
		Antigua	Br W Indies	250	/_	_	_	_
		Schwarzenburg	Switzerland	100/150	_	_		_
		Delhi	India	50	/			

MHz	Metres	Station	Country	kW	M	M	S	N
		Aligarh	India	250	_	_	_	_
		Armavir	USSR	240	∠	_	_	_
		Irkutsk	USSR	100/500	∠			_
		Bucuresti	Roumania	18/250	∠	_	_	_
		Noblejas	Spain	350	∠	_	_	_
		Jerusalem	Israel	20/300	_		_	_
		Carnarvon	Australia	100		_		
		Riyadh	Saudi Arabia	350	/	_	_	
		Santiago	Chile	100		_	_	
		Agana	Guam	100		_		
		Sackville	Canada	50/250	/		_	
11.780	25.47	London	UK	100/250	∠	_	_	_
		Limassol	Cyprus	100	/	_	_	_
		Masirah	Oman	100	∠	_	_	_
		Hurlingham	Argentina	7.5	∠	_	_	_
		Tokyo Yamata	Japan	100	∠	_	_	_
		Lvov	Ukraine	500		_		
		Krasnoiarsk	USSR	100			_	
		Serpukhov	USSR	100		_		
		Tbilisi	USSR	500	/			
		Scituate	USA	100	_			
		Allouis	France	100/500	_	_	_	_
		Brasilia	Brasil	250	∠	_	_	_
		Tinang	Philippines	50	/			_
		Schwarzenburg	Switzerland	100/500	∠	_	_	_
		Kamalabad	Iran	350		_		
		Kimjae	Korea (S)	100		_	_	_
		Lisbonne	Portugal	100	∠			
		Biblis	Germany (W)	100	/			
		Jerusalem	Israel	100/300	∠		_	_
		Agana	Guam	100	_	_		
		Abu Zaabal	Egypt	100			_	
		Ismaning	Germany (W)	100		_		
		Santiago	Chile	100	/			_
		Kavalla	Greece	250		_		
		Mt Carlo	Monaco	100	/			
		Salman Pack	Iraq					
11.782/5	25.46	Schwarzenburg*	Switzerland	30	/	_	_	_
11.785	25.46	London	UK	100	/		_	_
		Montserrat	Br W Indies	50	∠	_	_	_
		Antigua	Br W Indies	250	∠	_	_	_
		Omsk	USSR	100	∠	_		_
		Sverdlovsk	USSR	100	∠	_	_	_
		Kigali	Rwanda	250	∠	_	_	_
		K Wusterhausen	Germany (E)	50/100		_	_	
		Wertachtal	Germany (W)	500	∠	_	_	_
		Julich	Germany (W)	100	∠	_	_	_
		Pt Alegre	Brasil	7.5	∠	_	_	_
		Salman Pack	Iraq	100	_	_	_	_

MHz	*Metres*	*Station*	*Country*	*kW*	*M*	*M*	*S*	*N*
		Mahe	Seychelles	25/100	∠	_	_	_
		Cyclops	Malta	250	∠	_	_	_
		Abis	Egypt	250			_	
		Arganda	Spain	100	_	_	_	_
		Noblejas	Spain	350	/			
		Wavre	Belgium	100/250	_	_	_	_
		Beyrouth	Lebanon	100				_
11.790	25.45	London	UK	250	_	_	_	_
		Ascension	Ascension	250	/			_
		Greenville	USA	250	∠	_	_	_
		Cincinnati	USA	175	∠	_	_	_
		Riazan	USSR	240/500	∠	_	_	_
		Nikolaevskamur	USSR	240				_
		Blagovechtchen	USSR	240	∠			
		Shepparton	Australia	10	∠	_	_	_
		Lyndhurst	Australia					
		Bonaire Zuid	Neth Antilles	50	_			
		Bonaire Noord	Neth Antilles	300	∠	_	_	_
		Meyerton	South Africa	100	∠	_	_	_
		Bucuresti	Roumania	18/250	∠	_	_	_
		Hoerby	Sweden	350	_		_	_
		Jakarta	Indonesia	100				
		Wavre	Belgium	100	/			
		Kimjae	Korea (S)	250	∠	_	_	_
		Wien	Austria	100	_	_	_	_
		Abu Zaabal	Egypt	100			_	
		Allouis	France	100/500	∠	_	_	_
		Dacca	Bangladesh	250	/			_
		Bogota	Colombia					
		Mt Carlo	Monaco					
11.795	25.43	Antigua	UK	250	∠	_	_	_
		Tripoli	Libya	100	∠	_	_	
		Julich	Germany (W)	100	∠	_	_	_
		Wertachtal	Germany (W)	500		_		
		Kinshasa	Zaire	10	∠	_	_	_
		Kenga	USSR	100		_	_	
		Moskva	USSR	100	∠			_
		Tbilisi	USSR	50			_	
		Bogota	Colombia	25	∠	_	_	_
		Cyclops	Malta	250	∠	_	_	_
		K Wusterhausen	Germany (E)	50		_		
		Aligarh	India	250	/			
		Agana	Guam	100		_		
			Egypt					
		Baghdad	Iraq					
11.800	25.42	Ejura	Ghana	250	∠	_	_	_
		Ekala	Ceylon	100	∠	_	_	_
		Roma	Italy	60/100	∠	_	_	_
		Kiev	Ukraine	240/500	∠			_

MHz	Metres	Station	Country	kW	M	M	S	N
		Mt Carlo	Monaco	100	_			
		Santiago	Chile	100				
		Malolos	Philippines	50/100	_			
		Bonaire Zuid	Neth Antilles	50		_	_	_
		Meyerton	South Africa	250/ 500	_	_	_	_
		Mahe	Seychelles	25/100	_		_	
		Quito	Ecuador	100	L	_		
		Ankara	Turkey	250		_	_	
		Warszawa	Poland	1	L	_	_	_
		S Gabriel	Portugal	100	L	_	_	_
		Wellington	New Zealand	7.5	_	_	_	
		Carnarvon	Australia	250	L	_		
		Pori	Finland	250		_	_	_
		Hoerby	Sweden	350		_	_	
		Riyadh	Saudi Arabia	350				_
		Wavre	Belgium					
		Alger	Algeria					
		Okeechobee	USA					
11.805	25.41	London	UK	100/250	_		_	_
		Dixon	USA	100/250	L	_	_	_
		Greenville	USA	50/500	_	_	_	
		Okeechobee	USA	100				_
		Scituate	USA	50/100		_		_
		Kazan	USSR	50	L	_	_	_
		Tbilisi	USSR	240/500		_	_	
		Malolos	Philippines	100	L	_	_	_
		Tinang	Philippines	250	L	_	_	_
		Kavalla	Greece	250	L			
		Rio de Janeiro	Brasil	10	L	_	_	_
		Careysburg	Liberia	50	L	_	_	_
		Tanger	Morocco	10	/			
		Allouis	France	100/500	L	_	_	_
		Mahe	Seychelles	25/100	/	_	_	_
		Pt au Prince	Haiti	100				_
		Kabul	Afghanistan					
		Nauen	Germany (E)	100	/			_
		Ismaning	Germany (W)	100	/			
11.810	25.40	Antigua	Br W Indies	125/ 250	L	_	_	_
		Alger	Algeria	100	L	_	_	_
		Aligarh	India	100/250	_	_	_	_
		Delhi	India	100	/			
		Shepparton	Australia	50/100	_	_	_	_
		Simferopol	Ukraine	240		_	_	
		Serpukhov	USSR	100/240	L			_
		Tchita	USSR	100	L			
		Komsomolskamur	USSR	100			_	
		Julich	Germany (W)	100	L	_	_	_

MHz	*Metres*	*Station*	*Country*	*kW*	*M*	*M*	*S*	*N*
		Noblejas	Spain	350	∠	_	_	_
		Roma	Italy	60/100	∠	_	_	_
		Las Mesas	Canary Is	50	_	_		
		Bucuresti	Roumania	18/120	∠	_		_
		Jerusalem	Israel	50				_
		Agana	Guam	100	_	_	_	_
		Kavalla	Greece	250		_		
		Cyclops	Malta	250	/	_	_	_
		S M Galeria	Vatican	100	∠	_	_	_
		K Wusterhausen	Germany (E)	100	/			_
		Hoerby	Sweden	350	/			_
		Manzini	Swaziland	25				_
		Tinang	Philippines	250	/			_
11.815	25.39	Warszawa	Poland	100	∠	_	_	_
		Bonaire Zuid	Neth Antilles	50/ 250	_	_	_	_
		Goiana	Brasil	7.5	∠	_	_	_
		Biblis	Germany (W)	100	∠	_	_	_
		Lisbonne	Portugal	50	∠	_	_	_
		Tokyo Yamata	Japan	100	∠	_	_	_
		Scituate	USA	50/100				
		Okeechobee	USA	100	_			
		Tinang	Philippines	250	∠			
		Aligarh	India	250	∠	_	_	_
		Noblejas	Spain	350	∠	_	_	_
		Las Mesas	Canary Is	50	/	_	_	_
		Agana	Guam	100				
11.820	25.38	Antigua	Br W Indies	125/200		_		
		Limassol	Cyprus	100	_	_		
		Ascension	Ascension	125/250	∠		_	_
		Maputo	Mozambique	120	∠	_	_	_
		Krasnoirask	USSR	100		_		
		Frunze	USSR	240/500	∠			
		Khabarovsk	USSR	100		_	_	
		Tbilisi	USSR	240	/			_
		Voronej	USSR	100		_	_	
		Shepparton	Australia	50/100	/	_	_	_
		Bocaue	Philippines	50	_	_		
		Delano	USA	250	∠		_	_
		S M Galeria	Vatican	100				
		Wertachtal	Germany (W)	500	∠	_	_	_
		Quito	Ecuador	100	_			
		Sofia	Bulgaria	100	_			_
		Riyadh	Saudi Arabia	350		_		
		Kabul	Afghanistan	100				_
11.825	25.37	Bogota	Colombia	25	∠	_	_	_
		Shepparton	Australia	10/100	∠	_	_	_
		Papeete	Tahiti	20	∠	_	_	_
		Holzkirchen	Germany (W)	10	∠	_	_	_

MHz	Metres	Station	Country	kW	M	M	S	N
		Lisbonne	Portugal	50	/_	_	_	_
		Sines	Portugal	250			_	
		Recife	Brasil	10/15	/_	_	_	_
		Taipei	China Nat	25				
		Sackville	Canada	50/250	/_		_	
		Delhi	India	50		_	_	_
		Duchanbe	USSR	100	/_			
		Allouis	France	100	_	_	_	_
		Beyrouth	Lebanon	100	_			
		Malolos	Philippines	100				
		Pori	Finland	250			_	_
		Wien	Austria	100	/		_	_
		Nauen	Germany (E)	50	/			_
		Abu Ghraib	Iraq					
11.830	25.36	London	UK	100/250	_			_
		Bonaire Zuid	Neth Antilles	50/260	_	_		
		Greenville	USA	500	_			_
		Delano	USA	250			_	_
		Cincinnati	USA	250	/_		_	_
		Dixon	USA	250	/_	_		_
		Bombay	India	100	/_	_	_	_
		Moskva	USSR	240	/_	_	_	_
		Malolos	Philippines	100	_	_	_	_
		Tinang	Philippines	250				_
		Kavalla	Greece	250	/_	_	_	_
		Bucuresti	Roumania	18/250	/_	_	_	_
		Arganda	Spain	100	/_	_	_	_
		S M Galeria	Vatican	100	/_	_	_	_
		Quito	Ecuador	100				_
		V of Malayan Revln						
		Monrovia	Liberia	50	/_	_	_	_
11.835	25.35	London	UK	100				_
		Ekala	Ceylon	10/35	/_	_	_	_
		Omdurman	Sudan	120	_	_		_
		Serpukhov	USSR	100	_	_	_	_
		Krasnoiarsk	USSR	200	/_			_
		Tachkent	USSR		/			_
		Montevideo	Uruguay	5	/_	_	_	_
		Mt Carlo	Monaco	100	/_	_	_	
		Quito	Ecuador	100	/_	_	_	
		Monrovia	Liberia	50	/_	_	_	_
		Poro	Philippines	35/100	/	_	_	_
		Greenville	USA	250	/			
		Cincinnati	USA	175		_		
		Shepparton	Australia	10		_		_
		Carnarvon	Australia	100	/		_	
		Cap Haitien	Haiti					
		Franceville	Gabon Rep	500			_	

MHz	Metres	Station	Country	kW	M	M	S	N
		Delhi	India	100	/			
		Pori	Finland	250	/			
11.840	25.34	Nauen	Germany (E)	100		_	_	
		S Gabriel	Portugal	100	/_	_	_	_
		Warszawa	Poland	40/100	/_	_	_	_
		Kavalla	Greece	250	/_	_	_	_
		Greenville	USA	50/250	/_	_	_	_
		Tokyo Yamata	Japan	50/100	/_	_	_	_
		Poro	Philippines	35	_			_
		Bucuresti	Roumania	50/120	/	_	_	_
		Delhi	India	20	/_	_	_	_
		Quito	Ecuador	50	/_			
		Noblejas	Spain	350	/_	_	_	_
		Careysburg	Liberia	50/250		_		
		Biblis	Germany (W)	100				_
		Kimjae	Korea (S)	100			_	_
		Mahe	Seychelles	100	_	_	_	
		Allouis	France	100	/		_	_
		Wellington	New Zealand	7.5				_
		Jerusalem	Israel					
		Agana	Guam	100	/			_
11.845	25.33	London	UK	100/250				
		Moskva	USSR	250				_
		Serpukhov	USSR	100	_			
		Kazan	USSR	100	/_	_	_	_
		Montevideo	Uruguay	10	/_	_	_	_
		Allouis	France	100/500	/_	_	_	_
		S M Galeria	Vatican	100	/_	_	_	_
		Lopik	Netherlands	100	/_	_	_	_
		Tirane	Albania					
		Sackville	Canada	50/250	/_	_	_	_
		Hoerby	Sweden	350	/_	_	_	_
		Kimjae	Korea (S)	250		_		
		Riyadh	Saudi Arabia	350			_	
		Greenville	USA	250				_
		Okeechobee	USA	100				
		Jerusalem	Israel	300	/			
		Delhi	India	100	/			_
			Philippines	250				
		Athinai	Greece					
11.850	25.32	London	UK	100/250	_	_	_	_
		Kranji	Singapore	100	/			
		Limassol	Cyprus	100				
		Delano	USA	250	/	_		
		Scituate	USA	100	_			
		Okeechobee	USA	100				
		Delano	USA	250				_
		Dixon	USA	100/200	_		_	_
		Asuncion	Paraguay	5	/	_	_	_

MHz	*Metres*	*Station*	*Country*	*kW*	*M*	*M*	*S*	*N*
		Delhi	India	50/100	/	_	_	_
		Ejura	Ghana	250	/_	_	_	_
		Fredrikstad	Norway	100/250	/_	_	_	_
		Schwarzenburg	Switzerland	150	/_			_
		Vinnitsa	Ukraine	240		_	_	
		Konevo	USSR	240	/_	_	_	_
		Frunze	USSR	240	/_	_	_	
		Kazan	USSR	100	/_	_	_	
		Julich	Germany (W)	100	/_	_	_	_
		Wertachtal	Germany (W)	500	/_	_	_	_
		Kavalla	Greece	250	/_	_	_	
		Noblejas	Spain	350	_	_	_	_
		Wavre	Belgium	250	/_	_	_	_
		Nauen	Germany (E)	50	_			_
		Wellington	New Zealand	7.5	_		_	
		Karlsborg	Sweden	350	_			
		Hoerby	Sweden	350	_			
		Agana	Guam	100			_	_
		Sofia	Bulgaria	500		_	_	
		Allouis	France	500	/		_	_
		Kimjae	Korea (S)	250			_	_
11.855	25.31	London	UK	100/250	/	_	_	
		Bocaue	Philippines	50	/_	_	_	_
		Scituate	USA	50/100	_	_		
		Okeechobee	USA	100	/_	_	_	_
		Greenville	USA	100		_		
		Delhi	India	20/100	/_	_	_	_
		Jeddah	Saudi Arabia	50	/_	_	_	_
		Riyadh	Saudi Arabia	350		_	_	_
		Ulan Bator	Mongolian Rep					
		Sackville	Canada	50/250	_	_	_	_
		Praha	Czechoslovakia	120	/_	_	_	_
		Biblis	Germany (W)	50	_			
		Lampertheim	Germany (W)	50				_
		Lisbonne	Portugal	50/250	_	_	_	_
		Kavalla	Greece	250		_		
		Arganda	Spain	100	_			
		Santiago	Chile	100	_	_		
		Allouis	France	100/500			_	_
		Shepparton	Australia	50/100	/_	_	_	_
		Carnarvon	Australia	100			_	_
		Mt Carlo	Monaco	100	_	_	_	
		Tokyo Yamata	Japan	100	/_	_	_	
		Hoerby	Sweden	350	_			
		Mahe	Seychelles	50	/	_	_	_
		S M Galeria	Vatican	100				
11.860	25.30	Ascension	Ascension	125/250	/_	_	_	_
		Gorkii	USSR	240	/_	_	_	_
		Krasnoiarsk	USSR	100	_	_	_	

MHz	*Metres*	*Station*	*Country*	*kW*	*M*	*M*	*S*	*N*
		Iujnsakhalinsk	USSR	100	_			_
		Montevideo	Uruguay	10	/_	_	_	_
		Taipei	China Nat	50				
		Fredrikstad	Norway	100/250	/_	_	_	
		Kimjae	Korea (S)	250	/_	_		
		Allouis	France	100	/	_	_	_
		Monrovia	Liberia	50	/_	_	_	_
		Bucuresti	Roumania	250	_			
		Mahe	Seychelles	100	/_	_	_	_
		Sofia	Bulgaria	500	/			
		Sackville	Canada	250				
11.865	25.28	London	UK	100/250	/_	_	_	_
		Antigua	Br W Indies	250	/_	_	_	_
		Limassol	Cyprus	100	/			_
		Kranji	Singapore	250	/		_	_
		Recife	Brasil	1/7.5	/_	_	_	_
		Lubumbashi	Zaire	100	/_	_	_	_
		Tirane	Albania					
		Jayapura	Indonesia	25	/_	_	_	_
		Dixon	USA	100/250	_		_	_
		Delano	USA	100	/	_		
		Julich	Germany (W)	100	/_	_	_	_
		Wertachtal	Germany (W)	500	/_	_	_	_
		Sines	Portugal	250	/_	_	_	_
		Mahe	Seychelles	30/100	/_	_	_	_
		Cyclops	Malta	250	/_	_	_	_
		Habana	Cuba	10/100	_	_	_	
		Lampertheim	Germany (W)	100	_			
		Novosibirsk	USSR	100			_	
		Allouis	France	500	_	_		
		Kamalabad	Iran	100	_			
		Agana	Guam	100	_			
		Carnarvon	Australia	100				_
		Riyadh	Saudi Arabia	350				_
		Quito	Ecuador	100				
11.870	25.27	Khabarovsk	USSR	240		_	_	
		Serpukhov	USSR	240	/_	_	_	_
		Kaunas	USSR	240	/	_		
		Kenga	USSR	240	/_			_
		Bombay	India	100	/_	_	_	_
		Schwarzenburg	Switzerland	250	/_	_	_	_
		Wien	Austria	100	_	_	_	_
		Fredrikstad	Norway	100/250				_
		Riyadh	Saudi Arabia	350	/_	_	_	_
		Malolos	Philippines	100	/_	_	_	_
		Sofia	Bulgaria	50/500	/_	_	_	_
		Carnarvon	Australia	100				
		Lyndhurst	Australia	35				
		Ekala	Ceylon	100	/		_	_

MHz	*Metres*	*Station*	*Country*	*kW*	*M*	*M*	*S*	*N*
		Monrovia	Liberia					
11.875	25.26	Malolos	Philippines	100	∠	_	_	_
		Salvador	Brasil	10	∠	_	_	_
		Tokyo Yamata	Japan	100	∠	_	_	_
		Biblis	Germany (W)	100	∠			
		Lampertheim	Germany (W)	100	/	_	_	_
		Holzkirchen	Germany (W)	10		_		_
		Playa de Pals	Spain	50/500	∠	_	_	_
		Lisbonne	Portugal	250	/			
		S Gabriel	Portugal	100	∠	_	_	_
		Okeechobee	USA	100	_			
		K Wusterhausen	Germany (E)	100	∠			_
		S M Galeria	Vatican	100	_	_	_	_
		Kavalla	Greece	250	∠		_	_
		Managua	Nicaragua	100	∠	_	_	_
		Jerusalem	Israel	45/300	_		_	
		Roma	Italy	100	/	_	_	_
		Delhi	India	100	/			
11.880	25.25	Lusaka	Zambia	20/50	∠	_	_	_
		Lomas Mirador	Argentina	20	∠	_	_	_
		Lyndhurst	Australia	10	∠	_	_	_
		Shepparton	Australia	50	∠	_	_	_
		Mexico	Mexico	5	∠	_	_	_
		Armavir	USSR	240		_	_	
		Moskva	USSR	240	∠		_	
		Tchita	USSR	240		_		
		Ivanofrankovsk	Ukraine	100/240	∠			_
		Delhi	India	100	_	_	_	_
		Noblejas	Spain	350	∠	_	_	_
		Las Mesas	Canary Is	50	_	_	_	
		Riyadh	Saudi Arabia	350	_			
		Monrovia	Liberia	50		_	_	
		Wavre	Belgium					
11.885	25.24	Novosibirsk	USSR	50	/			_
		Bucuresti	Roumania	25/120	∠	_	_	_
		Karachi	Pakistan	50/120	∠	_	_	_
		Islamabad	Pakistan	100	_	_		_
		Montevideo	Uruguay	10	∠	_	_	_
		Lisbonne	Portugal	250		_		_
		Biblis	Germany (W)	100	_			
		Holzkirchen	Germany (W)	10		_		
		Lampertheim	Germany (W)	100	∠	_	_	_
		Playa de Pals	Spain	100	/	_	_	
		Rio de Janeiro	Brasil	10	∠	_	_	_
		Okeechobee	USA	100				_
		Meyerton	South Africa	100	∠	_	_	_
		Pyongyang	Korea (N)					
		Delhi	India	100				_
		Riyadh	Saudi Arabia					

MHz	Metres	Station	Country	kW	M	M	S	N
11.890	25.23	K Wusterhausen	Germany (E)	100		_	_	
		Nauen	Germany (E)	100	∠	_	_	_
		Bocaue	Philippines	50	∠	_	_	_
		Greenville	USA	250/500	∠	_	_	_
		Kenga	USSR	100	_			
		Frunze	USSR	240	∠	_		
		Erevan	USSR	500				_
		Riazan	USSR	240		_	_	_
		Tchita	USSR	240	/			
		Allouis	France	100/500	_	_	_	
		Riyadh	Saudi Arabia	350	/		_	_
		Seeb	Oman	50	∠	_	_	_
		Dacca	Bangladesh	100	∠	_	_	_
		Santiago	Chile	100	/	_	_	_
		Agana	Guam	100		_		
		Noblejas	Spain	350	/		_	_
		Kabul	Afghanistan					
11.895	25.22	Delhi	India	20/100	_	_	_	_
		Lisbonne	Portugal	50/250	∠	_	_	_
		Biblis	Germany (W)	100	/			_
		Lampertheim	Germany (W)	100		_	_	
		Playa de Pals	Spain	250			_	_
		Cincinnati	USA	250	∠	_	_	_
		Fredrikstad	Norway	250	_	_		_
		Jigulevsk	USSR	240	/			_
		Dakar	Senegal	100	/	_	_	_
		Wien	Austria	100	_	_	_	_
		Bata	Guinea	50	∠	_	_	_
		Dacca	Bangladesh	100	_			
		Carnarvon	Australia	250		_	_	_
11.900	25.21	Armavir	USSR	100		_	_	
		Komsomolskamur	USSR	100	∠			_
		Duchanbe	USSR	240		_	_	
		Lvov	Ukraine	240/500	∠			_
		Meyerton	South Africa	250/500	∠	_	_	_
		Kajang	Malaysia	100	∠	_	_	_
		Montevideo	Uruguay	20	∠	_	_	_
		Greenville	USA	50/250	∠	_	_	
		Dixon	USA	250				_
		Delano	USA	250	/			
		Ikorodu	Niger	100	∠	_	_	_
		Dacca	Bangladesh	100/250				_
		Bonaire Zuid	Neth Antilles	250	∠	_	_	_
		Agana	Guam	100	_	_	_	
		Quito	Ecuador	100	∠	_	_	_
		Riyadh	Saudi Arabia	350		_	_	_
11.905	25.20	London	UK	100/250	∠	_	_	_
		Taipei	China Rep	3				

MHz	Metres	Station	Country	kW	M	M	S	N
		Julich	Germany (W)	100	/_	_	_	_
		Wertachtal	Germany (W)	500	/_	_	_	_
		Roma	Italy	60/100	/_	_	_	_
		Frunze	USSR	100	/	_		_
		Greenville	USA	50/250	/_	_	_	_
		Red Lion	USA	50	_			
		Sines	Portugal	250	/_	_	_	_
		Patumthani	Thailand	100	/_	_	_	_
		Hoerby	Sweden	350	/	_	_	_
		Karlsborg	Sweden	350	/			
		Babel	Iraq	500	_	_		
		Quito	Ecuador	100				
		Sackville	Canada	250				_
		Riyadh	Saudia Arabia	350	/			
11.910	25.19	London	UK	100/250		_	_	_
		Montserrat	Br W Indies	15/50				
		Masirah	Oman	100/200	/_	_	_	_
		Kranji	Singapore	100	/			_
		Jaszbereny	Hungary	250	/_	_	_	_
		Diosd	Hungary	100	/_	_	_	_
		Moskva	USSR	240		_		
		Sverdlovsk	USSR	100	/	_	_	_
		Komsomolskamur	USSR	100	/_			
		Quito	Ecuador	100	/_	_	_	_
		Kimjae	Korea (S)	10	/			
		Noblejas	Spain	350	_	_	_	_
		Riyadh	Saudi Arabia	350	/	_	_	_
		Pori	Finland	250				_
		Manzini	Swaziland					
		Lampertheim	Germany (W)	100	/			
		Islamabad	Pakistan	250	/			
		Allouis	France	100/500	/			
11.915	25.18	London	UK	250	/	_		
		Abu Zaabal	Egypt	100			_	
		Concepcion	Paraguay	100	/_	_	_	_
		Petropavlo Kam	USSR	50/100		_	_	
		Pt Alegre	Brasil	7.5	/_	_	_	_
		Quito	Ecuador	50/100	/_	_	_	_
		Delano	USA	200	_	_		
		Greenville	USA	250	_		_	_
		Tanger	Morocco	35/100	/_		_	_
		Careysburg	Liberia	250	/		_	_
		Sines	Portugal	250	/_	_	_	_
		Playa de Pals	Spain	100		_		
		Holzkirchen	Germany (W)	10		_		
		Taipei	China Nat					
		Delhi	India	100	_	_	_	_
		Islamabad	Pakistan	250	/	_	_	_
		Karachi	Pakistan	50			_	

MHz	Metres	Station	Country	kW	M	M	S	N
		Hoerby	Sweden	350	_			
		Riyadh	Saudi Arabia	350	/	_	_	
		Sackville	Canada	250		_		
		Tirana	Albania					
		Munich	Germany (W)	100				
11.920	25.17	Kranji	Singapore	100	/			_
		Dixon	USA	100	_	_	_	_
		Greenville	USA					_
		Bocaue	Philippines	50	∠	_	_	_
		Iba	Philippines	100				_
		Arganda	Spain	100	/			
		Noblejas	Spain	350	_	_	_	_
		Moskva	USSR	240	∠	_		_
		Novosibirsk	USSR	100		_	_	
		Baku	USSR	240	∠	_	_	_
		Kazan	USSR	100	∠			
		Alma Ata	USSR	100	_			_
		Abidjan	Ivory Coast	100	∠	_	_	_
		Cyclops	Malta	250	_			
		Ismaning	Ismaning	100	∠			
		Nauen	Germany (E)	100	∠			_
			China Rep					
		Amman	Jordan	100	∠	_	_	_
		Tirana	Albania					
		Sofia	Bulgaria					
11.925	25.16	London	UK	100/250	/	_	_	_
		Kranji	Singapore	100	/			
		S Gabriel	Portugal	50	∠	_	_	_
		Lisbonne	Portugal	50		_	_	
		Lampertheim	Germany (W)	100	∠			
		Biblis	Germany (W)	100			_	
		Playa de Pals	Spain	250				_
		Arganda	Spain	100	_		_	_
		Noblejas	Spain	350	∠	_	_	_
		S Paulo	Brasil	10	∠	_	_	_
		Krasnoiarsk	USSR	240	_			
		Tachkent	USSR	50		_	_	
		Tchita	USSR	500	/			_
		Kiev	Ukraine	100	∠			_
		Kavalla	Greece	250	∠	_	_	_
		Bonaire Zuid	Neth Antilles	250	∠	_	_	_
		Ikorodu	Niger	100	_	_		
		Santiago	Chile	100	_			
		Tinang	Philippines	250		_		_
		Alger	Algeria	100	∠	_	_	_
		Allouis	France	100		_	_	
		Babel	Iraq	500			_	_
		Beyrouth	Lebanon	100			_	
		Okeechobee	USA	100				_

MHz	*Metres*	*Station*	*Country*	*kW*	*M*	*M*	*S*	*N*
		Scituate	USA	50				_
11.930	25.15	Habana	Cuba	50	∠	_	_	_
		Achkhabad	USSR	240/500		_	_	
		Moskva	USSR	240	∠	_	_	_
		Armavir	USSR	120	∠	_	_	_
		Tinang	Philippines	50/250	∠	_	_	_
		Poro	Philippines		/			_
		Lopik	Netherlands	100	∠	_	_	_
		Kamalabad	Iran	100		_	_	
		Noblejas	Spain	350	∠	_	_	_
		Arganda	Spain	100	/			_
		Allouis	France	100/500	∠	_	_	_
		Monrovia	Liberia	50	/			
11.935	25.14	London	UK	100/250	∠	_	_	_
		Wertachtal	Germany (W)	500	/	_	_	_
		Biblis	Germany (W)	100		_		
		Lampertheim	Germany (W)	100			_	
		Playa de Pals	Spain	250	∠		_	
		S Gabriel	Portugal	100	∠	_	_	_
		Meyerton	South Africa	100	∠	_	_	_
		Delhi	India	20	∠	_	_	_
		Greenville	USA	50/250				_
		Curitiba	Brasil	7.5/25	∠	_	_	_
		Tirane	Albania					
		Carnarvon	Australia	250	_	_	_	_
		Shepparton	Australia	100	_			
		Fredrikstad	Norway	100/250		_		_
		Ekala	Ceylon	35	∠	_	_	_
		Tula	USSR	120	_			_
		Mt Carlo	Monaco	100	_	_	_	
		Babel	Iraq	500				
		Lopik	Netherlands	100			_	_
		Hoerby	Sweden	350	X			_
11.940	25.13	Bucuresti	Roumania	120/250	∠	_	_	_
		Sulaibiyah	Kuwait	250	∠	_	_	_
		Blagovechtchen	USSR	240		_	_	
		Krasnoiarsk	USSR	240	_			_
		Sverdlovsk	USSR	100	∠			
		Singapore	Singapore	50	∠	_	_	_
		Tokyo Yamata	Japan	50	∠	_	_	_
		Encarnacion	Paraguay	5	∠	_	_	_
		Sackville	Canada	50	_	_	_	_
		Wavre	Belgium	100/250	∠	_	_	_
		Ekala	Ceylon	35	_		_	
11.945	25.12	London	UK	100/250	∠	_	_	_
		Encarnacion	Paraguay	5	∠	_	_	_
		Wertachtal	Germany (W)	500	∠	_	_	_
		Peking	China Rep					
		Monrovia	Liberia	50	_	_	_	_

MHz	*Metres*	*Station*	*Country*	*kW*	*M*	*M*	*S*	*N*
		Delhi	India	100	_	_	_	_
		Kimjae	Korea (S)	100	∠	_	_	_
		Mahe	Seychelles	100	∠	_	_	_
		Sackville	Canada	50/250	∠	_	_	_
		Noblejas	Spain	350/700	∠	_	_	_
		Playa de Pals	Spain	250				_
		Wellington	New Zealand	7.5				_
		Allouis	France	100				
		Quito	Ecuador	100				
11.950	25.11	Diriyya	Saudi Arabia	50	∠	_	_	_
		Alma Ata	USSR	100	∠	_	_	_
		Tula	Ukraine	100		_	_	_
		Kharkov	Ukraine	240	∠	_	_	
		Rio de Janeiro	Brasil	7.5/10	∠	_	_	_
		Lopik	Netherlands	100	∠	_	_	_
		Malolos	Philippines	50	_	_	_	_
		Tokyo Yamata	Japan	200		_	_	_
		Bucuresti	Roumania	120/250	∠	_	_	_
		Bonaire Zuid	Neth Antilles	50	/		_	_
		Tirane	Albania					
11.9525	25.10	Varberg*	Sweden	100				_
11.955	25.09	London	UK	250	∠	_	_	_
		Kranji	Singapore	125/250	/		_	_
		Limassol	Cyprus	100	/	_	_	_
		Masirah	Oman	100	_			_
		Ivanofrankovsk	Ukraine	240		_	_	
		Montevideo	Uruguay	10	∠	_	_	_
		Greenville	USA	250/500	∠	_	_	_
		Tanger	Morocco	35/100	_			_
		Hoerby	Sweden	350	/	_	_	_
		Malolos	Philippines					
		Ekala	Ceylon	10/100		_		
		Manzini	Swaziland	25	_	_		_
		Jerusalem	Israel	300			_	
		Lopik	Netherlands	100	_	_	_	_
		Athinai	Greece	100			_	_
		Allouis	France	100/500	_		_	
		Luanda	Angola	100	/		_	_
		K Wusterhausen	Germany (E)	100		_		
		Mt Carlo	Monaco	100			_	
		Ankara	Turkey	250	/			_
		Roma	Italy	60				_
		Meyerton	South Africa	250				_
		Islamabad	Pakistan	250				_
11.960	25.08	London	UK	100/250	∠	_	_	_
		Bamako	Malawi	50		_	_	
		Armavir	USSR	240	∠	_		
		Serpukhov	USSR	100	/			_
		Tachkent	USSR	100/500	∠	_	_	

MHz	Metres	Station	Country	kW	M	M	S	N
		Tbilisi	USSR	240				_
		Kenga	USSR	240		_	_	
		Krasnoiarsk	USSR	100	∠	_	_	_
		Kazan	USSR	100	/			
		Tokyo Yamata	Japan	200	/			_
		Jerusalem	Israel	300	∠	_	_	_
		Tanger	Morocco	35	_			
		Quito	Ecuador	100	∠	_	_	_
		Wellington	New Zealand	7.5	∠	_		_
		Sackville	Canada	250			_	
		Greenville	USA	50/250	∠		_	_
		Leipzig	Germany (E)	100	/			_
11.965	25.07	Kigali	Rwanda	250	∠	_	_	_
		Allouis	France	100	_	_	_	_
		Kazan	USSR	240		_	_	
		Tinang	Philippines	250	∠	_	_	_
		Poro	Philippines	100	_	_		
		S Paulo	Brasil	7.5	∠	_	_	_
		Conakry	Guinea	100	∠	_	_	_
		Noblejas	Spain	350	_	_	_	_
		Greenville	USA	50	∠			
		Riyadh	Saudi Arabia	350	_			
		Beyrouth	Lebanon	100		_		
		Kimjae	Korea (S)	250	/		_	_
		Tirane	Albania					
		Franceville	Gabon Rep	500			_	
		Ankara	Turkey	250	/			_
		Delhi	India	20	/			
		Tangier	Morocco	50				
11.970	25.06	Montserrat	Br W Indies	15	_	_	_	
		Sofia	Bulgaria	50	/	_	_	_
		Lisbonne	Portugal	50/100		_	_	_
		Biblis	Germany (W)	100			_	_
		Lampertheim	Germany (W)	250	∠	_	_	
		Playa de Pals	Spain	100/250	∠			_
		Sfax	Tunisia	50/100			_	_
		Habana	Cuba	10/50	∠	_	_	_
		Nauen	Germany (E)	50		_	_	
		K Wusterhausen	Germany (E)	100	∠	_	_	_
		Komsomolskamur	USSR	100		_	_	
		Frunze	USSR	240/500	∠			_
		Kharkov	Ukraine	240		_	_	
		Bucuresti	Roumania	120/250	∠	_	_	_
		Malolos	Philippines	100	∠	_	_	_
		Kimjae	Korea (S)	100	∠	_	_	_
		Kathmandu	Nepal	100	∠		_	_
		Islamabad	Pakistan	10	/			
11.975	25.05	Berlin	Germany (E)					
		Moskva	USSR					

MHz	*Metres*	*Station*	*Country*	*kW*
11.980	25.04	Bucurest	Roumania	
		Peking	China Rep	
		Moskva	USSR	
		V of Democratic Kampuchea		
11.985	25.03	Orcha	USSR	60
		Tirane	Albania	240
		Kabul	Afghanistan	
11.990	25.02	Prague	Czechoslovakia	100
			USSR	50
		Peking	China Rep	
		Sulaibiyah	Kuwait	
		Warszawa	Poland	
		V of Democratic Kampuchea		
11.995	25.01	Moskva	USSR	
12.000	25.00		USSR	100
12.005	24.99	Cairo	Egypt	100
		Minsk	USSR	
12.006	24.99	V NUFK		
12.010	24.98		USSR	
		Monrovia*	Liberia	
12.015	24.97	Wien	Austria	
		Peking	China Rep	
12.020	24.96		USSR	
		Monrovia*	Liberia	
12.025	24.95		USSR	
12.030	24.94		USSR	100
12.033	24.93	Hanoi	Vietnam	
12.035	24.93		USSR	
12.040	24.92	London	UK	
		Yerevan	USSR	100
12.045	24.91	Moskva	USSR	
		Cairo	Egypt	
12.050	24.90		USSR	
		Cairo	Egypt	
12.055	24.89	Peking	China Rep	
		Moskva	USSR	
12.060	24.88		USSR	
		Peking	China Rep	
12.070	24.86	Kiev	Ukraine	100
		Moskva*	USSR	
		Ulan Bator	Mongolian Rep	
		Cairo	Egypt	
12.075	24.84		USSR	
12.077	24.84	Jerusalem	Israel	
12.080	24.83	Peking	China Rep	
12.085	24.82		Kuwait	
12.095	24.80	London	UK	

MHz	*Metres*	*Station*	*Country*	*kW*
12.100	24.79		USSR*	
12.110	24.77	Peking	China	
12.113	24.76	Riyadh*	Saudi Arabia	
12.120	24.75	Peking	China Rep	
12.127	24.74	London*	UK	
12.140	24.71		USSR	
12.165	24.66		USSR*	
12.175	24.64		USSR*	
		Reykjavik*	Iceland	
12.179	24.63	London*	UK	
12.190	24.61		Australia	
12.200	24.59	Peking	China Rep	
12.205	24.58		USSR*	
12.2225	24.54		USSR*	
12.240	24.51	Magadan*	USSR	50
12.246	24.50	Liberty*		
12.250	24.49		USSR*	
12.25275	24.48		USSR*	
12.280	24.43		USSR	
12.290	24.41		Australia	
			Germany (W)*	
12.295	24.40	Mt Carlo*	Monaco	
12.420	24.15	Peking	China Rep	
12.450	24.10	Peking	China Rep	120
12.460	24.08	Paris	France	
13.272	22.60	New York	USA	
13.370	22.44		USSR*	
13.380	22.42		USSR*	
13.396	22.39	Riyadh	Saudi Arabia	
13.491	22.24	Greenville*	USA	
13.512	22.20	Julich/Wertachtal*	Germany (W)	
13.590	22.08		USSR*	
13.690	21.91	RFE*		
13.710	21.88		USSR*	
13.760	21.80		USSR*	
13.820	21.71		USSR*	
13.849	21.66	Bogota	Colombia	
13.860	21.65	Delano*	USA	
13.960	21.49		USSR*	
14.270	21.02	Tirane	Albania	
14.290	20.99		USSR	
14.440	20.78	R Free Portugal		
			USSR	
14.595	20.55		USSR	
14.670	20.45	'CHU' Time Sig	Canada	
14.700	20.41			
14.712	20.39	RFE/R Liberty*		
14.715	20.39	RFE/R Liberty*		
14.820	20.24	Peking	China Rep	

MHz	Metres	Station	Country	kW	M	M	S	N
14.850	20.20		USSR*					
14.990	20.01	Alger	Algeria					
14.996	20.01	Freq Standard	USSR					
15.000	20.00	Boulder Freq Std	USA					
		Honolulu Freq Std	Hawaii					
		Tokyo Freq Std	Japan					
15.009	19.99	Hanoi	Vietnam					
15.020	19.97	Peking	China Rep					
		Hanoi	Vietnam					
15.030	19.96	Peking	China Rep	120				
15.040	19.95	Peking*	China Rep					
15.045	19.94	Peking	China Rep	240				
			USSR					
15.050	19.93	Peking	China Rep					
		Grenada	Br W Indies					
15.060	19.92	Peking	China Rep	240				
			USSR					
		Riyadh	Saudi Arabia					
15.070	19.91	London	UK					
		Peking	China Rep					
		Conakry	Guinea					
		Jerusalem	Israel					
15.080	19.89	Delhi	India	100				
		Bombay	India					
		Peking	China Rep					
15.084	19.89	Teheran	Iran	250				
15.090	19.88	Monrovia	Liberia					
15.095	19.87	Peking	China Rep	240				
15.100	19.87	Jerusalem	Israel					
		Moskva	USSR					
		Berlin	Germany (E)					
		Tripoli	Libya					
15.105	19.86	London	UK	100	/		—	—
		Limassol	Cyprus	100	—	—		—
		Ascension	Ascension	250	∠	—	—	—
		Jerusalem	Israel	100/300	∠	—	—	—
		Tokyo Yamata	Japan	100	∠	—	—	—
		K Wusterhausen	Germany (E)	50/100	∠			—
		Rio de Janeiro	Brasil	7.5	∠	—	—	—
		Wien	Austria	100	—	—	—	—
		Pori	Finland	100	—	—	—	—
		Cyclops	Malta	250	/	—	—	—
		Julich	Germany (W)	100	/	—	—	
		Wertachtal	Germany (W)	500	/		—	—
		Shepparton	Australia	100				—
		Grenada	Br W Indies	10				
15.110	19.85	London	UK					
		Alma Ata	USSR	240				—
		Krasnoiarsk	USSR	240	∠			—

MHz	*Metres*	*Station*	*Country*	*kW*	*M*	*M*	*S*	*N*
		Kiev	Ukraine	500/240	/_	_	_	_
		Mexico	Mexico	5	/_	_	_	_
		Velkekostolany	Czechoslovakia	120	/_	_	_	_
		Scituate	USA	50/100	_	_	_	_
		Okeechobee	USA	100	_			
		Bata	Guinea	50	/_	_	_	_
		Santiago	Chile	100	/	_	_	_
		Delhi	India	100	/	_	_	_
		Wien	Austria	100		_		
		Riyadh	Saudi Arabia	350			_	
		Tinang	Philippines	50				_
		Carnarvon	Australia	250				_
15.115	19.85	Jeddah	Saudi Arabia	50	/_	_	_	_
		Riyadh	Saudi Arabia	350		_	_	_
		Quito	Ecuador	50/100	/_	_	_	_
		Lisbonne	Portugal	250	/_	_	_	_
		Playa de Pals	Spain	500		_		
		Khabarovsk	USSR	100		_		
		Jigulevsk	USSR	100			_	
		Tachkent	USSR	50		_	_	
		Islamabad	Pakistan	100/250	/_	_	_	_
		Nauen	Germany (E)	100				_
		Santiago	Chile	5	/_	_	_	_
		Malolos	Philippines	100/250	/	_	_	_
		Poro	Philippines	50				_
		Tinang	Philippines	250	/			
		Agana	Guam	100	_			
		Ekala	Ceylon	35	/		_	_
		Okeechobee	USA	100				_
		Peking	China Rep					
15.120	19.84	London	UK	100/250	_	_		
		Merauke	Indonesia	5	/_	_	_	_
		Ikorodu	Niger	100	/_	_	_	_
		S M Galeria	Vatican	100	/_	_	_	_
		Warszawa	Poland	15/100	/_	_	_	_
		Ekala	Ceylon	35/100	/_	_	_	_
		Hoerby	Sweden	350	/_			
		Jerusalem	Israel	50	_		_	
		Kavalla	Greece	250	/_	_	_	_
		Greenville	USA	50/250	/_	_	_	_
		Delhi	India	100		_	_	_
		Ismaning	Germany (W)	100				_
		Wertachtal	Germany (W)	500				
15.125	19.83	Limassol	Cyprus	20	_	_	_	_
		S Gabriel	Portugal	100	/_	_	_	_
		Lisbonne	Portugal	100	_		_	_
		Salvador	Brasil	10	/_	_	_	_
		Mexico	Mexico	10	/_	_	_	_

MHz	*Metres*	*Station*	*Country*	*kW*	*M*	*M*	*S*	*N*
		K Wusterhausen	Germany (E)	100		_	_	
		Meyerton	South Africa	100/	_	_	_	_
				250				
		Arganda	Spain	100	_	_	_	_
		Noblejas	Spain	350	/			
		Greenville	USA	50/500	_	_	_	_
		Bombay	India	100	∠	_	_	_
		Taipei	China Nat					
		Hoerby	Sweden	350	_	_	_	
		Santiago	Chile	100	/	_	_	_
		Ankara	Turkey	250	/			_
15.130	19.83	Simferopol	Ukraine	240/500	/	_	_	_
		Vladivostock	USSR	240	∠	_	_	_
		Lampertheim	Germany (W)	100	∠		_	
		Playa de Pals	Spain	100/500		_	_	_
		Scituate	USA	50/100	∠			_
		Okeechobee	USA	100			_	_
		Nauen	Germany (E)	100	∠	_		_
		Wellington	New Zealand	7.5	_	_	_	_
		Santiago	Chile	100	_		_	
		Kimjae	Korea (S)	250		_		
		Delhi	India	20/100	∠	_	_	_
		Hoerby	Sweden	350	∠			
		Kavalla	Greece	250	/			
		Carnarvon	Australia	250				_
15.135	19.82	Abu Zaabal	Egypt	100			_	
		Allouis	France	100/500	∠	_	_	_
		Wertachtal	Germany (W)	500	∠	_	_	_
		Julich	Germany (W)	100	∠	_	_	_
		S Paulo	Brasil	7.5	∠	_	_	_
		Scituate	USA	50/100		_		
		Cyclops	Malta	250	_		_	_
		Kamalabad	Iran	250		_		
		Malolos	Philippines	50	∠	_	_	_
		Fredrikstad	Norway	120	∠	_	_	
		Wien	Austria	100	_	_	_	_
		Agana	Guam	100	/	_	_	_
		Lvov	Ukraine	240		_		
		Vinnitsa	Ukraine	240	/			_
		Sofia	Bulgaria	500		_		
		Ankara	Turkey	250	/			_
15.140	19.82	London	UK	100/250	/	_	_	_
		Shepparton	Australia	10/50	_	_	_	_
		Riazan	USSR	240	∠	_	_	_
		Riga	USSR	100		_	_	
		Petropavlo Kam	USSR	240	∠			_
		Bombay	India	100	∠	_	_	_
		Schwarzenburg	Switzerland	100/250	_	_	_	_
		Cincinnati	USA	175	_	_	_	_

MHz	*Metres*	*Station*	*Country*	*kW*	*M*	*M*	*S*	*N*
		Greenville	USA	250	/		_	_
		Santiago	Chile	100	/	_	_	_
		S Gabriel	Portugal	100	_			
		Kabul	Afghanistan					
15.145	19.81	K Wusterhausen	Germany (E)	50		_	_	
		Nauen	Germany (E)	100	∠	_	_	_
		Biblis	Germany (W)	100		_		
		Lisbonne	Portugal	100	∠	_	_	_
		Recife	Brasil	10	∠	_	_	_
		Red Lion	USA	50	/	_	_	_
		Agana	Guam	100	_			
		Ankara	Turkey	250	/			_
15.150	19.80	Antigua	Br W Indies	250	∠	_	_	_
		Jakarta	Indonesia	7.5	∠	_	_	_
		Santiago	Chile	10/100	_	_	_	
		Julich	Germany (W)	10				_
		Wertachtal	Germany (W)	500	∠	_	_	_
		Greenville	USA	250	/	_	_	_
		Sulaibiyah	Kuwait	250		_	_	
		Islamabad	Pakistan	100	∠	_	_	_
		Sackville	Canada	50	∠		_	_
		Minsk	Bielorussia	240/500	∠	_	_	_
		Erevan	USSR	240		_	_	
		Serpukhov	USSR	240		_		
		Omsk	USSR	240/500	∠	_	_	_
		Tinang	Philippines	250		_	_	
15.155	19.80	Allouis	France	100/500	∠	_	_	_
		Tinang	Philippines	50/250	∠	_	_	_
		S Paulo	Brasil	25	∠	_	_	_
		Alma Ata	USSR	240			_	
		Duchanbe	USSR	500		_		
		Armavir	USSR	240		_		
		Meyerton	South Africa	250	∠	_	_	_
		Nauen	Germany (E)	50		_	_	
		K Wusterhausen	Germany (E)	100				
		Agana	Guam	100	_	_	_	_
		Greenville	USA	250		_		
		Riyadh	Saudi Arabia	350				
		Wien	Austria	100		_		
		Alger	Algier					
		Tanger	Morocco	50/100		_	_	_
		Rabat	Morocco	50/100	/			
15.160	19.79	London	UK	100/250	∠		_	
		Jaszbereny	Hungary	250	∠	_	_	_
		Szekesfehervar	Hungary	20	∠	_	_	_
		Mexico	Mexico	10	∠	_	_	_
		Greenville	USA	50/100	∠	_	_	_
		Cincinnati	USA	175/250	/	_	_	_
		Okeechobee	USA	100	_			

MHz	*Metres*	*Station*	*Country*	*kW*	*M*	*M*	*S*	*N*
		Scituate	USA	100	∠			
		Lyndhurst	Australia	10	∠	_	_	_
		Athinai	Greece	100	∠	_	_	_
		Kavalla	Greece	250	∠	_	_	_
		Alger	Algeria	50	∠	_	_	_
		Mahe	Seychelles	30/100	∠		_	_
		Quito	Ecuador	100	∠	_	_	_
		Lopik	Netherlands	100		_	_	
		Delhi	India	100	/	_	_	_
		Sines	Portugal	250	/	_	_	
		Tanger	Morocco	50/100				_
15.165	19.78	London	UK	250				_
		Limassol	Cyprus	100	_	_	_	
		Aligarh	India	250	/			
		Delhi	India	20/100	∠	_	_	_
		Forteleza	Brasil	5	∠	_	_	_
		S M Galeria	Vatican	100	_	_	_	_
		Kobenhavn	Denmark	50	∠	_	_	_
		Peking	China Rep					
		Talata Volon	Malagasy Rep	300	_	_		
		Manzini	Swaziland	25	_		_	_
		Lopik	Netherlands	100	_			
		Nauen	Germany (E)	100		_		
		Kavalla	Greece	250			_	
		Tinang	Philippines	250	/			_
15.170	19.78	Lisbonne	Portugal	10/100	∠	_	_	_
		Lampertheim	Germany (W)	100		_		
		Fredrikstad	Norway	100		_		
		Papeete	Tahiti	20	∠	_	_	_
		Leipzig	Germany (E)	100	∠	_	_	_
		Nauen	Germany (E)	500		_		
		Tinang	Philippines	250		_	_	
		Greenville	USA	250				_
		Kimjae	Korea (S)	100	∠	_	_	_
		Vinnitsa	Ukraine	50/500	∠			_
		Tula	USSR	50		_	_	
		Blagovechtchen	USSR	240		_		
		Schwarzenburg	Switzerland	250	/	_	_	
15.175	19.77	Alger	Algeria					
		Abu Zaabal	Egypt	100				
		Fredrikstad	Norway	100/250	∠	_	_	_
		Kazan	USSR	240				_
		Kenga	USSR	240	_	_		
		Irkutsk	USSR	240/500	/	_	_	_
		Lvov	Ukraine	100	_			
		Red Lion	USA	50	_			
		Agana	Guam	100	∠	_	_	_
		Riyadh	Saudi Arabia	350			_	_
		Santiago	Chile	100		_	_	

MHz	*Metres*	*Station*	*Country*	*kW*	*M*	*M*	*S*	*N*
15.180	19.76	London	UK	100/250	/_	_	_	_
		Talata Volon	Malagasy Rep	300	_			
		Armavir	USSR	240	/			_
		Leningrad	USSR	100		_	_	
		Petropavlo Kam	USSR	50/240	/	_	_	_
		Delhi	India	100	/_	_	_	_
		Carnarvon	Australia	100	/			
		Shepparton	Australia	100		_	_	_
		Bonaire Noord	Neth Antilles	300	/_	_	_	_
		Allouis	France	100/500	/		_	_
		Beyrouth	Lebanon	100			_	
15.185	19.76	Antigua	Br W Indies	250	/_	_	_	
		Ekala	Ceylon	35	_	_	_	_
		Jigulevsk	USSR	240		_	_	
		Serpukhov	USSR	100/240	/_	_		_
		Simferopol	Ukraine	240			_	
		Ikorodu	Niger	100	/_	_	_	_
		Wertachtal	Germany (W)	500	/_	_	_	_
		Red Lion	USA	50	/_	_	_	_
		Lopik	Netherlands	100	_	_	_	_
		Sines	Portugal	250	/_	_	_	_
		Delhi	India	100	/_	_	_	_
		Poro	Philippines	50	/	_	_	_
		Noblejas	Spain	350	/_	_	_	_
15.190	19.75	Brazzaville	Congo	50	/_	_	_	_
		Aligarh	India	250	/_	_	_	_
		Sackville	Canada	50/250	/_	_	_	
		Belo Horizonte	Brasil	25	/_	_	_	_
		Erevan	USSR	100		_	_	
		Tula	USSR	240	/_			_
		Tchita	USSR	240		_		
		Cincinnati	USA	175		_	_	
		Bata	Guinea	50	/_	_	_	_
		Riyadh	Saudi Arabia	350	_			
		Wavre	Belgium	100	/_	_	_	_
		Allouis	France	100	/_	_	_	_
		Schwarzenburg	Switzerland	150				_
15.1925	19.75	Varberg*	Sweden	100	/_	_	_	_
15.195	19.74	London	UK	250	/_	_	_	_
		Ascension	Ascension	250	/_	_	_	_
		Tokyo Yamata	Japan	100	/_	_	_	_
		Tanger	Morocco	35/100	/_	_	_	_
		Warszawa	Poland	1	_	_	_	
		Allouis	France	100/500	/		_	_
		Wien	Austria	100		_		
		Achkhabad	USSR	500		_		
		Irkutsk	USSR	100	/			
		Kavalla	Greece		/		_	
		Greenville	USA	250	/		_	_

MHz	*Metres*	*Station*	*Country*	*kW*	*M*	*M*	*S*	*N*
		Okeechobee	USA	100			_	
		Peking	China Rep					
15.200	19.74	Singapore	Singapore	50	∠	_	_	_
		Kalatch	USSR	240	∠	_	_	_
		S M Galeria	Vatican	100	_	_		
		Allouis	France	100	∠	_	_	_
		Kathmandu	Nepal	100	∠		_	_
		Beyrouth	Lebanon	100	_			
		S Gabriel	Portugal	100	/	_	_	_
		Pori	Finland	100				_
		Jerusalem	Israel	50				_
		Talata Volon	Malagasy Rep	300				_
		Tinang	Philippines	250				_
		Meyerton	South Africa	100	/			
15.205	19.73	London	UK	100/250	∠		_	_
		Simferopol	Ukraine	240	_			_
		Greenville	USA	50/500	∠			_
		Delano	USA	100/250	∠			_
		Aligarh	India		/			
		Delhi	India	100/250	_	_	_	_
		Tanger	Morocco	35/100	∠	_	_	_
		Kavalla	Greece	250		_		_
		Shepparton	Australia	100	_	_	_	_
		Carnarvon	Australia	250	/	_		
		Wavre	Belgium	100				_
		Poro	Philippines	50		_	_	
		Kimjae	Korea (S)	250			_	_
		Ismaning	Germany (W)	100		_		
		Hoerby	Sweden	350		_		
		Karachi	Pakistan	50			_	
		Arganda	Spain	100	X	_	_	_
15.210	19.72	Asuncion	Paraguay	100	∠	_	_	_
		Moskva	USSR	240	/			_
		Armavir	USSR	240/500		_		
		Alma Ata	USSR	240	_			
		Khabarovsk	USSR	100			_	
		Petropavlo Kam	USSR	240		_		
		Irkutsk	USSR	240	∠	_	_	_
		Abis	Egypt	100/250			_	
		Wavre	Belgium	100	∠	_	_	
		Allouis	France	100	∠	_	_	_
		Tinang	Philippines	250	/			_
		Pori	Finland	100			_	_
		Delhi	India	20	/			
		S M Galeria	Vatican	100			_	_
		Jerusalem	Israel	50				_
15.215	19.72	London	UK	100	∠	_	_	_
		Okeechobee	USA	100	∠	_	_	_
		Lisbonne	Portugal	100	∠	_	_	_

MHz	*Metres*	*Station*	*Country*	*kW*	*M*	*M*	*S*	*N*
		S Luiz	Brasil	2.5	∠	_	_	_
		Malolos	Philippines	50	∠	_	_	_
		Tinang	Philippines	250	∠	_	_	_
		Alger	Algeria	100	∠	_	_	_
		Agana	Guam	100				_
15.220	19.71	Meyerton	South Africa	/500	∠	_	_	_
				250				
		Duchanbe	USSR	100/500		_	_	
		Riga	USSR	240	∠			_
		Lopik	Netherlands	100	∠	_	_	_
		Szekesfehervar	Hungary	20	∠			_
		Jaszbereny	Hungary	250		_	_	
		Bonaire Noord	Neth Antilles	300	∠	_	_	_
		Talata Volon	Malagasy Rep	300	∠	_	_	
		Kamalabad	Iran	250	_			
		Riyadh	Saudi Arabia	350	_			
		Peking	China Rep					
		Allouis	France	100	/		_	_
		Arganda	Spain	100	/			_
15.225	19.70	Limassol	Cyprus	20	/			_
		Lisbonne	Portugal	100		_		
		Wertachtal	Germany (E)	500	_	_	_	
		Sfax	Tunisia	100			_	_
		Szekesfehervar	Hungary	20	_	_	_	
		Jaszbereny	Hungary	250	_	_	_	
		Cyclops	Malta	250	∠	_	_	_
		Delano	USA	250	_	_	_	
		Greenville	USA	250				_
		Wavre	Belgium	100	_			
		Taipei	China Nat					
		Careysburg	Liberia	250				_
		Agana	Guam	100	/			
15.230	19.70	Habana	Cuba	100	∠	_	_	_
		Voronej	USSR	240	∠	_	_	
		Alma Ata	USSR	240	∠	_	_	_
		Lyndhurst	Australia	10	∠	_	_	_
		Melo	Uruguay	5	∠	_	_	_
		Roma	Italy	100	_	_		
		Jerusalem	Israel	50/300	∠	_	_	_
		Kabul	Afghanistan	100				
		Schwarzenburg	Switzerland	250	_	_		
		Wavre	Belgium	100/250			_	_
		Alger	Algeria					
		Tinang	Philippines	250				_
		Riyadh	Saudi Arabia	350	/			
		Karachi	Pakistan	10	/			
		Agana	Guam	100	/			
15.235	19.69	London	UK	100	∠	_	_	_
		Greenville	USA	250/500	∠	_	_	_

MHz	Metres	Station	Country	kW	M	M	S	N
		Delhi	India	100	∠	_	_	_
		Tokyo Yamata	Japan	100	∠	_	_	_
		Careysburg	Liberia	250	∠	_	_	_
		Bocaue	Philippines	50	_	_	_	_
		Malolos	Philippines	50	∠	_	_	_
		Tanger	Morocco	35	/			_
		Kavalla	Greece	250	∠		_	
		Lopik	Netherlands	100	∠	_	_	_
		Agana	Guam	100			_	_
		Schwarzenburg	Switzerland	150	/			_
		Kampala	Uganda	250				_
15.240	19.69	Armavir	USSR	240	∠			
		Baku	USSR	500				_
		Novosibirsk	USSR	100	∠			
		Nikolaevskamur	USSR	240	/			_
		Kiev	Ukraine	240				_
		Belgrade	Yugoslavia	100	∠	_	_	_
		K Wusterhausen	Germany (E)	100	∠	_	_	_
		Nauen	Germany (E)	500	∠	_	_	_
		Lyndhurst	Australia	10	∠	_	_	_
		Karlsborg	Sweden	350	_			
		Hoerby	Sweden	350	∠	_	_	_
		Greenville	USA	250/500	∠	_	_	_
		Careysburg	Liberia	250	_			
		Jerusalem	Israel	20/300	_	_	_	
		Wavre	Belgium	100	_	_		
		Noblejas	Spain	350	∠	_	_	_
		Velkekostolany	Czechoslovakia	120	/			_
		Santiago	Chile					
15.245	19.68	London	UK	100	/		_	_
		Wertachtal	Germany (W)	500	∠	_	_	_
		Julich	Germany (W)	100	∠	_	_	_
		Sines	Portugal	250	∠	_	_	_
		Brasilia	Brasil	10	/			_
		Nikolaevskamur	USSR	120		_	_	
		Simferopol	Ukraine	240		_		
		Kinshasa	Zaire	100	∠	_	_	_
		Wien	Austria	100		_		
		Tanger	Morocco	35	/			_
		Quito	Ecuador	100			_	
15.250	19.67	London	UK	250			_	
		Limassol	Cyprus	20		_		
		Bucuresti	Roumania	120/250	∠	_	_	_
		Cincinnati	USA	250	/			_
		Dixon	USA	250	_	_	_	_
		Greenville	USA	250	∠	_	_	_
		Tinang	Philippines	250	∠	_	_	_
		Poro	Philippines	50	/		_	_
		K Wusterhausen	Germany (E)	100		_	_	

MHz	*Metres*	*Station*	*Country*	*kW*	*M*	*M*	*S*	*N*
		Delhi	India	100	L	_	_	_
		Kampala	Uganda	250				_
		Careysburg	Liberia	250		_	_	_
		Roma	Italy	60/100		_	_	
		Wavre	Belgium	250			_	
		Franceville	Gabon Rep	500			_	
		Noblejas	Spain	350	/			_
		Peking	China Rep					
15.255	19.67	K Wusterhausen	Germany (E)	100				_
		Lisbonne	Portugal	250	L	_	_	_
		Khabarovsk	USSR	120		_		
		Kenga	USSR	50/500	L			_
		Tachkent	USSR	100		_	_	
		Abu Zaabal	Egypt	100			_	
		Bonaire Zuid	Neth Antilles	50	/	_	_	_
		Bucuresti	Roumania	250	/	_	_	_
15.260	19.66	London	UK	100/250	_	_	_	
		Ascension	Ascension	250	L	_	_	_
		Kazan	USSR	120		_	_	
		Baku	USSR	240/500		_	_	_
		Simferopol	Ukraine	240	_			
		Tokyo	Japan	1	L	_	_	_
		Momote	Japan	10	_			
		Greenville	USA	50/250	L	_	_	
		Scituate	USA	50			_	
		Kavalla	Greece	250	L	_	_	_
		Careysburg	Liberia	250	_			_
		Malolos	Philippines	50	L	_	_	_
		Tinang	Philippines	250				_
		Kamalabad	Iran	350	_			_
		Sackville	Canada	50/250	L	_	_	_
		Shepparton	Australia	50/100	_	_	_	_
15.265	19.65	London	UK	100			_	_
		Kenga	USSR	100	/	_	_	
		Armavir	USSR	100	_	_		_
		Frunze	USSR	240/500		_		
		Gorkii	USSR	100	_			_
		Tachkent	USSR	500	/			
		S Paulo	Brasil	7.5/10	L	_	_	_
		Pori	Finland	100	L	_	_	_
		Jerusalem	Israel	100/300			_	
		Peking	China Rep					
15.270	19.65	London	UK	100/250	L	_	_	_
		Kranji	Singapore	100	/			
		Limassol	Cyprus	100				
		Peking	China Rep					
		Tanger	Morocco	100	L	_	_	_
		Delano	USA	250	_	_	_	_
		Scituate	USA	50			_	

MHz	*Metres*	*Station*	*Country*	*kW*	*M*	*M*	*S*	*N*
		Tokyo Yamata	Japan	200	/_	_	_	_
		Pori	Finland	100/250	/_	_	_	_
		Wien	Austria	100	_	_	_	_
		Kavalla	Greece	250	/_	_	_	_
		Islamabad	Pakistan	10	/			
		Delhi	India	100	/			
		V of Democratic Kampuchea						
15.275	19.64	Julich	Germany (W)	100	/_	_	_	
		Wertachtal	Germany (W)	500	/_	_	_	_
		Montevideo	Uruguay	10	/_	_	_	_
		Riazan	USSR	100	/_	_	_	_
		Bonaire Zuid	Neth Antilles	50/80	/_	_	_	_
		Warszawa	Poland	100	/_	_	_	_
		Malolos	Philippines	50	/			
		Delhi	India	50				_
15.280	19.63	London	UK	100	/			
		Kranji	Singapore	100	/			_
		Greenville	USA	250/500	/_	_	_	_
		Redwood City	USA	50	/_	_	_	_
		Brazilia	Brasil					
		Armavir	USSR	240		_	_	
		Krasnoiarsk	USSR	240	/	_		
		Khabarovsk	USSR		/			
		Delhi	India	50	_	_	_	_
		Malolos	Philippines	50/100	_	_	_	_
		Wavre	Belgium	250		_		
		Wellington	New Zealand	7.5			_	_
		Lopik	Netherlands	100	/			_
		Allouis	France	100				
15.285	19.63	Tema	Ghana	100	/_	_	_	_
		Ejura	Ghana	250	/_	_	_	_
		K Wusterhausen	Germany (E)	100		_		_
		Tanger	Morocco	35		_		
		Szekesfehervar	Hungary	20	_	_	_	
		Jaszbereney	Hungary	250	_	_	_	
		Malolos	Philippines	50	/_	_	_	_
		Dacca	Bangladesh	100	/_	_	_	_
		Irkutsk	USSR	50				
		Meyerton	South Africa	100	/			
		Wavre	Belgium					
15.290	19.62	Shepparton	Australia	100	/_	_	_	_
		S Fernando	Argentina	10	/_	_	_	_
		Tinang	Philippines	35/250	/_	_	_	_
		Poro	Philippines	100	/			
		Noblejas	Spain	350	/_	_	_	_
		Playa de Pals	Spain	500	/_	_	_	
		Lisbonne	Portugal	50	/_			

MHz	Metres	Station	Country	kW	M	M	S	N
		Lampertheim	Germany (W)	50/100		_	_	_
		Holzkirchen	Germany (W)	10	/			
		Biblis	Germany (W)	100	/			_
		Santiago	Chile	100	_			
		Allouis	France	500		_		
15.295	19.61	Kajang	Malaysia					
		Kazan	USSR	100		_	_	
		Voronej	USSR	240	_	_	_	_
		Khabarovsk	USSR	100	∠			_
		Maputo	Mozambique	100	∠	_		_
		Kajang	Malaysia	100	∠	_	_	_
		Quito	Ecuador	100	∠	_	_	_
		Wien	Austria	100	_			
		Kabul	Afghanistan					
15.300	19.61	Bocaue	Philippines	50	∠	_	_	_
		Tokyo Yamata	Japan	100	∠	_	_	_
		Quito	Ecuador	100	_	_	_	_
		Allouis	France	500	∠	_	_	_
		Santiago	Chile	100	_			
		Novosibirsk	USSR	100		_	_	
		Jerusalem	Israel	300			_	_
		Peking	China Rep					
		Tripoli	Libya					
15.305	19.60	Schwarzenburg	Switzerland	150/250	∠	_	_	_
		Sottens	Switzerland	500	∠	_	_	_
		Sverdlovsk	USSR	100		_		
		Tbilisi	USSR	100	∠			_
		Voronej	USSR	100		_	_	
		Tanger	Morocco	35/100	∠		_	_
		Karlsborg	Sweden	350	_			
		Sines	Portugal	250		_		
		Red Lion	USA	50	_			
		Peking	China Rep					
		Kavalla	Greece	250		_		
		Jerusalem	Israel	300			_	
15.310	19.60	London	UK	100	/		_	_
		Kranji	Singapore	100/250				_
		Masirah	Oman	100	∠	_	_	_
		Delhi	India	50	∠	_	_	_
		Sofia	Bulgaria	50/145	∠	_	_	_
		Conakry	Guinea	100	∠	_	_	_
		Tanger	Morocco	35/100	_	_		_
		Novosibirsk	USSR	100	∠			_
		Tokyo Yamata	Japan	100/200	/	_	_	_
		Quito	Ecuador	100	∠	_		_
		Malolos	Philippines	50/100	_	_		
		Bonaire Noord	Neth Antilles	300	_	_		
		Allouis	France	100	_	_	_	
		Shepparton	Australia					

MHz	Metres	Station	Country	kW	M	M	S	N
15.315	19.59	London	UK	100		_	_	_
		Sines	Portugal	250	∟			_
		Careysburg	Liberia	250	∟	_	_	_
		Allouis	France	100/500	∟	_	_	_
		Kamalabad	Iran	100		_	_	
		Bonaire Noord	Neth Antilles	300	∟	_	_	_
		Roma	Italy	100		_		
		Greenville	USA	50	/		_	_
		Arganda	Spain	100	/	_	_	_
15.320	19.58	London	UK	100/250	∟	_	_	_
		Armavir	USSR	100	∟			_
		Kazan	USSR	100		_	_	
		Shepparton	Australia	50	∟	_	_	_
		Wertachtal	Germany (W)	500	∟	_	_	_
		Julich	Germany (W)	100	∟	_	_	_
		Bonaire Noord	North Antilles	50/ 300	_	_	_	_
		Careysburg	Liberia	250	/	_		_
		Leipzig	Germany (E)	100		_		
		Nauen	Germany (E)	100	/		_	_
		Wien	Austria	100	∟	_	_	_
		Santiago	Chile	100			_	
		Sines	Portugal	250				_
		Tinang	Philippines	250	/			_
15.325	19.58	Karachi	Pakistan	50	∟		_	_
		Islamabad	Pakistan	250	∟	_	_	_
		Simferopol	Ukraine	120		_	_	
		Sackville	Canada	50/250	∟	_	_	_
		Mahe	Seychelles	50	∟	_	_	_
		S Paulo	Brasil	1	∟	_	_	_
		Tokyo Yamata	Japan	100	_			
		Lopik	Netherlands	100	∟	_	_	_
		Kampala	Uganda	250				
		Athinai	Greece	100	/			
15.330	19.57	Cincinnati	USA	175/250	∟	_	_	_
		Dixon	USA	100	∟	_	_	_
		Ivanofrankovsk	Ukraine	240		_		
		Starobelsk	Ukraine	240		_	_	
		Tachkent	USSR	240	∟			_
		Kursk	USSR	240			_	
		Sverdlovsk	USSR	100		_	_	
		Roma	Italy	100	∟	_	_	_
		Jerusalem	Israel	20/300	_			_
		Tanger	Morocco	100	/			_
		Allouis	France	100	_		_	
		Pori	Finland	250	_			
		Kavalla	Greece	250	∟	_	_	_
		Careysburg	Liberia	250	/	_		_
		Kimjae	Korea (S)	250		_		

MHz	*Metres*	*Station*	*Country*	*kW*	*M*	*M*	*S*	*N*
		Sofia	Bulgaria	50		_		
		Riyadh	Saudi Arabia	350				_
		Arganda	Spain	100	/			_
15.335	19.56	Madras	India	100	∠	_	_	_
		Aligarh	India	250	/			
		Delhi	India	100/250	_	_	_	_
		Kimjae	Korea (S)	250	∠		_	_
		Wien	Austria	100	∠	_	_	_
		Bogota	Colombia	25	∠	_	_	_
		Abu Zaabal	Egypt	100			_	
		Bucuresti	Roumania	250	∠	_	_	_
		Dacca	Bangladesh	100			_	_
		Arganda	Spain	100	X	_	_	_
		Tanger	Morocco	100				_
		Novosibirsk	USSR	100	/			
15.340	19.56	Biblis	Germany (W)	100			_	
		Holzkirchen	Germany (W)	10	_			
		Lampertheim	Germany (W)	100	∠	_	_	
		Playa de Pals	Spain	250/500		_	_	_
		Lisbonne	Portugal	250				
		S Gabriel	Portugal	100		_		
		Habana	Cuba	50	∠	_	_	_
		K Wusterhausen	Germany (E)	100	∠			_
		Bucuresti	Roumania	120	∠	_	_	_
		Roma	Italy	100	/	_	_	_
		Beyrouth	Lebanon	100				_
15.345	19.55	Gral Pacheo	Argentina	50	∠	_	_	_
		Sebaa Aioun	Morocco	50	∠	_	_	_
		Tinang	Philippines	250	∠	_	_	_
		Jayapura	Indonesia	5	∠	_	_	_
		Sulaibiyah	Kuwait	250	∠	_	_	_
		Fredrikstad	Norway	250	∠		_	_
		Delano	USA	250	/			_
		Dixon	USA	250		_		
		Athinai	Greece	100	∠	_	_	_
		Taipei	China Nat					
		Bucuresti	Roumania	120	∠	_	_	_
		Roma	Italy	100	∠			
		Wellington	New Zealand	7.5	/		_	_
15.350	19.54	London	UK	250	_	_	_	
		Junglinster	Luxembourg	6	∠	_	_	_
		Armavir	USSR	100				_
		Vologda	USSR	240	∠	_	_	_
		Komsomolskamur	USSR	240	_	_	_	_
		Greenville	USA	250	_	_	_	
		Kinshasa	Zaire	100	∠		_	_
		Schwarzenburg	Switzerland	100/250	∠	_	_	_
		Riyadh	Saudi Arabia	350		_		
		Beyrouth	Lebanon	100				_

MHz	*Metres*	*Station*	*Country*	*kW*	*M*	*M*	*S*	*N*
		Scituate	USA	50				_
		Delhi	India	50	/			
15.355	19.54	London	UK	250	/			
		Holzkirchen	Germany (W)	10		_		
		Montevideo	Uruguay	10	∠	_	_	_
		Lisbonne	Portugal	100	∠	_	_	_
		Shepparton	Australia	50/100	∠	_	_	_
		Redwood City	USA	250	_	_	_	_
		Lyndhurst	Australia	10				
15.360	19.53	Kranji	Singapore	250	/			_
		Rabat	Morocco	50	/			
		Tanger	Morocco	50	_	_	_	_
		Moskva	USSR	100/240	∠	_	_	_
		Irkutsk	USSR	50	_			
		Allouis	France	500	∠	_	_	_
		Kimjae	Korea (S)	250	/		_	_
		Tinang	Philippines	50/250				_
		Quito	Ecuador					
15.365	19.52	Las Mesas	Canary Is	50	∠	_	_	_
		Delano	USA	240	_	_	_	_
		Bucuresti	Roumania	18/120	∠	_	_	_
		Sackville	Canada	50	_	_	_	_
		Tinang	Philippines	50/250	_	_	_	_
		Kabul	Afghanistan	100	_	_	_	
		Noblejas	Spain	350		_		
		Riyadh	Saudi Arabia	350		_		
15.370	19.52	Limassol	Cyprus	20/100			_	
		Lisbonne	Portugal	25	/			
		Lampertheim	Germany (W)	100		_	_	
		Holzkirchen	Germany (W)	10	/			_
		Playa de Pals	Spain	250/500	/	_	_	_
		Rio de Janeiro	Brasil	10	∠	_	_	_
		Tula	USSR	100		_		
		Riyadh	Saudi Arabia	350	/			
		Delhi	India	100	/			
		Grenada	Br W Indies					
15.375	19.51	Armavir	USSR	240		_	_	
		Jigulevsk	USSR	100				_
		Kazan	USSR	100	∠			
		Leningrad	USSR	220	_			
		Moskva	USSR	240	/			
		Kenga	USSR	100	∠			_
		Quito	Ecuador	50/100	∠	_	_	_
		Abu Zaabal	Egypt	100				
		Greenville	USA	500	/	_	_	_
		Noblejas	Spain	350	/			_
15.380	19.51	Limassol	Cyprus	20	/	_	_	
		Kranji	Singapore	100	/			
		Bucuresti	Roumania	15/250	∠	_	_	_

MHz	*Metres*	*Station*	*Country*	*kW*	*M*	*M*	*S*	*N*
		Biblis	Germany (W)	100	/	_	_	
		Lampertheim	Germany (W)	100	_		_	_
		Lisbonne	Portugal	100	/			
		Playa de Pals	Spain	250/500	_	_		
		Wellington	New Zealand	7.5	_	_	_	
		Wien	Austria	100	/		_	_
		Quito	Ecuador					
		Meyerton	South Africa	100	/			
15.385	19.50	Bocaue	Philippines	35/50	_	_		
		Roma	Italy	60/100	L	_	_	_
		Armavir	USSR	240	L			
		Tchita	USSR	120	L	_		_
		Gorkii	USSR	240		_	_	
		Novosibirsk	USSR	240		_		
		Mexico	Mexico	10/50				
		Cyclops	Malta	250	_			
		Sofia	Bulgaria	145/500	L	_	_	_
		Lopik	Netherlands	100				_
		Talata Volon	Malagasy Rep	300				_
		Wien	Austria	100		_		
		Schwarzenburg	Switzerland	150	/			_
15.390	19.49	London	UK	100/250	L	_	_	_
		Ascension	Ascension		/			_
		Bocaue	Philippines	50	L	_	_	_
		Hoerby	Sweden	350	L	_	_	_
		Caracas	Venezuela	10	L	_	_	_
		Bucuresti	Roumania	18/120	/	_		_
		Kabul	Afghanistan	100	_	_	_	
		K Wusterhausen	Germany (E)	100		_		
		Leipzig	Germany (E)	100	/	_	_	_
		Scituate	USA	100		_		
		Noblejas	Spain	350	/			
15.395	19.49	Arganda	Spain	100	/			_
		Noblejas	Spain	300				
		Greenville	USA	50/500	L	_	_	_
		Tinang	Philippines	250	L	_	_	_
		Armavir	USSR	100	/			
		Tachkent	USSR	240	L	_	_	_
		Velkekostolany	Czechoslovakia	120	L	_	_	_
		Ekala	Ceylon	35	L	_	_	_
15.400	19.48	Ascension	Ascension	125/250	L	_	_	_
		Limassol	Cyprus	20/100	_	_		
		Greenville	USA	250/500	L	_	_	_
		Roma	Italy	100	L	_	_	_
		Kazan	USSR	100	_			_
		Novosibirsk	USSR	50	/			
		Careysburg	Liberia	250	L		_	_
		Dacca	Bangladesh	100	L	_	_	_
		Karachi	Pakistan	50	_			

MHz	Metres	Station	Country	kW	M	M	S	N
		S M Galeria	Vatican	100	∠	_	_	
		Pori	Finland	100	/			
15.405	19.47	Khabarovsk	USSR	240		_	_	
		Armavir	USSR	240	∠	_	_	
		Sverdlovsk	USSR	100		_	_	
		Tirane	Albania					
		Jerusalem	Israel	50/300	∠	_	_	_
		Cyclops	Malta	250	∠	_	_	_
		Noblejas	Spain	350	∠	_	_	_
		Quito	Ecuador	100	_		_	
		Peking	China Rep					
		Mahe	Seychelles	100	/			
		Wertachtal	Germany (W)	500		_		
		Ekala	Ceylon	35				_
15.410	19.47	Greenville	USA	250	∠	_	_	_
		Tinang	Philippines	250	∠	_	_	_
		Malolos	Philippines	50	_			
		Wien	Austria	100	∠	_	_	_
		Kigali	Rwanda	250	∠	_	_	_
		Moskva	USSR	50	∠			_
		Dacca	Bangladesh	100	_	_		
		Shepparton	Australia	100	∠	_	_	_
		Julich	Germany (W)	100		_		
15.415	19.46	Greenville	USA	50/250	∠	_	_	_
		Ribeirao Preto	Brasil	1/7.5	∠	_	_	_
		Baku	USSR	100		_	_	
		Kazan	USSR	100	_	_	_	_
		Nikolaevskamur	USSR	100		_	_	
		Bonaire Zuid	Neth Antilles	50/250	_	_	_	_
		Jerusalem	Israel	20/300	∠	_	_	_
		Shepparton	Australia	25	/	_	_	_
		Hoerby	Sweden	350		_		
		Malolos	Philippines	50				_
		Quito	Ecuador					
15.420	19.46	London	UK	250	/	_	_	_
		Ascension	Ascension	250				_
		Antigua	Br W Indies	250	_	_		
		Limassol	Cyprus	20/100	∠	_	_	_
		Tokyo Yamata	Japan	20/100	∠	_	_	_
		Islamabad	Pakistan	100/250	∠	_	_	_
		Karachi	Pakistan	50	∠	_	_	_
		Serpukhov	USSR	100	_			
		Tchita	USSR	500				_
		Khabarovsk	USSR	100	/			
		Irkutsk	USSR	100	_			
		Krasnoiarsk	USSR	100	/			
		Peking	China Rep					
		Ribeiro Preto	Brasil	1/7.5				
		Bonaire Zuid	Neth Antilles	50	/			

MHz	*Metres*	*Station*	*Country*	*kW*	*M*	*M*	*S*	*N*
15.425	19.45	London	UK	250		_		
		Petropavlo Kam	USSR	100		_	_	
		Kenga	USSR	240		_	_	
		Simferopol	Ukraine	100		_	_	
		Perth	Australia	10/50	∠	_	_	_
		Wertachtal	Germany (W)	500	∠	_	_	_
		Julich	Germany (W)	100	∠	_	_	
		Ekala	Ceylon	100	∠	_	_	_
		Allouis	France	100/500	∠	_	_	_
		Jerusalem	Israel	50/300	_		_	_
		Taipei	China Nat					
		Wien	Austria	100	_			
		Tinang	Philippines	250	/			_
		Greenville	USA	50				_
15.430	19.44	Limassol	Cyprus	100	/		_	_
		Schwarzenburg	Switzerland	100/250	_	_	_	_
		Sottens	Switzerland	500	_	_	_	_
		Greenville	USA	50/250	∠	_	_	_
		Cincinnati	USA	175	∠	_	_	_
		Delhi	India	100	_	_	_	_
		Ismaning	Germany (W)	100		_		
		Mexico	Mexico	50	∠	_	_	_
		Kimjae	Korea (S)	100	∠	_	_	_
		Mahe	Seychelles	100	∠	_	_	_
		Nauen	Germany (E)	500		_		
		Careysburg	Liberia	200	/			_
		Pori	Finland	100	/			
15.435	19.44	London	UK	100/250	∠	_	_	_
		Kranji	Singapore	100	/			_
		Dar es Salaam	Tanzania	50	_	_	_	_
		Erevan	USSR	100	/			
		Frunze	USSR	100/500	∠	_	_	
		Tula	USSR	120				_
		Vinnitsa	Ukraine	100	∠			
		Wien	Austria	100	_	_	_	_
		Quito	Ecuador	100	/	_		_
		Kamalabad	Iran	350		_		
		Allouis	France	100			_	_
		Hoerby	Sweden	350		_		
15.440	19.43	London	UK	100/250		_		
		Scituate	USA	50/100	∠	_	_	_
		Okeechobee	USA	100	∠	_	_	_
		Bocaue	Philippines	25/50	∠	_	_	_
		Riazan	USSR	120	∠	_	_	_
		Allouis	France	100	_	_	_	_
		Bonaire Zuid	Neth Antilles	50/250	∠	_	_	_
		Beyrouth	Lebanon	100	_			
		Wien	Austria	100		_		
		Careysburg	Liberia	250		_		

MHz	Metres	Station	Country	kW	M	M	S	N
		Franceville	Gabon Rep	500			—	
15.445	19.42	London	UK	100/250	—	—	—	
		Careysburg	Liberia	250	∠	—	—	—
		Holzkirchen	Germany (W)	10			—	
		Lampertheim	Germany (W)	100	∠	—		—
		Playa de Pals	Spain	250/500	/		—	—
		Lisbonne	Portugal	100				
		Brasilia	Brasil	10	∠	—	—	—
		Novosibirsk	USSR	240		—	—	
15.450	19.42	Serpukhov	USSR	100				
		Manila	Philippines					
		Peking	China Rep					
15.455	19.41	Moskva	USSR	240				
15.460	19.40		USSR	100				
15.465	19.40	Jerusalem	Israel					
			USSR					
		Karachi	Pakistan					
15.470	19.39		USSR	50				
15.475	19.39	Cairo	Egypt	50				
15.480	19.38		USSR	50				
		Peking	China Rep					
15.485	19.37	Jerusalem	Israel					
			USSR					
15.490	19.37		USSR*	50				
15.497	19.36		USSR*					
15.500	19.35		USSR					
			China Rep					
15.505	19.35	Sverdlovsk	USSR	50				
15.510	19.34	Peking	China Rep					
			USSR					
15.512	19.34	Jerusalem	Israel					
15.515	19.34		USSR					
		Karachi	Pakistan					
15.520	19.33	Peking	China Rep	120				
		Dacca	Bangladesh	100				
		Karachi	Pakistan					
			USSR					
15.525	19.32		USSR					
15.530	19.32		USSR					
		Jerusalem	Israel					
15.535	19.31		USSR					
15.540	19.31		USSR*					
15.545	19.30	Jerusalem*	Israel					
			USSR					
15.550	19.29	Peking	China Rep					
15.570	19.27	Seoul	Korea (S)					
15.583	19.25	Monte Carlo*	Monaco					
15.587	19.24	Peking	China					
15.589	19.24	London*	UK					

MHz	Metres	Station	Country	kW	M	M	S	N
15.590	19.24	Peking*	China Rep	50				
15.595	19.24		Germany (E)*					
15.600	19.23		USSR*					
			China Rep					
		V of Democratic Kampuchea						
15.645	19.18	Kuwait	Kuwait					
15.650	19.17	Greenville*	USA					
15.660	19.16	Monte Carlo*	Monaco					
15.665	19.15	Riyadh*	Saudi Arabia					
15.670	19.14	London*	UK					
15.685	19.13	Peking	China Rep					
		Karachi	Pakistan					
15.710	19.10	Peking	China Rep					
15.730	19.07	RFE/R Liberty*						
15.735	19.07		China Rep					
15.752	19.04	Greenville*	USA					
15.770	19.02	Greenville*	USA					
15.773	19.02	Algiers	Algeria					
15.775	19.02	RFE/R Liberty*						
15.780	19.01		USSR*					
15.790	19.00	V Malayan Revln						
15.849	18.93	London*	UK					
15.850	18.93		USSR*					
15.870	18.90		USSR					
		Lyndhurst*	Australia					
15.880	18.89	Peking	China Rep					
15.910	18.86	London*	UK					
15.920	18.84	Monrovia*	Liberia					
16.030	18.71		USSR*					
16.065	18.67	RFE/R Liberty*						
16.140	18.59		USSR*					
16.190	18.53		USSR*					
16.222	18.49	Bethany*	USA					
16.240	18.47	RFE/R Liberty*						
16.247	18.46	Tangier*	Morocco					
16.250	18.46		USSR*					
16.330	18.37		USSR*					
16.430	18.26	Greenville*	USA					
16.870	17.78		USSR					
17.135	17.51		USSR					
17.387	17.25	Delhi	India	100				
17.422	17.21	Hilversum*	Netherlands					
17.445	17.20	RFE/R Liberty*						
17.450	17.19	Peking	China Rep					
17.490	17.15	Peking	China Rep					
17.515	17.13	Peking	China Rep					
17.530	17.11	Peking	China Rep					
17.535	17.11	Peking	China Rep					

MHz	Metres	Station	Country	kW	M	M	S	N
17.560	17.08		USSR					
17.580	17.06		USSR*					
17.593	17.05	Baghdad	Iraq					
17.605	17.04	Peking	China Rep					
17.630	17.02	Jerusalem	Israel					
17.635	17.01	Peking	China Rep					
17.640	17.01	Santiago	Chile					
		Karachi	Pakistan					
17.642	17.00	London*	UK					
17.650	17.00	Peking	China Rep	240				
17.665	16.98	Karachi	Pakistan					
17.670	16.98	Cairo	Egypt	100				
17.680	16.97	Peking	China Rep	240				
17.685	16.96	Jerusalem	Israel					
17.690	16.96	Cairo	Egypt					
			USSR					
17.695	16.95	London	UK					
17.700	16.95	Berlin	Germany (E)					
			USSR					
		Lopik	Netherlands					
		Jerusalem	Israel					
17.705	16.94	London	UK	100/250	∠	_	_	_
		Delhi	India	50/100	∠	_	_	_
		Aligarh	India	250	∠	_	_	_
		K Wusterhausen	Germany (E)	100	∠	_		_
		Nauen	Germany (E)	500		_	_	
		Nikolaevskamur	USSR	100	_			_
		Habana	Cuba	50	∠	_	_	_
		S M Galeria	Vatican	100	_			
		Tanger	Morocco	20/100	∠	_	_	_
		Praha	Czechoslovakia	120	∠	_	_	_
		Lampertheim	Germany (W)	100		_		
		Peking	China Rep					
		V of Democratic Kampuchea						
17.710	16.94	Cincinnati	USA	250	∠	_	_	_
		Moskva	USSR	240	_	_	_	_
		Tachkent	USSR	100	/			_
		Jaszbereney	Hungary	250	_	_	_	
		Szekesfehervar	Hungary	20	_	_	_	
		Jerusalem	Israel	100/300	∠	_	_	_
		Wellington	New Zealand	7.5		_		
		Wien	Austria	100	_	_		
		Tanger	Morocco	35		_		
		Malolos	Philippines	100	/			
		Poro	Philippines	50				_
		Beyrouth	Lebanon	100			_	
		Tokyo	Japan	100				
17.715	16.93	London	UK	100/250	∠	_	_	_

MHz	Metres	Station	Country	kW	M	M	S	N
		Kranji	Singapore	100	/			
		Jaszbereny	Hungary	250	_	_	_	
		Szekesfehervar	Hungary	20	_	_	_	
		Wertachtal	Germany (W)	500	∠	_	_	_
		Julich	Germany (W)	100	∠	_	_	_
		Fredrikstad	Norway	100				_
		Santiago	Chile	45/100	∠	_	_	_
		Wavre	Belgium	100/250	∠	_	_	
		Schwarzenburg	Switzerland	100		_		
		Sottens	Switzerland	500	/		_	_
		Allouis	France	500				_
		Mahe	Seychelles	100	/		_	_
		Careysburg	Liberia	250	/			
		Beirut	Lebanon					
17.720	16.93	Allouis	France	500	∠	_	_	_
		Red Lion	USA	50	_	_	_	_
		Dixon	USA	250	/			
		Delano	USA	250				_
		Kazan	USSR	100	∠	_	_	_
		Khabarovsk	USSR	100	_	_		
		Petropavlo Kam	USSR	100	/	_		_
		Taipei	China Nat	50				
		Wien	Austria	100	_	_	_	_
		Dacca	Bangladesh	100	_	_		
		Bucuresti	Roumania	120/250				_
		Noblejas	Spain	350	/			_
17.725	16.93	Kajang	Malaysia	500		_		
		Biblis	Germany (W)	100				_
		Holzkirchen	Germany (W)	10		_		_
		Lampertheim	Germany (W)	100	∠		_	
		Playa de Pals	Spain	250		_	_	
		Lisbonne	Portugal	50	∠	_	_	_
		Tokyo Yamata	Japan	100	∠	_	_	_
		Abu Zaabal	Egypt	100			_	
		Wien	Austria	100	_	_	_	_
		Shepparton	Australia	100	∠	_	_	_
		Delhi	India	100	_	_	_	_
		Praha	Czechoslovakia	120	/	_	_	_
		Santiago	Chile	100			_	
		Franceville	Gabon Rep	500			_	
		Kuwait	Kuwait					
17.730	16.92	Scituate	USA	50/100	∠			
		Greenville	USA	250	∠	_	_	_
		Red Lion	USA	50	/			
		Okeechobee	USA	100		_	_	_
		Armavir	USSR	100/500		_	_	
		Serpukhov	USSR	240	∠	_		_
		Irkutsk	USSR	240		_	_	
		Kamalabad	Iran	350		_	_	

MHz	Metres	Station	Country	kW	M	M	S	N
		Bucuresti	Roumania	50/120	∠	_	_	_
		Wavre	Belgium					
		Julich	Germany (W)	100	_	_	_	_
		Wertachtal	Germany (W)	500	/			_
		Aligarh	India	250	/			
		Dacca	Bangladesh	100				_
		Leipzig	Germany (E)	100				_
17.735	16.92	Peking	China Rep					
		Lisbonne	Portugal	50/250	∠	_	_	_
		Hoerby	Sweden	30/350	_	_	_	
		Karlsborg	Sweden	350	_	_	_	_
		Noblejas	Spain	350	∠	_	_	_
		Wavre	Belgium					
		Erevan	USSR	100	∠			
		Schwarzenburg	Switzerland	150	/			_
		Allouis	France	500				_
			Philippines					
		Ismaning	Germany (W)					
17.740	16.91	London	UK	100/250	∠	_	_	_
		Tinang	Philippines	250	∠	_	_	_
		Poro	Philippines	100		_	_	_
		Wavre	Belgium	20/250	∠	_	_	_
		Leipzig	Germany (E)	500	∠	_		_
		K Wusterhausen	Germany (E)	100				_
		Greenville	USA	250/500	∠	_	_	_
		Lopik	Netherlands	100	_	_	_	
		Fredrikstad	Norway	250			_	
		Schwarzenburg	Switzerland	150		_	_	_
		Achkhabad	USSR	100		_	_	
		Wien	Austria	100			_	
		Sulaibiyah	Kuwait	250	/			
		Careysburg	Liberia					
17.7425	16.91	Schwarzenburg*	Switzerland	30		_	_	
17.745	16.91	Kursk	USSR	240	∠	_	_	_
		Frunze	USSR	240/500	∠	_		_
		Vladivostock	USSR	50		_		
		Abis	Egypt	250			_	
		Manzini	Swaziland	25	_	_	_	
		Wien	Austria	100		_	_	_
		Quito	Ecuador	100			_	
17.750	16.90	Holzkirchen	Germany (W)	10			_	
		Biblis	Germany (W)	100		_	_	
		Lampertheim	Germany (W)	100	∠	_		
		Lisbonne	Portugal	10/50	/			_
		Playa de Pals	Spain	250	∠			_
		Noblejas	Spain	250	∠	_	_	_
		Poro	Philippines	100	_			
		Tinang	Philippines	250	/	_		_
		Delano	USA	250	/			

MHz	*Metres*	*Station*	*Country*	*kW*	*M*	*M*	*S*	*N*
		Dixon	USA	250	_	_	_	_
		Islamabad	Pakistan	100	∠	_	_	_
		Karachi	Pakistan	50	∠	_	_	_
		Habana	Cuba	50	∠	_	_	_
		Bucuresti	Roumania	120	∠	_	_	_
		Sackville	Canada	250	_		_	
		Sulaibiyah	Kuwait	250	_			_
17.755	16.90	K Wusterhausen	Germany (E)	100	∠	_	_	_
		Vinnitsa	Ukraine	120	/		_	
		Erevan	USSR	500	/			
		Tachkent	USSR	100/500	∠	_	_	_
		Vladivostock	USSR	100		_		
		Roma	Italy	100	∠	_	_	_
		Quito	Ecuador	100		_		_
		Fredrikstad	Norway	100	∠			
		Riyadh	Saudi Arabia	350	∠	_	_	
		Karachi	Pakistan	50	_			
		Santiago	Chile	100		_	_	
		Tokyo Yamata	Japan	100	/	_	_	_
		Pori	Finland	250			_	
		Greenville	USA					
		Shepparton	Australia	100	/			
17.760	16.89	London	UK					_
		Biblis	Germany (W)	100	_			
		Lampertheim	Germany (W)	100	/	_	_	_
		Playa de Pals	Spain	250		_	_	_
		Lisbonne	Portugal	10			_	
		Lvov	Ukraine	240		_	_	
		Sackville	Canada	50/250	_			_
		Riyadh	Saudi Arabia	350		_	_	
		Wien	Austria	100			_	
		Quito	Ecuador	100			_	
		Tanger	Morocco					
		Delhi	India	100	/			
17.765	16.89	Nikolaevskamur	USSR	100		_	_	
		Tula	USSR	100	∠	_	_	_
		Duchanbe	USSR	50		_	_	
		Dixon	USA	100	∠	_	_	_
		Kigali	Rwanda	250	∠	_	_	_
		Wertachtal	Germany (W)	500	∠	_	_	_
		Julich	Germany (W)	100	_			
		Poro	Philippines	100	∠	_	_	_
		Wavre	Belgium	100	∠	_		
		Alger	Algeria					
		Allouis	France	100	_	_	_	
		Beyrouth	Lebanon	100		_		
		Mexico City	Mexico					
17.770	16.88	London	UK	100		_		
		Masirah	Oman	100	/		_	_

MHz	Metres	Station	Country	kW	M	M	S	N
		Lisbonne	Portugal	50				_
		Playa de Pals	Spain	100/250	∟	_	_	
		Lampertheim	Germany (W)	100		_		
		Roma	Italy	60/100	∟	_	_	_
		Wien	Austria	100	∟	_	_	_
		Abu Ghraib	Iraq	100	_	_	_	_
		Hoerby	Sweden	350	_	_		
		Karlsborg	Sweden	350	_			
		Wavre	Belgium	250			_	_
		Riyadh	Saudi Arabia	350	∟	_	_	
		Julich	Germany	100		_		
		Carnarvon	Australia	250	_			
		Sottens	Switzerland	500	/			_
		Schwarzenburg	Switzerland	100				_
		Wellington	New Zealand					
17.775	16.88	Vinnitsa	Ukraine	500	/			
		Armavir	USSR	100			_	
		Riazan	USSR	240	∟			_
		Frunze	USSR	240/500	_			_
		Khabarovsk	USSR	240				_
		Krasnoiarsk	USSR	500				_
		Greenville	USA					
		Allouis	France	500	∟	_	_	_
		Jaszbereny	Hungary	250	_			
		Praha	Czechoslovakia	120	/			_
		Delhi	India	50	/			
		Malolos	Philippines	100	/			
17.7775	16.88	Varberg	Sweden	100	_		_	_
17.780	16.87	London	UK	100	∟	_	_	_
		Lvov	Ukraine	120	∟			_
		Aligarh	India	250	∟	_	_	_
		Szekesfehervar	Hungary	20	_	_	_	
		Meyerton	South Africa	500	∟	_	_	_
		Julich	Germany (W)	100	∟	_	_	_
		Wertachtal	Germany (W)	500	∟	_	_	_
		Delano	USA	200/250	∟	_	_	_
		Careysburg	Liberia	250	_		_	
		Lampertheim	Germany (W)	100	_			
		Sackville	Canada	50/250		_		
		Poro	Philippines	50	/			_
		Tinang	Philippines	50/250	/			_
		Roma	Italy	60/100	∟	_	_	_
		Kavalla	Greece	250	/	_	_	_
17.785	16.87	Abis	Egypt	250			_	
		Frunze	USSR	50	∟			_
		Ivanofrankovsk	Ukraine	240	_	_	_	
		Greenville	USA	250/500	∟	_	_	_
		Okeechobee	USA	100		_	_	
		Scituate	USA	100		_		_

MHz	Metres	Station	Country	kW	M	M	S	N
		Jaszbereny	Hungary	250	_	_	_	
		Szekesfehervar	Hungary	20	_	_	_	
		Bucuresti	Roumania	250	_			_
		Pori	Finland	250	_			
		Leipzig	Germany (E)	100		_		
		Santiago	Chile	100			_	
		S M Galeria	Vatican	100			_	
		Kavalla	Greece	250	/			
		Allouis	France	500	/			_
17.7875	16.87	Varberg	Sweden	100	_	_	_	_
17.790	16.86	London	UK	100/250	∠	_	_	_
		Ascension	Ascension	250	_			
		Krasnoiarsk	USSR	50				
		Tinang	Philippines	250	∠	_	_	_
		Bucuresti	Roumania	120/250	/	_		_
		Hoerby	Sweden	30/350	_	_	_	_
		Careysburg	Liberia	250	/	_	_	_
		Allouis	France	100		_		
		Santiago	Chile	100		_	_	
		Schwarzenburg	Switzerland	150	X			_
		Vatican	Vatican					
		Quito	Ecuador	100				
17.795	16.86	London	UK	1000			_	
		Serpukhov	USSR	100	∠	_	_	_
		Roma	Italy	100	∠	_	_	_
		Fredrikstad	Norway	100	/	_	_	
		Tokyo Yamata	Japan	200	∠	_	_	_
		Shepparton	Australia	50	∠	_	_	_
		Careysburg	Liberia	250	_	_		
		Allouis	France	100	/			_
		Jerusalem	Israel	50/300	∠	_	_	_
		Wertachtal	Germany (W)	500	/	_	_	
		Julich	Germany (W)	100		_	_	_
17.800	16.85	London	UK	100	∠	_	_	_
		Karachi	Pakistan	50	_	_	_	_
		Islamabad	Pakistan	250			_	
		Greenville	USA	50/250	∠	_	_	_
		Julich	Germany (W)	100	∠	_	_	_
		Wertachtal	Germany (W)	500	∠	_	_	_
		K Wusterhausen	Germany (E)	100		_	_	
		Leipzig	Germany (E)	100		_		
		Fredrikstad	Norway	100	_		_	
		Tinang	Philippines	250	_	_		
		Baku	USSR	100	_			
		Allouis	France	100	∠	_	_	_
		Taipei	China Nat					
		Tripoli	Libya	500				
		Kigali	Rwanda	250	∠	_	_	_
		Santiago	Chile	100	∠	_	_	_

MHz	Metres	Station	Country	kW	M	M	S	N
		Kathmandu	Nepal	100	/_		_	_
17.805	16.85	Tbilisi	USSR	100/500	/_	_	_	_
		Tallin	USSR	100				_
		Khabarovsk	USSR	100	/			_
		Irkutsk	USSR	240	_			
		Tula	USSR	240				_
		Lisbonne	Portugal	50/250	/_	_	_	_
		Bucuresti	Roumania	250	/_	_	_	_
		Okeechobee	USA	100	/	_	_	
		Meyerton	South Africa	250				_
17.810	16.84	London	UK	250	/_	_	_	_
		Bocaue	Philippines	50	/_	_	_	_
		Lopik	Netherlands	100	/_	_	_	_
		Bonaire Noord	Neth Antilles	300	/_	_	_	_
		Talata Volon	Malagasy Rep	300	/_	_	_	_
		Warszawa	Poland	1	_	_	_	
		Tokyo	Japan					
17.815	16.84	London	UK	100			_	
		Roma	Italy	60/100	/_	_	_	_
		S Paulo	Brasil	10	/_	_	_	_
		Tchita	USSR	500	/			_
		Frunze	USSR	50	_			_
		Simferopol	Ukraine	240		_	_	
		Bucuresti	Roumania	18/250	/_	_	_	_
		Jerusalem	Israel	50/300	/_	_	_	_
		Wien	Austria	100	_	_		
		Talata Volon	Malagasy Rep	300	_	_	_	_
		K Wusterhausen	Germany (E)	100			_	
		Santiago	Chile	100	/			
		Sottens	Switzerland	500	X			
		Greenville	USA					
17.820	16.84	Kiev	Ukraine	240	_	_	_	
		Kazan	USSR	240	/			_
		Sackville	Canada	50/250	/_	_	_	_
		Dixon	USA	200	_	_	_	_
		Greenville	USA	250	/			_
		Sofia	Bulgaria	50	_	_	_	_
		Karachi	Pakistan	50	/		_	_
		Sines	Portugal	250			_	
		Tinang	Philippines	250	/			_
		Sulaibiyah	Kuwait	250				_
17.825	16.83	Tokyo Yamata	Japan	100/200	/_	_	_	_
		Baku	USSR	240/500		_	_	
		Frunze	USSR	100/500	/_	_	_	_
		Sofia	Bulgaria	50	/_	_	_	_
		Julich	Germany (W)	100	/_	_	_	_
		Wertachtal	Germany (W)	500	/_	_	_	_
		Cyclops	Malta	250	/_	_	_	_
		S M Galeria	Vatican	100	/_	_	_	_

MHz	Metres	Station	Country	kW	M	M	S	N
		Bucuresti	Roumania	120/250	∠	_	_	_
		Pori	Finland	250	/			
17.830		Ascension	Ascension	250	∠	_	_	_
		Antigua	Br W Indies	250	/		_	_
		Schwarzenburg	Switzerland	150/250	∠	_	_	_
		Islamabad	Pakistan	250	∠	_	_	_
		Athinai	Greece	100	∠	_	_	_
		Agana	Guam	100	∠	_	_	_
		Bucuresti	Roumania	120	/		_	
		Sackville	Canada	250	/			_
17.835	16.82	Lisbonne	Portugal	50	∠	_	_	_
		Vinnitsa	Ukraine	240		_	_	
		Moskva	USSR	240	∠			_
		Krasnoiarsk	USSR	100	_			
		Bucuresti	Roumania	120	∠	_	_	_
		Peking	China Rep					
17.840	16.82	Antigua	Br W Indies	250		_		
		Ascension	Ascension	250	_	_	_	
		Armavir	USSR		/			
		Frunze	USSR	240/500	∠			_
		Dixon	USA	200/250	/	_	_	_
		Delano	USA	200/250	∠	_	_	_
		Velkekostolany	Czechoslovakia	120	∠	_	_	_
		S M Galeria	Vatican	100	∠	_	_	_
		Bucuresti	Roumania	120	/			_
		Schwarzenburg	Switzerland	150		_	_	
		Wien	Austria	100	_	_	_	_
		Fredrikstad	Norway	120	∠	_	_	_
		Pori	Finland	250		_	_	
		Sackville	Canada	50			_	
17.845	16.81	Scituate	USA	50/100	∠			_
		Okeechobee	USA	100	∠	_	_	_
		Wertachtal	Germany (W)	500	_	_	_	_
		Nikolaevskamur	USSR	240			_	
		Sverdlovsk	USSR	100		_	_	
		Lvov	Ukraine	240		_	_	
		Lampertheim	Germany (W)	100				_
		Islamabad	Pakistan	250	∠	_	_	_
		Karachi	Pakistan	50	∠	_	_	_
		Kavalla	Greece	250	/	_		
		Allouis	France	500	/		_	_
		Jerusalem	Israel	50				_
		S M Galeria	Vatican	100	/		_	
17.850	16.81	Armavir	USSR	240	_	_		_
		Alma Ata	USSR	100	_	_	_	_
		Khabarovsk	USSR	100	/			_
		Allouis	France	500	∠	_	_	_
		Bucuresti	Roumania	50/120	/	_	_	_
		Ekala	Ceylon	35	∠	_	_	_

MHz	*Metres*	*Station*	*Country*	*kW*	*M*	*M*	*S*	*N*
		Jerusalem	Israel	50				_
		Wertachtal	Germany (W)	500				
		Tinang	Philippines	250	/			_
17.855	16.80	London	UK	250	∠	_	_	_
		Talata Volon	Malagasy	300	∠	_	_	_
		Careysburg	Liberia	250	∠	_	_	_
		Greenville	USA	50/250	∠	_	_	_
		Simferopol	Ukraine	240	∠			_
		Jerusalem	Israel	50/300	∠	_	_	
		Kavalla	Greece	250	/			_
		Agana	Guam	100	∠	_	_	_
		Peking	China Rep					
		Allouis	France	100/500		_		_
		Habana	Cuba	100	/		_	_
		Wien	Austria	100		_		
17.860	16.80	London	UK	100/250	/	_	_	_
		Allouis	France	100/500	∠	_	_	_
		Armavir	USSR	240		_		
		Kursk	USSR	100	∠			_
		Khabarovsk	USSR	240	/			
		Tachkent	USSR	240/500		_	_	
		Vladivostock	USSR	100	_			_
		Kiev	Ukraine		∠			_
		Greenville	USA	250	_	_		
		Delhi	India	100	/			
		Aligarh	India	250	_	_	_	_
		Careysburg	Liberia	250	∠	_		_
		Talata Volon	Malagasy Rep	300	_	_	_	_
		Wellington	New Zealand	7.5	/		_	_
17.865	16.79	Scituate	USA	50/100	_			
		Warszawa	Poland	100	∠	_	_	
		Komsomolskamur	USSR	120		_		
		Delhi	India	50	_	_	_	_
		Allouis	France	100/500	∠	_	_	_
		Noblejas	Spain	250	/			_
		Playa de Pals	Spain	250	/			
		Biblis	Germany (W)	100				_
		Lisbonne	Portugal	10/50	_	_	_	_
		Quito	Ecuador	100				
		Kavalla	Greece	250	∠	_	_	_
		Poro	Philippines	100	/			_
		Wien	Austria	100	/			_
		Careysburg	Liberia	250	/			
		Sackville	Canada	250	/			
17.870	16.79	London	UK	100/250	∠	_	_	_
		Ascension	Ascension	250	/		_	_
		Riazan	USSR	240	_	_		
		Khabarovsk	USSR	240	∠			_
		Tachkent	USSR	100	/			

MHz	Metres	Station	Country	kW	M	M	S	N
		Taichet	USSR	100				_
		Carnarvon	Australia	250	∠	_	_	_
		Shepparton	Australia	100	/	_		
		Montevideo	Uruguay	25	∠	_	_	_
		Careysburg	Liberia	250	∠	_	_	_
		Bucuresti	Roumania	120	∠	_	_	_
		Pori	Finland	250	_	_	_	
		Okeechobee	USA	100			_	
		Allouis	France	100				_
		Noblejas	Spain	350	/			
		Delhi	India	50	/			
17.875	16.78	Julich	Germany (W)	100	∠	_	_	_
		Wertachtal	Germany (W)	500	∠	_	_	_
		Rio de Janeiro	Brasil	7.5	∠	_	_	_
		Tachkent	USSR	240/500		_	_	
		Scituate	USA	50/100	_	_	_	_
		Okeechobee	USA	100	/		_	
		Allouis	France	500	_	_		
		Jerusalem	Israel	300	/			_
		Lisbonne	Portugal	50	_			
		Pori	Finland	250		_		
		Quito	Ecuador	100				
		Fredrikstad	Norway	100	/			
		Sackville	Canada	250	/			
		Delhi	India	100				_
17.880	16.78	Ascension	Ascension	125	/		_	
		Kranji	Singapore	100	/			_
		Leipzig	Germany (E)	100	_			
		Nauen	Germany (E)	500	_			
		K Wusterhausen	Germany (E)	100	∠			_
		S Gabriel	Portugal	100	∠	_	_	_
		Petropavlo Kam	USSR	240		_	_	
		Irkutsk	USSR	240	∠			_
		Tula	USSR	100		_	_	
		Kiev	Ukraine	240/500	∠			_
		Tokyo Yamata	Japan	100	∠	_	_	_
		Santiago	Chile	100	_			
		Allouis	France	100	_	_		
		Red Lion	USA	50	_			
		Peking	China Rep					
		Jerusalem	Israel	100		_		
17.885	16.77	Ascension	Ascension	125/250			_	_
		Limassol	Cyprus	20/100	∠	_	_	_
		Habana	Cuba	100	_	_	_	
		Vinnitsa	Ukraine	240		_		
		Tbilisi	USSR	100			_	
		Bogota	Colombia	25	∠	_	_	_
		Playa de Pals	Spain	250	_	_	_	
		Hoerby	Sweden	350	_			

MHz	*Metres*	*Station*	*Country*	*kW*	*M*	*M*	*S*	*N*
		Quito	Ecuador	100	/		_	
		Agana	Guam	100				_
		Lampertheim	Germany (W)	100	/			
		Dacca	Bangladesh	100	/			
		Allouis	France	500	/			
17.890	16.77	Tchita	USSR	240		_	_	
		Sverdlovsk	USSR	100	∠			_
		Lvov	Ukraine	500				_
		Taiwan	China Nat	50				
		Dacca	Bangladesh	10	∠	_	_	_
		Bucuresti	Roumania	120	∠			_
		Quito	Ecuador	100	_		_	
		Shepparton	Australia	10/100	/			_
		Noblejas	Spain	350	/			_
		Agana	Guam	100	/			_
		Islamabad	Pakistan	250	/			
		Karachi	Pakistan	50	/			
17.895	16.76	Holzkirchen	Germany (W)	10		_		
		Lampertheim	Germany (W)	100	∠		_	_
		Playa de Pals	Spain	250	∠	_	_	_
		S Gabriel	Portugal	100	∠	_	_	_
		Greenville	USA	250/500			_	_
		Delano	USA	250	∠	_	_	_
		Lusaka	Zambia	50	∠	_	_	_
		S M Galeria	Vatican	100	∠	_	_	_
17.900	16.76	Moskva	USSR	100				
		Vatican	Vatican	100				
17.920	16.74	Cairo	Egypt	100				
18.015	16.65		USSR					
18.080	16.59	London	UK					
18.135	16.54	Dixon*	USA	50				
18.195	16.49		USSR*					
18.275	16.42	Greenville*	USA					
18.285	16.41		USSR*					
18.310	16.38		USSR*					
18.370	16.33		USSR*					
18.420	16.29		USSR					
18.460	16.25		USSR*					
18.653	16.08		USSR*					
18.782	15.97	Greenville*	USA					
18.830	15.93		USSR*					
19.130	15.68		USSR*					
19.210	15.62		USSR					
19.261	15.58	Bethany*	USA					
19.455	15.43	London	UK					
19.480	15.40	Dixon*	USA					
19.505	15.38	Greenville*	USA					
19.710	15.22	Hilversum*	Netherlands					
19.721	15.21	Greenville*	USA					

MHz	Metres	Station	Country	kW	M	M	S	N
19.725	15.21		USSR*					
19.833	15.13		USSR*					
19.845	15.12		USSR*					
19.912	15.07	Delano*	USA					
19.915	15.06	Tanger*	Morocco					
20.000	15.00	Boulder Freq Std	USA					
20.060	14.96	Greenville*	USA					
			USSR					
20.125	14.91	Greenville*	USA					
20.215	14.84	RFE*						
20.250	14.81		USSR*					
20.345	14.75	London*	UK					
20.605	14.56		USSR*					
20.710	14.49	RFE*						
21.450	13.99	Moskva	USSR					
21.455	13.98	Holskirchen	Germany (W)	10		_		
		Lampertheim	Germany (W)	20	∠	_	_	_
		Lisbonne	Portugal	50		_		_
		Shepparton	Australia	100	_	_		
		Karachi	Pakistan	50	/			
21.460	13.98	Tula	USSR	100	/			
		Wavre	Belgium	250	∠	_	_	_
		Delano	USA	100	∠	_	_	_
		Dacca	Bangladesh	100	_	_		
21.465	13.98	Abis	Egypt	250			_	
		Nauen	Germany (E)	100/ 500	∠	_		_
		Leipzig	Germany (E)	100		_	_	
21.470	13.97	London	UK	100/250	∠	_	_	_
		Greenville	USA	50		_		
		Cincinnati	USA	250	/			_
		Wien	Austria	100			_	_
21.475	13.97	Wavre	Belgium	100	∠	_	_	_
		Pori	Finland					
21.480	13.97	Lopik	Netherlands	100	_	_	_	_
		Talata Volon	Malagasy Rep	300	∠	_	_	_
		Riyadh	Saudi Arabia	350		_		
		Quito	Ecuador	100	/		_	_
21.485	13.96	Cincinnati	USA	250	∠	_	_	_
		S M Galeria	Vatican	100	∠	_	_	_
		K Wusterhausen	Germany (E)	100		_		
		Pori	Finland	250	/			
		Islamabad	Pakistan	250	/			
21.490	13.96	Tula	USSR	100	∠	_	_	_
		Baku	USSR	240	∠	_		_
		Bucuresti	Roumania	100/250	∠	_	_	_
		Quito	Ecuador	100		_		_
		Poro	Philippines	35	/		_	_

MHz	*Metres*	*Station*	*Country*	*kW*	*M*	*M*	*S*	*N*
21.495	13.96	S Gabriel	Portugal	100	∠	_	_	_
			USSR					
		Pori	Finland	250	∠	_	_	_
		Jerusalem	Israel	300	/		_	_
21.500	13.95	London	UK	250	/			
		Cincinnati	USA	175	_	_	_	_
		Greenville	USA	50/250	∠		_	_
		Wertachtal	Germany (W)	500	∠	_	_	_
		Jerusalem	Israel	300	_	_	_	
		Wien	Austria	100			_	
		Riyadh	Saudi Arabia	350	/			
21.505	13.95	Duchanbe	USSR	240	∠			_
		Erevan	USSR	240/500		_	_	
		Khabarovsk	USSR	100	∠			_
		Riyadh	Saudi Arabia	350				_
		Hoerby	Sweden	350	X			_
21.510	13.95	Armavir	USSR	240		_	_	
		Bogota	Colombia	25	∠	_	_	_
		Roma	Italy	100	∠	_	_	_
		Lisbonne	Portugal	50	∠	_	_	
		Poro	Philippines	50				_
21.515	13.94	Bocaue	Philippines	2	∠	_	_	_
		Simferopol	Ukraine	240/500	∠	_	_	_
		Kiev	Ukraine	240	∠			_
		Frunze	USSR	50			_	
21.520	13.94	Schwarzenburg	Switzerland	150/250	∠	_	_	_
		Riyadh	Saudi Arabia	350	∠	_		_
		Kavalla	Greece	250	/		_	_
21.525	13.94	Jaszbereny	Hungary	250	_	_	_	
		Szekesfehervar	Hungary	20	_	_	_	
		Scituate	USA	50/100	∠	_	_	_
		Okeechobee	USA	100	/			_
		Shepparton	Australia	100	∠	_	_	_
		Pori	Finland	250	/			
21.530	13.93	Armavir	USSR	120	∠			_
		Frunze	USSR	100	∠	_		
		Riyadh	Saudi Arabia	350			_	_
		Lopik	Netherlands	100			_	_
		Wien	Austria	100	/			_
21.535	13.93	Meyerton	South Africa	250	∠	_	_	_
		Greenville	USA	50		_	_	
		Tokyo Yamata	Japan	100		_	_	_
		Noblejas	Spain	350	/			
21.540	13.93	Kursk	USSR	100	_			
		Nauen	Germany (E)	500		_	_	
		K Wusterhausen	Germany (E)	100	_			
		Leipzig	Germany (E)	100	/	_	_	_
		Kigali	Rwanda	250	∠	_	_	_

MHz	Metres	Station	Country	kW	M	M	S	N
		Lopik	Netherlands	100		_	_	_
		Greenville	USA	250	/		_	_
21.545	13.92	Tema	Ghana	100	∠	_	_	_
		Sottens	Switzerland	500	/			_
		Schwarzenburg	Switzerland	150/250	/			_
21.5475	13.92	Schwarzenburg*	Switzerland	30	/			_
21.550	13.92	London	UK	100/250	∠	_	_	_
		Ascension	Ascension		/			_
		Riyadh	Saudi Arabia	350	_	_	_	_
		Sackville	Canada	50		_		
		Greenville	USA		/			_
		Sines	Portugal	250	/		_	
21.5525	13.92	Varberg*	Sweden	100	/	_	_	_
21.555	13.92	Roma	Italy	60/100	∠			_
		Riyadh	Saudi Arabia	350	/	_		
		Wien	Austria	100	/		_	_
		Armavir	USSR	100	/			_
		Meyerton	South Africa	250	/			
21.5575	13.92	Varberg*	Sweden	100	∠	_	_	_
21.560	13.91	Roma	Italy	60/100	/	_	_	
		Julich	Germany (W)	100	∠	_	_	_
		Wertachtal	Germany (W)	500	/			
21.565	13.91	Lvov	Ukraine	100	∠	_	_	_
		Talata Volon	Malagasy Rep	300	/			
		Jerusalem	Israel	300	/	_	_	_
		Red Lion	USA	50				
21.570	13.91	London	UK	100/250	/	_	_	_
		Ascension	Ascension	250				
		Sottens	Switzerland	500	/	_	_	_
		Schwarzenburg	Switzerland	150	/			_
		Karachi	Pakistan	50			_	
		Carnarvon	Australia	100	∠	_	_	_
		Kavalla	Greece	250		_		
21.575	13.90	Moskva	USSR	100/240	∠	_	_	_
		Jerusalem	Israel					
21.580	13 90	Julich	Germany (W)	100	_	_		
		Holzkirchen	Germany (W)	10				
		Allouis	France	500	∠	_	_	_
		Jerusalem	Israel	300		_	_	_
		Rabat	Morocco	50	/			
21.585	13.90	Tachkent	USSR	100	/			
		Duchanbe	USSR	100		_	_	
		Minsk	Bielorussia	240	_	_		
		Schwarzenburg	Switzerland	150	∠	_	_	_
21.590	13.90	London	UK	100/250	∠	_	_	_
		Ascension	Ascension	250			_	_
		Karachi	Pakistan	50	∠	_	_	_
		Islamabad	Pakistan	250	∠	_	_	_
		Riyadh	Saudi Arabia	350	_		_	

MHz	*Metres*	*Station*	*Country*	*kW*	*M*	*M*	*S*	*N*
		Serpukhov	USSR	240/500	∠			_
		Cyclops	Malta	250	/	_	_	_
		Greenville	USA	250	/	_	_	_
		Julich	Germany (W)	100	/	_		_
		Wien	Austria	100	/			
21.595	13.89	Allouis	France	100	/			_
		Islamabad	Pakistan	100/250	/			_
		Monrovia	Liberia					
21.600	13.89	Leningrad	USSR	240	∠	_	_	_
		Julich	Germany (W)	100	∠	_	_	_
		Wertachtal	Germany (W)	500	∠	_	_	_
		Careysburg	Liberia	50	_	_	_	_
		Karachi	Pakistan	50	/			
21.605	13.89	Sulaibiyah	Kuwait	250	∠	_	_	_
		Poro	Philippines	50				_
21.610	13.88	London	UK	100/250	_	_	_	_
		Dixon	USA	50/250	∠	_	_	_
		Scituate	USA	100			_	
		Greenville	USA	250	/		_	_
		Tanger	Morocco	35		_		
		Hoerby	Sweden	350				_
		Tokyo	Japan	100				
21.615	13.88	Riga	USSR	100	∠	_	_	_
		Okeechobee	USA	100	/			_
		Hoerby	Sweden	350	/			_
21.620	13.88	Allouis	France	100/500	∠	_	_	_
		Noblejas	Spain	350	/			
21.625	13.87	Frunze	USSR	100	∠	_		_
		Jerusalem	Israel	300	∠	_	_	
		Islamabad	Pakistan	250	/		_	_
21.630	13.87	London	UK	250	/	_	_	_
		Schwarzenburg	Switzerland	150	/	_	_	_
		Tinang	Philippines	250	/			_
21.635	13.87	Kalinin	USSR	240	∠	_	_	_
		Armavir	USSR	240	_			_
21.640	13.86	London	UK	250	∠	_	_	_
		Lopik	Netherlands	100	∠	_	_	_
		Bonaire Noord	Neth Antilles	300	∠	_	_	_
		Talata Volon	Malagasy Rep	300	_	_		
		Tokyo Yamata	Japan	100	/		_	_
21.645	13.86	Armavir	USSR	240	∠	_	_	_
		Allouis	France	100/500	∠	_	_	_
21.650	13.86	London	UK	250	/			
		Julich	Germany (W)	100	∠	_	_	_
		Cyclops	Malta	250	∠	_	_	_
		Karlsborg	Sweden	350	_			_
		Jerusalem	Israel	300		_	_	
		Tanger	Morocco	35			_	_
		Dixon	USA	100	/			_

MHz	Metres	Station	Country	kW	M	M	S	N
		Tripoli	Libya					
21.655	13.85	Roma	Italy	30	∠	_	_	_
		Allouis	France	100	_	_		
		Fredrikstad	Norway	120	/			
		Karachi	Pakistan					
21.660	13.85	London	UK	250	/			
		Ascension	Ascension	250	/			
		Limassol	Cyprus	50/100	_	_	_	_
		Lisbonne	Portugal	50	∠	_		
		Velkekostolany	Czechoslovakia	120	/	_	_	_
		Hoerby	Sweden	350		_		
		Cincinnati	USA	175	/			
21.665	13.85	Lisbonne	Portugal	50				
			USSR					
21.670	13.84	Greenville	USA	250/500	∠	_	_	_
		Poro	Philippines	35	/	_	_	_
		Dacca	Bangladesh	100	/		_	_
21.675	13.84	Allouis	France	100/500	∠	_	_	_
		Delhi	India	100	_	_	_	_
		Islamabad	Pakistan	250			_	_
21.680	13.84	Baku	USSR	100		_	_	
		Shepparton	Australia	100	/		_	_
		Julich	Germany (W)	100	/	_	_	_
		Poro	Philippines	50	/			
21.685	13.83	Sulaibiyah	Kuwait	250	∠	_	_	_
		Dacca	Bangladesh	100	∠	_	_	_
		Talata Volon	Malagasy Rep	300	∠	_	_	_
		Lopik	Netherlands	100	/	_	_	
		Kavalla	Greece					
21.690	13.83	Hoerby	Sweden	350	∠	_	_	_
		Karlsborg	Sweden	350	/	_	_	_
		Roma	Italy	100	∠	_	_	_
		Careysburg	Liberia	50			_	
		Greenville	USA	250		_		
21.695	13.83	London	UK	250	/	_	_	_
		Sackville	Canada	250	/		_	_
		Delano	USA	100/250	/			_
		Meyerton	South Africa	250	/			
21.700	13.82	S Gabriel	Portugal	100	∠	_	_	_
		Velkekostolany	Czechoslovakia	100	∠	_	_	_
		Karlsborg	Sweden	350	/		_	_
21.705	13.82	Armavir	USSR	240/500	∠			_
		Allouis	France	100	/			_
21.710	13.82	London	UK	100/250	∠	_	_	_
21.715	13.82	Wien	Austria	100	/		_	_
		Bucuresti	Roumania					
21.720	13.81	Ejura	Ghana	250	∠	_	_	_
		Lisbonne	Portugal	50	∠	_	_	_
21.725	13.81	Tachkent	USSR	240		_	_	

MHz	Metres	Station	Country	kW	M	M	S	N
21.730	13.81	Fredrikstad	Norway	100	∠	_	_	_
		Islamabad	Pakistan	250	/	_	_	_
		Karachi	Pakistan	50	∠	_	_	_
		Allouis	France	100	∠	_	_	_
		Jerusalem	Israel	300	/			
		Rabat	Morocco					
21.735	13.80	S Gabriel	Portugal	100	∠	_	_	_
		Lisbonne	Portugal	50	/			
		Tanger	Morocco	50		_		_
		Jerusalem	Israel	50			_	
		Islamabad	Pakistan	250	/			
		Allouis	France	500	/			
21.740	13.88	Shepparton	Australia	10/50	/			_
		Delhi	India	100	/			
		Islamabad	Pakistan	100	/			
21.745	13.80	Lisbonne	Portugal	50	∠	_	_	_
		Lampertheim	Germany (W)	20		_		
		Armavir	USSR	100	∠			_
		Frunze	USSR	240/500		_	_	
		Simferopol	Ukraine	240	∠			_
		Delano	USA	100	_	_	_	
		Dixon	USA	50/250	/			_
		Karachi	Pakistan	50	/			
21.750	13.79	Lisbon	Portugal					
21.760	13.79	Careysburg	Liberia					
21.917	13.69	Jerusalem	Israel					
22.205	13.51		USSR					
22.770	13.18		USSR*					
22.930	13.08	London	UK					
22.970	13.06	RFE*						
23.191	12.94	London*	UK					
25.350	11.83	RFE*						
25.605	11.72	Jerusalem*	Israel	50	∠	_	_	_
25.620	11.71	Allouis	France	100	X		_	_
		Dixon	USA	100	/			
25.640	11.70	Jerusalem	Israel					
25.645	11.70	Kuwait	Kuwait					
25.650	11.70	Talata Volon	Malagasy Rep					
25.690	11.68	Lisbonne	Portugal	10	/	_	_	_
25.750	11.65	Bogota	Colombia	25	∠	_	_	_
25.760	11.65	Jerusalem	Israel	50				_
25.790	11.63	Meyerton	South Africa	250	/			_
25.800	11.63	Allouis	France	100	/			
25.820	11.62	Allouis	France	100				
25.880	11.59	Bethany	USA					
25.900	11.58	Allouis	France	100	/			
25.990	11.54	Delano	USA	100	/			_
26.000	11.54	Poro	Philippines	35				_
26.040	11.52	Greenville	USA	50	∠	_	_	_

MHz	*Metres*	*Station*	*Country*	*kW*	*M*	*M*	*S*	*N*
26.095	11.50	Dixon	USA	250	/		_	_
		Jerusalem	Israel	300	/			
29.705	10.10	Jerusalem	Israel					

GEOGRAPHICAL LIST OF SHORT WAVE STATIONS OF THE WORLD

	MHz
ADEN	
Aden	5.06
	5.97
	7.19
AFARS & ISSAS	
Djibuti	4.78
AFGHANISTAN	
Kabul	3.39
	4
	4.775
	6
	6.23
	7.2
	11.77
	11.805
	11.82
	11.89
	11.985
	15.14
	15.23
	15.295
	15.365
	15.39
ALBANIA	
Tirane	5.057
	5.945
	5.96
	6.18
	6.185
	6.19
	6.2
	6.21
	7.065
	7.075
	7.08
	7.09
	7.12
	7.275
	7.28
	7.29
	7.3
	9.375
	9.43
	9.48
Tirane	9.485
	9.5
	9.515
	9.75
	9.76
	9.78
	9.79
	11.845
	11.865
	11.915
	11.92
	11.935
	11.95
	11.965
	11.985
	14.27
	15.405
ALGERIA	
Alger	6.08
	6.145
	6.16
	7.055
	7.145
	7.195
	7.245
	8.063
	9.51
	9.61
	9.615
	9.64
	9.68
	9.685
	9.705
	9.76
	9.86
	11.74
	11.77
	11.8
	11.81
	11.925
	14.99
	15.155
	15.16
	15.175
	15.215
	15.23
Alger	15.773
	17.765
ANDORRA	
Andorra	6.215
ANGOLA	
Bassacongo	4.72
Huambo	4.78
	7.16
Luanda	3.345
	3.355
	3.375
	4.82
	4.985
	5.61
	5.96
	6.175
	7.235
	7.245
	7.265
	9.535
	9.66
	11.955
Saurimo	4.86
Silva Porto	4.896
Unknown	4.79
ARGENTINA	
Buenos Aires	5.882
Hurlingham	6.09
	9.76
	11.78
Gral Pacheo	6.06
	9.69
	11.71
	15.345
Lomas Mirador	5.985
	9.74
	11.88
Malargue	6.16
Mendoza	6.18
S Fernando	6.12
	9.71
	11.755
	15.29

	MHz
ASCENSION	
Ascension	6.005
	6.055
	6.14
	7.105
	7.15
	7.27
	9.58
	9.6
	9.7
	9.76
	9.765
	9.77
	11.75
	11.77
	11.79
	11.82
	11.86
	15.105
	15.195
	15.26
	15.39
	15.4
	15.42
	17.79
	17.83
	17.84
	17.87
	17.88
	17.885
	21.55
	21.57
	21.59
	21.66
AUSTRALIA	
Brisbane	4.92
	9.66
Carnarvon	6.005
	6.035
	7.24
	9.56
	9.64
	9.7
	11.705
	11.715
	11.76
	11.775
	11.8
	11.835
Carnarvon	11.855
	11.865
	11.895
	11.935
	15.11
	15.13
	15.18
	15.205
	17.77
	17.87
	21.57
Lyndhurst	5.995
	6.045
	6.15
	7.24
	7.5
	9.54
	9.68
	11.87
	11.88
	15.16
	15.23
	15.24
	15.355
	15.87
Perth	6.14
	9.61
	15.425
Shepparton	5.955
	5.97
	6.06
	6.08
	6.12
	6.865
	7.24
	9.505
	9.55
	9.56
	9.57
	9.58
	9.6
	9.67
	9.76
	9.77
	11.705
	11.72
	11.725
	11.74
	11.76
	11.79
Shepparton	11.81
	11.82
	11.825
	11.835
	11.855
	11.88
	11.935
	15.105
	15.14
	15.18
	15.205
	15.26
	15.29
	15.31
	15.32
	15.355
	15.41
	15.415
	17.725
	17.755
	17.795
	17.87
	17.89
	21.455
	21.525
	21.68
	21.74
Sydney	6.09
Unknown	9.645
	11.87
	12.19
	12.29
AUSTRIA	
Innsbruck	6
Wien	5.925
	5.945
	5.96
	6.015
	6.155
	6.221
	7.17
	9.585
	9.605
	9.62
	9.65
	9.66
	9.725
	9.765
	9.77

	MHz
Austria – contd	
Wien	11.72
	11.79
	11.825
	11.87
	11.895
	12.015
	15.105
	15.11
	15.135
	15.155
	15.195
	15.245
	15.27
	15.295
	15.32
	15.335
	15.38
	15.385
	15.41
	15.425
	15.435
	15.44
	17.71
	17.72
	17.725
	17.74
	17.745
	17.76
	17.77
	17.815
	17.84
	17.855
	17.865
	21.47
	21.5
	21.53
	21.555
	21.59
	21.715

BANGLADESH

	MHz
Dacca	4.88
	4.89
	6.145
	6.155
	6.18
	7.15
	7.245
	7.28
Dacca	7.29
	9.5
	9.53
	9.535
	9.54
	9.55
	9.585
	9.605
	9.615
	9.68
	9.72
	9.75
	11.65
	11.725
	11.765
	11.79
	11.89
	11.895
	11.9
	15.285
	15.335
	15.4
	15.41
	15.52
	17.72
	17.73
	17.885
	17.89
	21.46
	21.67
	21.685

BELGIUM

	MHz
Wavre	5.965
	6.01
	6.015
	6.065
	6.08
	6.16
	9.61
	9.615
	9.645
	9.655
	9.68
	9.685
	9.73
	9.755
	11.735
	11.785
	11.79
Wavre	11.8
	11.85
	11.88
	11.94
	15.19
	15.205
	15.21
	15.225
	15.23
	15.24
	15.25
	15.28
	15.285
	17.715
	17.73
	17.735
	17.74
	17.765
	17.77
	21.46
	21.475

BENIN

	MHz
Cotonou	4.87

BIELORUSSIA

	MHz
Minsk	5.975
	7.26
	9.54
	11.745
	15.15
	21.585
Orcha	6.015
	7.105
	9.655
	9.715
	11.715

BOLIVIA

	MHz
Animas	6.015
Catavi	6.08
Choraya	4.99
Cochabamba	5.975
Huanuni	5.965
La Paz	4.845
	5.005
	6.005
	6.025
	6.035
	6.055

	MHz
Bolivia – contd	
La Paz	6.125
	6.155
	6.185
	6.195
	9.505
	9.555
Llallagua	5.955
Oruro	6.07
Potosi	4.965
R. Nueva America	4.795
S Cruz	6.135
	6.175
	9.605
S Cruz del Sur	4.875
Sucre	5.995
	9.715
Tarija	6.145
Yacuiba	4.805
BOTSWANA	
Gaberone	3.355
	4.845
BRASIL	
Aparacida	9.635
Aquidauana	4.795
	5.025
Aranquara	3.365
Belem	4.865
Belo Horizonte	6
	6.175
	15.19
Boa Vista	4.835
Braganca	4.825
Brasilia	6.065
	9.605
	9.665
	11.78
	15.245
	15.28
	15.445
Cuiba	4.775
	5.055
Curityba	6.045
	9.545
	11.935
Florianapolis	5.975
	9.675
Fortaleza	6.105
	15.165
Goiana	4.985
	4.995
	9.755
	11.735
	11.815
Lins	3.225
Londrina	4.815
Macapa	4.915
Maceio	3.325
Manaos	4.805
	9.695
Marilia	3.235
Pernambuco	3.285
Pocas de Caldas	4.885
	4.945
	9.645
P. Alegre	5.965
	6.135
	11.875
	11.915
Prudente	5.045
Recife	6.015
	6.085
	9.565
	11.825
	11.865
	15.145
Ribeirao	15.415
Preto	15.42
Rio de	4.875
Janeiro	4.905
	5.015
	5.99
	6.035
	6.065
	6.115
	6.195
	9.515
	9.61
	9.705
	9.77
	11.805
	11.885
	11.95
	15.105
	15.37
	17.875
Salvador	6.155
Salvador	9.595
	11.875
	15.125
Sao Luiz	4.785
	4.976
	15.215
Sao Paulo	5.035
	5.955
	6.055
	6.095
	6.125
	6.165
	6.185
	9.505
	9.585
	9.645
	9.685
	9.745
	11.765
	11.925
	11.965
	15.135
	15.155
	15.265
	15.325
	17.815
Taubate	4.855
Uberaba	4.965
Uberlandia	3.345
Unknown	3.265
	4.925
BRUNEI	
Berakas	4.865
	7.215
BRITISH WEST INDIES	
Antigua	5.96
	6.04
	6.055
	6.085
	6.175
	6.195
	9.51
	9.545
	9.64
	9.69
	9.735
	9.765
	11.775

	MHz
British West Indies – contd	11.785
Antigua	11.795
	11.81
	11.82
	11.865
	15.15
	15.185
	15.42
	17.83
	17.84
Grenada	15.05
	15.105
	15.37
Montserrat	6
	6.135
	6.145
	9.545
	9.59
	9.64
	11.705
	11.785
	11.91
	11.97
BULGARIA	
Sofia	5.915
	5.94
	6.07
	6.085
	6.16
	7.115
	7.15
	7.215
	7.255
	7.27
	7.67
	9.53
	9.56
	9.685
	9.7
	9.705
	9.745
	9.765
	11.72
	11.735
	11.75
	11.765
	11.82
	11.85
Sofia	11.86
	11.87
	11.92
	11.97
	15.135
	15.31
	15.33
	15.385
	17.82
	17.825
BURMA	
Rangoon	4.725
	5.189
	5.965
	5.985
	7.185
	9.73
Unknown	4.98
BURUNDI	
Bujumbari	3.3
	6.14
Cordac	4.9
CAMBODIA	
Phnom Penh	4
	4.908
	6.095
	9.695
CAMEROON	
Bertoua	4.75
Buea	3.97
	6.005
Garoura	5.01
	7.24
Yaounde	4.85
	4.925
	4.972
	7.205
	9.745
CANADA	
Calgary	6.03
CHU time signal	14.67
Halifax	6.13
Montreal	6.005
Ottawa	7.335
Sackville	5.96
	5.99
	6.045
	6.065
	6.085
	6.105
	6.12
	6.14
	6.15
	6.175
	6.195
	9.51
	9.53
	9.535
	9.575
	9.58
	9.59
	9.605
	9.625
	9.635
	9.655
	9.715
	9.73
	9.755
	11.705
	11.72
	11.735
	11.775
	11.825
	11.845
	11.855
	11.86
	11.905
	11.915
	11.94
	11.945
	11.96
	15.15
	15.19
	15.26
	15.325
	15.365
	17.75
	17.76
	17.78
	17.82
	17.83
	17.84
	17.865
	17.875

	MHz
Canada – contd	
Sackville	21.55
	21.695
St Johns	6.16
Toronto	6.07
Vancouver	6.08
	6.16
CANARY ISLANDS	
Las Mesas	6.045
	6.09
	9.515
	9.525
	11.81
	11.815
	11.88
	15.365
CAPE VERDE ISLANDS	
Cape Verde Islands	3.885
Barlavento	3.93
Sao Vicente	4.72
CENTRAL AFRICAN REPUBLIC	
Bangui	4.999
	7.22
CEYLON	
Colombo	3.383
	3.395
	4.94
Ekala	4.87
	4.9
	4.945
	6.005
	6.075
	6.13
	6.15
	6.185
	7.105
	7.11
	7.19
	9.72
	11.745
	11.755
	11.76
	11.8

	MHz
Ekala	11.835
	11.87
	11.935
	11.94
	11.955
	15.115
	15.12
	15.185
	15.395
	15.405
	15.425
	17.85
CHAD	
Fort Lamy	4.905
N Djamena	6.165
	7.12
CHILE	
Concepcion	6.135
Santiago	5.955
	6.135
	6.15
	6.19
	9.51
	9.52
	9.53
	9.55
	9.57
	9.63
	9.65
	9.675
	9.685
	9.69
	9.715
	9.75
	9.755
	9.765
	11.705
	11.715
	11.72
	11.755
	11.76
	11.765
	11.775
	11.78
	11.8
	11.855
	11.89
	11.925

	MHz
Santiago	15.11
	15.115
	15.125
	15.13
	15.14
	15.15
	15.175
	15.24
	15.29
	15.3
	15.32
	17.64
	17.715
	17.725
	17.755
	17.785
	17.79
	17.8
	17.815
CHINA (NATIONALIST)	
Taipei	5.98
	6.405
	9.51
	9.575
	9.6
	9.685
	9.765
	11.745
	11.825
	11.86
	11.905
	11.915
	15.125
	15.225
	15.345
	15.425
	17.72
	17.8
	17.89
CHINA (REPUBLIC)	
Changsha	4.99
Foochow	4.975
Fukien Front Station	3.0
	3.2
	3.4
	3.535

	MHz
China (Republic) — contd	
Fukien Front Station	3.64
	4.045
	4.33
	4.38
	5.24
	5.265
	5.9
	6.765
	7.85
Heeilunkiang	4.84
Kweiyang	3.26
Lanchow	4.865
	7.325
Nanning	4.915
Peking	3.22
	3.27
	3.29
	3.36
	3.39
	3.45
	3.5
	3.66
	3.7
	3.83
	3.92
	3.95
	3.952
	3.96
	3.985
	4.02
	4.13
	4.18
	4.2
	4.25
	4.46
	4.62
	4.63
	4.76
	4.77
	4.8
	4.815
	4.85
	4.872
	4.88
	4.885
	4.895
	4.905
	4.94
Peking	4.96
	4.98
	5.075
	5.09
	5.125
	5.135
	5.145
	5.163
	5.22
	5.25
	5.295
	5.32
	5.42
	5.84
	5.85
	5.86
	5.88
	5.915
	5.935
	5.95
	5.96
	5.975
	5.99
	6.175
	6.205
	6.21
	6.225
	6.26
	6.27
	6.28
	6.29
	6.32
	6.33
	6.345
	6.41
	6.43
	6.495
	6.52
	6.54
	6.55
	6.555
	6.56
	6.585
	6.59
	6.645
	6.665
	6.75
	6.79
	6.81
	6.825
Peking	6.86
	6.88
	6.89
	6.935
	6.955
	6.995
	7.01
	7.025
	7.03
	7.035
	7.04
	7.045
	7.055
	7.06
	7.065
	7.08
	7.095
	7.19
	7.315
	7.33
	7.36
	7.375
	7.38
	7.385
	7.44
	7.47
	7.48
	7.504
	7.55
	7.59
	7.62
	7.66
	7.7
	7.77
	7.775
	7.78
	7.8
	7.815
	7.82
	7.827
	7.855
	7.935
	8.007
	8.24
	8.26
	8.3
	8.32
	8.345
	8.425
	8.45

	MHz		MHz		MHz
China (Republic) –		Peking	11.33	Peking	15.195
contd	8.49		11.375		15.22
Peking	8.565		11.445		15.25
	8.6		11.455		15.265
	8.66		11.5		15.27
	9.02		11.505		15.3
	9.03		11.515		15.305
	9.064		11.53		15.405
	9.08		11.533		15.42
	9.17		11.6		15.45
	9.29		11.61		15.48
	9.336		11.63		15.51
	9.34		11.635		15.52
	9.365		11.65		15.55
	9.38		11.66		15.587
	9.39		11.665		15.59
	9.4		11.675		15.67
	9.417		11.685		15.71
	9.44		11.69		15.88
	9.455		11.695		17.45
	9.46		11.705		17.49
	9.47		11.71		17.515
	9.48		11.715		17.53
	9.49		11.72		17.535
	9.515		11.725		17.605
	9.525		11.73		17.635
	9.53		11.735		17.65
	9.535		11.945		17.68
	9.54		11.98		17.705
	9.55		12.015		17.735
	9.625		12.095		17.835
	9.67		12.06		17.855
	9.82		12.08	Urumchi	3.99
	9.85		12.11		4.11
	9.86		12.12		4.19
	9.88		12.2		4.22
	9.893		12.42		4.5
	9.9		12.45		4.97
	9.92		14.82		5.03
	9.94		15.02		5.06
	9.945		15.03		5.44
	9.965		15.04		5.925
	10.245		15.045		7.05
	10.26		15.05	Wunan	3.94
	10.865		15.06	Yunnan	4.785
	11.000		15.07	Unknown	3.925
	11.04		15.08		4.41
	11.1		15.095		4.968
	11.29		15.115		6.937
	11.3		15.165		9.895

	MHz
China (Republic) – contd	11.92
Unknown	15.5
	15.6
	15.735

COLOMBIA

	MHz
Aranca	4.865
	4.925
Bogota	4.755
	4.955
	4.965
	5.075
	5.96
	5.97
	6.02
	6.03
	6.065
	6.075
	6.125
	6.14
	6.16
	6.18
	9.635
	9.655
	11.79
	11.795
	11.825
	13.849
	15.335
	17.885
	21.51
	25.75
Bucramanga	4.845
Cali	6.055
	6.195
Espinal	6.095
Florencia	5.035
	6.17
Gutapuri	4.915
Ibaque	4.785
	4.875
	6.04
Medellin	4.875
	5.98
	6.105
Neiva	4.945
	6.15
Pereira	5.995
	6.01
Popayan	6.145
Sutatenza	5.095
Tumaco	6.015
Tunja	5.985
Vallendupar	4.815
Villavicencio	4.885
	5.04
	5.955
	6.115

COMORO ISLANDS

	MHz
Dzaudazi	3.33

CONGO REPUBLIC

	MHz
Brazzaville	3.232
	4.765
	4.795
	6.115
	7.105
	7.175
	9.61
	9.715
	15.19
Point Noire	4.843

COOK ISLANDS

	MHz
Rarotonga	5.045
	9.695
	11.76

COSTA RICA

	MHz
Faro del Caribe	6.175
Pt Limon	5.955
San Jose	4.832
	5.054
	6.005
	6.04
	6.05
	6.15
	9.615
	9.645

CUBA

	MHz
Havana	6.06
	9.525
	9.55
	9.655
	9.685
	9.765
	9.77
Havana	11.725
	11.76
	11.865
	11.93
	11.97
	15.23
	15.34
	17.705
	17.75
	17.855
	17.885

CYPRUS

	MHz
Limassol	3.99
	5.99
	6.01
	6.02
	6.05
	6.07
	6.085
	6.12
	6.125
	6.13
	6.14
	6.15
	6.155
	6.18
	6.195
	7.11
	7.14
	7.19
	7.21
	7.23
	7.25
	7.26
	7.27
	7.275
	9.53
	9.54
	9.58
	9.59
	9.6
	9.605
	9.61
	9.615
	9.62
	9.64
	9.69
	9.695
	9.705

	MHz
Cyprus – contd	
Limassol	9.715
	9.75
	9.77
	11.705
	11.71
	11.72
	11.735
	11.74
	11.75
	11.76
	11.78
	11.82
	11.85
	11.865
	11.955
	15.105
	15.125
	15.165
	15.225
	15.25
	15.27
	15.37
	15.38
	15.4
	15.42
	15.43
	17.885
	21.66
R Bayrak	6.285
CZECHOSLOVAKIA	
Prague	5.93
	6.055
	7.245
	7.345
	9.6
	9.605
	11.855
	11.98
	17.705
	17.725
	17.775
Velkeko-stolany	6.015
	6.055
	9.505
	9.54
	9.63
	9.74
	15.11

	MHz
Velkeko-stolany	15.24
	15.395
	17.84
	21.66
	21.7
DENMARK	
Koebenhavn	6.175
	9.71
	15.165
DOMINICAN REPUBLIC	
Pt Plata	6.19
S Domingo	4.85
	5.01
	5.965
	6.09
	6.11
	6.13
	9.505
	9.59
	11.7
S Pedromacoris	6.025
Santiago	4.807
	6.06
	6.075
	6.12
EGYPT	
Abis	9.705
	9.75
	11.715
	11.785
	15.21
	17.745
	17.785
	21.465
Abu Zaabal	6.17
	7.225
	9.52
	9.62
	9.675
	9.77
	11.78
	11.79
	11.915
	15.135
	15.175

	MHz
Abu Zaabal	15.255
	15.335
	15.375
	17.725
Cairo	6.23
	7.05
	7.075
	9.455
	9.475
	9.495
	9.76
	9.805
	9.85
	9.995
	11.63
	12.005
	12.045
	12.05
	12.07
	15.475
	17.67
	17.69
	17.92
Mokattam	9.755
Unknown	11.745
	11.795
ECUADOR	
Caracas	4.79
Esmeraldas	3.38
Espejo	4.679
Guayaquil	4.765
Quito	4.91
	4.92
	4.923
	4.93
	5.061
	6.01
	6.05
	6.07
	6.075
	6.095
	6.13
	9.56
	9.585
	9.605
	9.62
	9.635
	9.65
	9.665

	MHz
Ecuador – cont	
Quito	9.68
	9.685
	9.705
	9.715
	9.73
	9.745
	9.76
	9.765
	11.715
	11.73
	11.745
	11.8
	11.82
	11.83
	11.835
	11.84
	11.865
	11.9
	11.905
	11.91
	11.915
	11.945
	11.96
	15.115
	15.16
	15.245
	15.295
	15.3
	15.31
	15.36
	15.375
	15.38
	15.405
	15.415
	15.435
	17.745
	17.755
	17.76
	17.79
	17.865
	17.875
	17.885
	17.89
	21.48
	21.49
Riobamba	3.985
S Domingo	3.39
Sucua	4.96

	MHz
ETHIOPIA	
Gedja	4.905
	5.99
	6.015
	6.185
	7.11
	7.115
	7.165
	7.18
	9.61
	9.615
	9.705
FINLAND	
Pori	6.12
	9.55
	9.565
	9.575
	9.585
	9.645
	9.66
	9.755
	11.735
	11.755
	11.8
	11.825
	11.835
	11.91
	15.105
	15.2
	15.21
	15.265
	15.27
	15.33
	15.4
	15.43
	17.755
	17.785
	17.825
	17.84
	17.87
	17.875
	21.475
	21.485
	21.495
	21.525
FRANCE	
Allouis	3.965

	MHz
Allouis	5.955
	5.985
	5.99
	5.995
	6.01
	6.04
	6.145
	6.175
	7.16
	7.28
	7.285
	9.505
	9.51
	9.525
	9.535
	9.54
	9.55
	9.56
	9.605
	9.61
	9.615
	9.62
	9.66
	9.665
	9.69
	9.695
	9.705
	9.715
	9.75
	9.755
	10.356
	11.705
	11.73
	11.735
	11.745
	11.755
	11.77
	11.78
	11.79
	11.805
	11.825
	11.84
	11.845
	11.85
	11.855
	11.86
	11.865
	11.89
	11.91

	MHz
France -- contd	
Allouis	11.925
	11.93
	11.945
	11.955
	11.965
	15.135
	15.155
	15.18
	15.19
	15.195
	15.2
	15.21
	15.22
	15.28
	15.29
	15.3
	15.31
	15.315
	15.33
	15.36
	15.425
	15.435
	15.44
	17.715
	17.72
	17.735
	17.765
	17.775
	17.785
	17.79
	17.795
	17.8
	17.845
	17.85
	17.855
	17.86
	17.865
	17.87
	17.875
	17.885
	21.58
	21.595
	21.62
	21.645
	21.655
	21.675
	21.705
	21.73
	21.735

	MHz
Allouis	25.62
	25.8
	25.82
	25.9
Paris	6.21
	6.745
	6.876
	11.033
	12.46
GABON REPUBLIC	
Franceville	3.35
	4.83
	5.955
	6.03
	7.27
	9.53
	9.585
	9.695
	11.835
	11.965
	15.25
	15.44
	17.725
Libreville	3.3
	4.775
GERMANY (EAST)	
Berlin	7.3
	9.5
	11.7
	11.975
	15.1
	17.7
Leipzig	6.07
	6.195
	9.62
	9.665
	9.73
	11.72
	11.96
	15.17
	15.32
	15.39
	17.73
	17.74
	17.785
	17.8
	17.88
	21.465

	MHz
Leipzig	21.54
Nauen	5.955
	6.01
	6.04
	6.07
	6.08
	6.195
	7.26
	9.645
	9.665
	9.73
	9.77
	11.705
	11.805
	11.825
	11.84
	11.85
	11.89
	11.92
	11.97
	15.115
	15.13
	15.145
	15.155
	15.165
	15.17
	15.24
	15.32
	15.43
	17.705
	17.88
	21.465
	21.54
K Wuster-hausen	6.01
	6.07
	6.08
	6.115
	7.185
	7.225
	9.505
	9.51
	9.535
	9.6
	9.665
	9.755
	11.705
	11.785
	11.795
	11.81
	11.875

	MHz
Germany (East) — contd K Wusterhausen	11.89
	11.955
	11.97
	15.105
	15.125
	15.145
	15.155
	15.24
	15.25
	15.255
	15.285
	15.34
	15.39
	17.705
	17.74
	17.755
	17.8
	17.815
	17.88
	21.485
	21.54
Standard Time Signal	4.525
	15.595
GERMANY (WEST)	
Berlin (RIAS)	6.005
	3.985
	3.96
	3.99
	5.97
	5.985
	5.955
	6.01
	6.07
	6.105
	6.115
	6.135
	6.17
	7.115
	7.165
	7.18
	7.19
	7.215
	7.22
	7.245
	7.255
	7.295
	9.505
	9.52
Berlin (RIAS)	9.555
	9.595
	9.625
	9.66
	9.68
	9.695
	9.705
	9.725
	9.751
	11.725
	11.77
	11.78
	11.815
	11.84
	11.855
	11.875
	11.885
	11.895
	11.925
	11.935
	11.97
	15.145
	15.291
	15.34
	15.38
	17.725
	17.75
	17.76
	17.865
Bremen	6.19
Holzkirchen	3.96
	3.97
	3.99
	5.97
	5.955
	5.985
	6.01
	6.105
	6.155
	6.135
	6.17
	7.155
	7.165
	7.19
	7.2
	7.22
	7.245
	7.255
	7.295
	9.505
Holzkirchen	9.52
	9.565
	9.68
	9.695
	9.705
	9.725
	9.75
	11.825
	11.875
	11.885
	11.915
	15.29
	15.34
	15.355
	15.37
	15.445
	17.725
	17.75
	17.895
	21.455
	21.58
Ismaning	3.98
	5.955
	5.965
	6.005
	6.04
	6.06
	6.085
	6.095
	6.11
	6.15
	6.195
	7.15
	7.155
	7.205
	7.275
	7.29
	7.727
	9.53
	9.54
	9.62
	9.65
	9.68
	9.715
	11.78
	11.805
	11.92
	15.12
	15.205
	15.43

Germany (West) – contd

	MHz
Ismaning	17.735
Julich	3.995
	5.96
	5.995
	6.04
	6.065
	6.075
	6.12
	6.13
	6.145
	6.185
	6.975
	7.105
	7.15
	7.16
	7.21
	7.235
	7.275
	7.285
	7.675
	7.767
	9.545
	9.565
	9.59
	9.605
	9.61
	9.615
	9.63
	9.64
	9.65
	9.69
	9.7
	9.765
	9.875
	11.705
	11.765
	11.785
	11.795
	11.81
	11.85
	11.865
	11.905
	13.512
	15.105
	15.135
	15.15
	15.245
	15.275
	15.32
Julich	15.41
	15.425
	17.715
	17.73
	17.765
	17.77
	17.78
	17.785
	17.8
	17.825
	17.875
	21.56
	21.58
	21.59
	21.6
	21.65
	21.68
Lampertheim	3.96
	3.985
	3.99
	5.985
	5.955
	6.105
	6.17
	7.145
	7.155
	7.165
	7.18
	7.19
	7.2
	7.22
	7.245
	7.255
	7.295
	9.505
	9.52
	9.54
	9.555
	9.625
	9.66
	9.68
	9.705
	9.715
	9.725
	9.75
	9.77
	11.725
	11.77
	11.855
	11.865
Lampertheim	11.875
	11.885
	11.895
	11.91
	11.925
	11.935
	11.97
	15.13
	15.17
	15.29
	15.34
	15.37
	15.38
	15.445
	17.705
	17.75
	17.76
	17.77
	17.725
	17.78
	17.845
	17.885
	17.895
	21.455
	21.745
Muehlacker	6.03
Munich	11.915
Rohrdorf	7.265
Wertachtal	5.96
	6.065
	6.075
	6.085
	6.1
	6.12
	6.13
	6.145
	6.185
	7.105
	7.15
	7.175
	7.21
	7.235
	7.275
	7.285
	9.51
	9.545
	9.59
	9.605
	9.61
	9.64

	MHz
Germany (West)—contd	
Wertachtal	9.735
	9.765
	11.72
	11.765
	11.785
	11.795
	11.82
	11.85
	11.865
	11.905
	11.935
	11.945
	15.105
	15.12
	15.135
	15.15
	15.185
	15.225
	15.245
	15.275
	15.32
	15.405
	15.425
	17.715
	17.73
	17.765
	17.78
	17.795
	17.8
	17.825
	17.845
	17.85
	17.875
	21.5
	21.56
	21.6
Unknown	4.882
	5.83
	10.761
	10.922
	12.29
GHANA	
Accra	3.295
	4.915
	7.295
Ejura	5.99
	11.8
	11.85

	MHz
Ejura	15.285
	21.72
Tema	3.35
	3.366
	4.98
	6.07
	6.13
	9.545
	15.285
	21.545
GREECE	
Athinai	5.955
	6.045
	6.14
	6.185
	7.125
	7.205
	7.215
	9.515
	9.53
	9.61
	9.64
	9.655
	9.675
	9.76
	11.76
	11.845
	11.955
	15.16
	15.325
	15.345
	17.83
Kavalla	5.955
	5.985
	6.06
	6.09
	6.095
	6.14
	6.15
	6.17
	6.19
	7.125
	7.145
	7.15
	7.16
	7.17
	7.18
	7.205
	7.215

	MHz
Kavalla	7.24
	7.27
	7.275
	9.53
	9.54
	9.545
	9.555
	9.585
	9.615
	9.64
	9.655
	9.66
	9.68
	9.69
	9.7
	9.735
	9.76
	9.77
	11.73
	11.74
	11.76
	11.78
	11.805
	11.81
	11.83
	11.84
	11.85
	11.855
	11.875
	11.925
	15.12
	15.13
	15.16
	15.165
	15.195
	15.205
	15.235
	15.26
	15.27
	15.305
	15.33
	17.78
	17.785
	17.845
	17.855
	17.865
	21.52
	21.57
	21.685
Rhodos	5.965

	MHz
Greece – contd	
Rhodos	6.015
	6.07
	6.941
	7.11
	7.205
Thessaloniki	7.28
	9.655
	9.71
GREENLAND	
Godthaab	3.999
	5.96
	5.98
	9.575
	11.745
GUAM	
Agana	11.73
	11.75
	11.775
	11.78
	11.795
	11.81
	11.815
	11.84
	11.85
	11.865
	11.89
	11.9
	15.115
	15.135
	15.145
	15.155
	15.175
	15.215
	15.225
	15.23
	15.235
	17.83
	17.855
	17.885
	17.89
GUATEMALA	
Guatemala City	3.3
	6.18
GUINEA	
Bata	4.925
	6.12
	7.15
	7.19
	9.555
	9.585
	11.715
	11.895
	15.11
	15.19
Bissau	5.04
Conakry	4.91
	6.155
	7.125
	9.65
	11.965
	15.07
	15.31
Malabo	6.25
GUYANA	
Georgetown	3.265
Sparendum	5.98
GUYANA (FR)	
Cayenne	3.385
	4.972
	6.17
HAITI	
Cap Haitien	11.835
Port au Prince	6.195
	9.69
	11.805
HAWAII	
Honolulu Standard Freq.	5.0
	10.0
	15.0
HONDURAS (BR)	
Belmopan	3.3
HONDURAS REP	
Comayaguela	6.11
El Progreso	4.92
La Ceiba	6.135
	6.195
S Barbara	6.075
S Pedro Sula	5.965
	5.995
S Pedros Sula	6.125
	6.185
S Rosa Copan	5.96
Tegucigalpa	4.82
	6.02
	6.035
	6.05
	6.06
	6.085
	6.095
	6.165
HUNGARY	
Budapest	8.5
	9.833
Diosd	5.955
	5.965
	5.97
	5.98
	6.025
	6.08
	6.105
	6.11
	7.2
	9.585
	9.655
	11.91
Jaszbereny	5.96
	5.98
	6.0
	6.025
	6.04
	6.06
	6.08
	6.105
	6.11
	6.125
	6.16
	6.165
	7.155
	7.215
	7.275
	9.585
	9.655
	11.91
	15.16
	15.22
	15.225
	15.285
	17.71

	MHz
Hungary – contd	
Jaszbereny	17.715
	17.775
	17.785
	21.525
Juticalpa	6.145
Szekesfeher-var	5.98
	6.0
	6.04
	6.105
	6.11
	6.115
	7.155
ICELAND	
Reykjavik	12.175
INDIA	
Aligarh	7.125
	7.225
	7.28
	9.525
	9.535
	9.565
	9.59
	9.615
	9.625
	9.675
	9.705
	11.725
	11.74
	11.745
	11.77
	11.775
	11.795
	11.81
	11.815
	15.165
	15.19
	15.205
	15.335
	17.705
	17.73
	17.78
	17.86
Bhopal	3.315
	5.99
	7.18
Bombay	4.84
	6.035
Bombay	7.24
	7.26
	9.525
	9.55
	9.575
	11.765
	11.83
	11.87
	15.08
	15.125
	15.14
Calcutta	4.82
	6.01
	7.21
	9.53
Delhi	3.295
	3.365
	3.905
	3.925
	4.86
	4.96
	5.96
	6.015
	6.02
	6.05
	6.075
	6.085
	6.105
	6.12
	6.14
	6.145
	6.15
	6.16
	6.19
	7.105
	7.11
	7.12
	7.125
	7.145
	7.165
	7.195
	7.215
	7.225
	7.235
	7.26
	7.27
	7.28
	7.29
	7.412
	9.525
Delhi	9.535
	9.565
	9.575
	9.59
	9.615
	9.625
	9.645
	9.675
	9.705
	9.73
	9.755
	9.63
	9.912
	9.95
	10.335
	11.62
	11.715
	11.725
	11.735
	11.74
	11.745
	11.755
	11.76
	11.765
	11.77
	11.775
	11.81
	11.825
	11.835
	11.84
	11.845
	11.85
	11.855
	11.875
	11.88
	11.885
	11.895
	11.935
	11.95
	11.945
	11.965
	15.08
	15.11
	15.12
	15.13
	15.16
	15.165
	15.18
	15.185
	15.205

	MHz
India – contd	
Delhi	15.21
	15.235
	15.25
	15.27
	15.275
	15.28
	15.31
	15.335
	15.35
	15.37
	15.43
	17.387
	17.705
	17.725
	17.76
	17.775
	17.86
	17.865
	17.87
	17.875
	21.74
	21.675
Gauhati	3.235
	3.375
	4.775
	5.97
	6.13
	7.15
	7.28
Hyderabad	4.8
	6.12
	7.14
	9.72
Jammu	5.96
Kohima	6.065
	7.105
Kurseong	3.355
	4.895
	6.1
	7.23
Lucknow	3.205
	6.17
	7.25
Madras	4.92
	6.085
	7.16
	7.26
	9.51
	9.515

	MHz
Madras	9.59
	9.715
	9.75
	15.335
Ranchi	6.14
	7.17
Sibolga	4.775
Simla	3.223
	6.02
Srinagar	6.11
Unknown	3.277
	4.85
	17.855
INDONESIA	
Amboina	7.14
Bandjarmasin	5.97
Biak	7.21
Bukittinggi	4.885
Denpassar	3.945
	7.12
Djambi	4.927
Jakarta	3.925
	4.07
	4.72
	4.755
	4.77
	4.805
	4.875
	4.92
	4.93
	4.932
	4.96
	5.05
	5.88
	5.885
	5.98
	6.045
	7.11
	7.22
	7.27
	9.68
	9.77
	11.715
	11.77
	11.79
	15.15
Jajapura	6.07
	7.165
	7.19

	MHz
Jajapura	9.745
	11.865
	15.345
Jermate	3.95
Makassar	9.55
	11.75
Manokwari	6.185
Medan	3.421
	4.764
	7.24
Menado	5.99
	7.295
Merauke	7.185
	15.12
Padang	3.905
	3.955
	3.96
	6.19
	6.2
	9.51
Pakanbaru	5.955
Palembang	4.855
Pontianak	3.965
	3.995
	4.005
Semarinda	6.135
Semarang	3.935
Sorong	9.6
	9.66
Surabaya	3.98
	4.7
	6.12
Surakarta	3.91
Yogyakarta	7.105
Unknown	3.94
IRAN	
Kamalabad	7.205
	7.215
	7.23
	7.27
	9.535
	9.575
	9.72
	9.765
	9.77
	11.735
	11.745
	11.77
	11.78

	MHz
Iran – contd	
Kamalabad	11.865
	11.93
	15.135
	15.22
	15.26
	15.315
	15.435
	17.73
Teheran	9.022
	15.084
IRAQ	
Abu Ghraib	7.18
	11.725
	11.825
	17.77
Babel	7.18
	9.635
	9.745
	11.905
	11.925
	11.935
Baghdad	3.195
	3.242
	3.95
	11.795
	17.593
Salman Pack	6.095
	6.155
	7.22
	7.24
	9.555
	9.745
	11.78
	11.785
ISRAEL	
Jerusalem	5.882
	5.9
	5.915
	7.125
	7.2
	7.395
	7.412
	7.465
	9.009
	9.355
	9.425
	9.435

	MHz
Jerusalem	9.445
	9.495
	9.54
	9.63
	9.65
	9.74
	9.815
	9.82
	9.833
	11.602
	11.62
	11.625
	11.635
	11.64
	11.655
	11.7
	11.705
	11.715
	11.735
	11.76
	11.775
	11.78
	11.81
	11.84
	11.845
	11.875
	11.955
	11.96
	12.077
	15.07
	15.1
	15.105
	15.12
	15.2
	15.21
	15.23
	15.24
	15.265
	15.3
	15.305
	15.33
	15.405
	15.415
	15.425
	15.465
	15.485
	15.512
	15.53
	15.545
	17.63

	MHz
Jerusalem	17.685
	17.7
	17.71
	17.795
	17.815
	17.845
	17.85
	17.855
	17.875
	21.495
	21.5
	21.565
	21.575
	21.58
	21.625
	21.65
	21.73
	21.735
	21.917
	25.605
	25.64
	25.76
	26.095
	29.705
ITALY	
Caltanissetta	6.06
	7.175
	9.515
Milan	6.208
Roma	3.995
	5.0
	5.99
	6.01
	6.05
	6.06
	6.075
	7.235
	7.275
	7.29
	9.575
	9.58
	9.63
	9.71
	11.8
	11.81
	11.875
	11.905
	11.955
	15.23

	MHz
Italy – contd	
Roma	15.25
	15.315
	15.33
	15.34
	15.345
	15.385
	15.4
	17.755
	17.77
	17.78
	17.795
	17.815
	21.51
	21.555
	21.56
	21.655
	21.69
Turin	5.0
IVORY COAST	
Abidjan	4.94
	6.015
	7.215
	11.92
JAPAN	
Hiroshima	6.175
Hokkaido	3.945
Kumamoto	6.13
Matsuyama	6.005
Momote	15.26
Nagoya	9.535
Osaka	6.19
Tokyo	3.91
	3.925
	3.93
	6.155
	11.75
	15.0
	15.26
	17.71
	17.81
	21.61
Tokyo Kawagu	9.55
Tokyo Nagara	6.055
	6.115
	9.595
	9.76
Tokyo Yamata	5.99
Tokyo Yamata	6.03
	7.14
	7.195
	7.205
	7.25
	9.505
	9.525
	9.53
	9.585
	9.605
	9.67
	9.7
	9.705
	11.705
	11.725
	11.78
	11.815
	11.84
	11.855
	11.875
	11.94
	11.95
	11.96
	15.105
	15.195
	15.235
	15.27
	15.3
	15.31
	15.325
	15.42
	17.725
	17.755
	17.825
	21.535
	21.64
JORDAN	
Amman	7.155
	9.53
	9.56
	11.92
KENYA	
Langata	4.915
Mombasa	4.885
Nairobi	4.804
	4.934
	4.95
	7.125
Nairobi	7.14
	7.15
	7.21
	7.24
	7.295
	9.665
KOREA (NORTH)	
Hamhung	2.775
Kanggye	4.03
Pyongyang	2.765
	2.85
	3.015
	3.32
	3.56
	3.695
	3.89
	4.77
	5.88
	6.25
	6.29
	6.338
	6.4
	6.576
	6.6
	6.77
	9.42
	9.977
	11.35
	11.535
	11.568
	11.885
Sariwon	2.670
Sinuiju	2.745
Wonsan	3.03
KOREA (SOUTH)	
Kimjae	6.03
	6.17
	7.115
	7.24
	7.425
	7.25
	7.29
	9.51
	9.525
	9.535
	9.54
	9.555
	9.565

	MHz
Korea (South) – contd	9.575
Kimjae	9.595
	9.61
	9.65
	9.66
	9.675
	9.71
	9.72
	9.765
	11.74
	11.78
	11.79
	11.84
	11.845
	11.85
	11.86
	11.91
	11.945
	11.965
	11.97
	15.13
	15.17
	15.205
	15.33
	15.335
	15.36
	15.43
Kyong San	5.975
Seoul	6.24
	7.55
	9.87
	15.57
Suwon	6.135
	7.15
	7.275
	9.58
	9.64
	9.665
Unknown	3.93
KUWAIT	
Kuwait	9.84
	15.645
	17.725
Magwa	9.65
Sulaibiyah	6.055
	7.12
	9.505
	9.575
Sulaibiyah	11.94
	11.98
	15.15
	15.345
	17.74
	17.75
	17.82
	21.605
	21.685
	25.645
Unknown	12.085
LAOS	
Houa Phan	4.65
	4.665
	6.2
Laos	7.095
Luang Prabang	4.7
	6.975
	8.395
Pakse	6.6
Savannakhet	7.385
Udomsai	6.91
Vientiane	3.9
	4.32
	4.645
	5.16
	6.13
	6.21
	7.48
Xieng Khouang	4.603
	6.675
Unknown	8.66
LEBANON	
Beyrouth	5.98
	9.545
	11.785
	11.825
	11.925
	11.965
	15.18
	15.2
	15.34
	15.35
	15.35
	15.44
	17.71
	17.715
	17.765
LESOTHO	
Maseru	4.8
LIBERIA	
Careysburg	6.035
	6.045
	6.18
	7.175
	7.195
	7.28
	9.54
	9.615
	9.63
	9.67
	9.7
	9.715
	9.74
	11.705
	11.71
	11.715
	11.74
	11.805
	11.84
	11.915
	15.225
	15.235
	15.24
	15.25
	15.26
	15.315
	15.32
	15.33
	15.4
	15.43
	15.44
	15.445
	17.715
	17.74
	17.78
	17.79
	17.795
	17.855
	17.86
	17.865
	17.87
	21.6
	21.69
	21.76
Monrovia	3.227
	3.255

	MHz
Liberia – contd	
Monrovia	3.99
	4.77
	6.075
	7.442
	9.55
	10.88
	11.83
	11.835
	11.86
	11.87
	11.88
	11.93
	11.945
	12.01
	12.02
	15.09
	15.92
	21.595
LIBYA	
Tripoli	6.1
	6.185
	6.2
	7.165
	7.2
	9.5
	9.565
	9.65
	9.655
	9.66
	11.7
	11.755
	11.795
	15.1
	15.3
	17.8
	21.65
LUXEMBOURG	
Junglinster	6.09
	15.35
MALAGASY REP	
Talata Volon	6.02
	7.285
	11.73
	11.735
	11.74
	15.165
Talata Volon	15.18
	15.2
	15.22
	15.385
	17.81
	17.815
	17.855
	17.86
	21.48
	21.565
	21.64
	21.685
	25.65
Tananarive	3.232
	3.288
	3.37
	4.985
	5.01
	6.135
	6.17
	7.105
	7.155
	7.23
MALAWI	
Blantyre	3.38
Limbe	5.995
MALAYSIA	
Kajang	5.965
	6.025
	6.1
	6.175
	7.295
	9.515
	9.665
	9.75
	11.9
	15.295
	17.725
Koya Kinabalu	4.972
Kuala Lumpur	4.845
Kuching	4.835
	4.895
	4.955
	5.03
Mira	6.06
Penang	4.985
	7.2
Penang	9.71
Sabah	4.975
Sibu	6.05
Stapok	7.145
	7.16
	7.27
	9.535
	9.605
Tebrau	3.915
	6.195
MALDIVE ISLANDS	
Male	6.15
	7.225
	9.55
MALI	
Bamako	4.783
	4.825
	4.875
	5.995
	7.11
	7.285
	9.635
	11.96
MALTA	
Cyclops	5.96
	5.98
	5.99
	6.0
	6.025
	6.1
	7.12
	7.16
	7.265
	7.275
	9.53
	9.565
	9.605
	9.61
	9.65
	9.67
	9.735
	9.77
	11.765
	11.785
	11.795
	11.81
	11.92

	MHz
Malta – contd	
Cyclops	15.105
	15.135
	15.225
	15.385
	15.405
	17.825
	21.59
	21.65
MARTINIQUE	
Fort de France	3.315
	5.995
MAURETANIA	
Nouakchott	4.845
	7.245
	9.61
MAURITIUS	
Malherbes	9.71
Mauritius	9.71
P Louis	4.85
MEXICO	
Cd Mante	6.09
Chihuahua	6.14
Hermosillo	6.115
Linares	5.98
Merida	6.105
Mexico City	5.985
	6.01
	6.165
	6.185
	9.515
	9.555
	9.6
	9.705
	11.74
	11.77
	11.88
	15.11
	15.125
	15.16
	15.385
	15.43
	17.765
S Luis Potosi	6.045

	MHz
Sisoguichi	5.96
Tapachula	6.12
Texmelucan	6.065
Tlaxiaco	6.145
Vera Cruz	6.02
MONACO	
Monte Carlo	5.945
	5.965
	6.035
	6.215
	7.105
	7.125
	7.145
	7.215
	7.23
	7.245
	7.26
	7.275
	9.51
	9.525
	9.545
	9.575
	9.59
	9.61
	9.625
	9.63
	9.64
	9.65
	9.655
	9.66
	9.675
	9.725
	9.73
	9.735
	11.705
	11.71
	11.735
	11.74
	11.75
	11.78
	11.79
	11.8
	11.835
	11.855
	11.935
	11.955
	12.295
	15.583
	15.66

	MHz
MONGOLIAN REPUBLIC	
Hailar	3.9
	6.08
Huhetot	3.93
	3.97
	4.068
	4.895
	6.84
Ikechao/ Tungsheng	4.525
Silinhot	4.952
	7.2
Tungsheng	6.045
Ulan Bator	4.08
	4.763
	4.79
	4.83
	4.85
	4.995
	5.055
	6.385
	7.26
	11.74
	11.855
	12.07
MOROCCO	
Rabat	6.1
	7.365
	9.615
	15.155
	15.36
	21.58
Sebaa Aioun	6.19
	7.225
	15.345
Tangier	5.46
	5.965
	6.015
	6.06
	6.08
	6.095
	6.1
	6.11
	6.13
	6.17
	6.18
	7.17
	7.18

	MHz
Morocco – contd	
Tangier	7.19
	7.21
	7.22
	7.23
	7.24
	7.265
	7.275
	7.295
	9.51
	9.53
	9.54
	9.58
	9.615
	9.63
	9.65
	9.66
	9.68
	9.715
	9.735
	9.76
	9.77
	10.972
	11.71
	11.715
	11.735
	11.76
	11.77
	11.805
	11.915
	11.955
	11.96
	11.965
	15.155
	15.16
	15.195
	15.205
	15.235
	15.245
	15.27
	15.285
	15.305
	15.31
	15.33
	15.335
	15.36
	16.247
	17.705
	17.71
	17.76

	MHz
Tangier	19.915
	21.61
	21.65
	21.735
MOZAMBIQUE	
Beira	3.37
	4.895
	6.025
	6.09
	9.635
L Marques	3.265
	4.855
Maputo	3.21
	3.338
	4.865
	4.925
	6.05
	6.115
	7.11
	7.24
	9.62
	11.82
	15.295
Nampula	3.343
	7.14
	7.255
Pemba	7.15
Quelimane	7.145
NEPAL	
Kathmandu	6.1
	7.105
	7.165
	9.59
	10.010
	11.97
	15.2
	17.8
Khumaltar	3.425
Lalitpur	5.005
NETHERLANDS	
Hilversum	9.895
	17.422
	19.71
Lopik	5.955
	6.02
	6.045
	6,085

	MHz
Lopik	7.21
	7.24
	9.63
	9.66
	9.715
	9.77
	11.72
	11.725
	11.73
	11.74
	11.845
	11.93
	11.935
	11.95
	11.955
	15.16
	15.165
	15.185
	15.22
	15.235
	15.28
	15.325
	15.385
	17.7
	17.74
	17.81
	21.48
	21.53
	21.54
	21.64
NETHERLANDS ANTILLES	
Bonaire Noord	6.02
	6.165
	9.59
	9.63
	9.715
	9.77
	11.79
	15.18
	15.22
	15.31
	15.315
	15.32
	17.81
	21.64
Bonaire Zuid	6.19
	9.535
	9.55

	MHz
Netherlands – cont	
Antilles	
Bonaire Zuid	9.57
	9.61
	9.62
	9.755
	11.705
	11.71
	11.79
	11.8
	11.815
	11.83
	11.9
	11.925
	11.95
	15.255
	15.275
	15.415
	15.42
	15.44
NEW CALEDONIA	
Noumea	3.355
	7.17
	9.505
	11.71
NEW HEBRIDES	
Port Vila	7.26
NEW ZEALAND	
Wellington	6.105
	9.62
	9.77
	11.705
	11.8
	11.84
	11.85
	11.945
	11.96
	15.28
	15.345
	15.38
	17.71
	17.77
	17.86
NICARAGUA	
Bluefields	5.955
	6.12
Bonanza	9.52
Managua	5.95
	6.01
	11.875
Nueva Segovia	6.1
Ocotal	6.1
NIGER	
Niamey	3.26
	5.02
NIGERIA	
Benin	4.932
	5.965
Calabar	6.145
Enugu	6.025
	7.235
Ibadan	3.205
	6.05
	7.285
Ikorodu	7.255
	7.275
	11.77
	11.9
	11.925
	15.12
	15.185
Jaji	6.09
	9.57
Jos	5.965
Kaduna	3.396
	6.175
Lagos	4.99
Maidguri	6.1
Sogunle	7.255
Sokoto	6.195
NORWAY	
Fredrikstad	6.005
	6.015
	6.085
	6.18
	9.55
	9.59
	9.605
	9.61
	9.645
	11.735
	11.85
	11.86
Fredrikstad	11.87
	11.895
	11.935
	15.135
	15.17
	15.175
	15.345
	17.715
	17.74
	17.755
	17.795
	17.8
	17.84
	17.875
	21.655
	21.73
OMAN	
Masirah	6.03
	6.14
	6.195
	7.14
	7.23
	7.25
	7.275
	9.59
	11.74
	11.78
	11.91
	11.955
	15.31
	17.77
Seeb	6.175
	11.89
PAKISTAN	
Islamabad	3.24
	3.915
	4.02
	5.01
	5.061
	6.01
	6.03
	6.18
	7.165
	7.195
	7.265
	7.29
	9.505
	9.545

	MHz
Pakistan – contd	
Islamabad	9.59
	9.645
	9.66
	9.69
	11.705
	11.75
	11.885
	11.91
	11.915
	11.955
	11.97
	15.115
	15.15
	15.27
	15.325
	15.42
	17.75
	17.8
	17.83
	17.845
	17.89
	21.485
	21.59
	21.595
	21.625
	21.675
	21.73
	21.735
	21.74
Karachi	3.86
	3.89
	3.965
	4.72
	4.735
	6.115
	6.235
	7.085
	7.165
	7.195
	7.21
	7.265
	7.375
	9.46
	9.465
	9.645
	9.66
	9.69
	9.79
	11.64
Karachi	11.672
	11.68
	11.705
	11.75
	11.885
	11.915
	15.205
	15.23
	15.325
	15.4
	15.42
	15.465
	15.515
	15.52
	15.685
	17.64
	17.665
	17.75
	17.755
	17.8
	17.82
	17.845
	17.89
	21.455
	21.57
	21.59
	21.6
	21.655
	21.73
	21.745
Peshawar	3.155
	3.33
	6.08
Quetta	3.273
	5.98
	7.105
	7.15
	9.735
Rawalpindi	3.215
	4.06
	4.935
	6.13
	9.77
Unknown	3.396
	3.925
	5.112
PANAMA	
David	6.045
Panama	5.995
Panama	9.685
PAPUA NEW GUINEA	
Alotau	6.04
Daru	3.305
	6.08
Milne Bay	3.36
Port Moresby	3.925
	4.89
	9.52
	9.575
Rabaul	3.385
	5.985
Wewak	3.335
	6.14
PARAGUAY	
Ascuncion	6.015
	6.025
	6.11
	9.735
	11.85
	15.21
Concepcion	11.915
Encarnacion	11.94
	11.945
PJ Caballero	5.995
Villarrica	5.975
PERU	
Andahuaylas	5.025
Andina	4.995
Iquitos	4.79
Lima	3.24
	4.975
Tarapoto	4.935
PHILIPPINES	
Bocaue	6.03
	6.12
	7.225
	9.505
	9.715
	9.755
	9.765
	11.765
	11.82
	11.855
	11.89
	11.92

	MHz
Philippines – contd	
Bocaue	15.235
	15.3
	15.385
	15.39
	15.44
	17.81
	21.515
Iba	9.715
	9.755
	11.92
Malolos	5.965
	9.515
	9.535
	9.54
	9.57
	9.58
	9.59
	9.605
	9.64
	9.69
	9.7
	9.74
	11.725
	11.8
	11.805
	11.825
	11.83
	11.87
	11.875
	11.95
	11.955
	11.97
	15.115
	15.135
	15.215
	15.235
	15.26
	15.275
	15.28
	15.285
	15.31
	15.41
	15.415
	17.71
	17.775
Manila	3.285
	3.345
	15.45
Marulas	6.17
	9.615
Poro	2.6
	5.995
	6.185
	7.155
	7.165
	7.19
	7.235
	7.24
	7.255
	7.275
	7.285
	9.53
	9.65
	9.77
	11.715
	11.73
	11.835
	11.84
	11.93
	11.965
	15.115
	15.185
	15.205
	15.25
	15.29
	17.71
	17.74
	17.75
	17.765
	17.78
	17.865
	21.49
	21.51
	21.67
	21.68
Tinang	5.955
	6.01
	6.11
	7.16
	7.275
	9.545
	9.555
	9.56
	9.585
	9.605
	9.615
	9.625
Tinang	9.63
	9.645
	9.65
	9.67
	9.725
	9.76
	11.715
	11.73
	11.75
	11.755
	11.76
	11.78
	11.805
	11.81
	11.815
	11.83
	11.925
	11.93
	11.965
	15.11
	15.115
	15.15
	15.155
	15.165
	15.17
	15.2
	15.21
	15.215
	15.23
	15.25
	15.26
	15.29
	15.32
	15.345
	15.36
	15.365
	15.395
	15.41
	15.425
	17.74
	17.75
	17.78
	17.79
	17.8
	17.82
	17.85
	21.63
Unknown	9.575
	11.845

	MHz
Philippines – contd	
Unknown	15.195
POLAND	
Warsaw	3.955
	5.96
	5.995
	6.035
	6.095
	6.135
	6.155
	6.195
	7.11
	7.125
	7.145
	7.18
	7.205
	7.27
	7.285
	9.525
	9.54
	9.57
	9.675
	9.755
	11.8
	11.815
	11.84
	11.98
	15.12
	15.195
	15.275
	17.81
	17.865
PORTUGAL	
Lisbon	5.97
	5.985
	6.07
	6.105
	6.115
	6.135
	7.115
	7.145
	7.165
	7.19
	7.2
	7.215
	7.245
	7.255
	7.295

	MHz
Lisbon	9.505
	9.555
	9.565
	9.595
	9.695
	9.705
	9.725
	9.75
	10.905
	11.725
	11.77
	11.78
	11.815
	11.825
	11.855
	11.875
	11.885
	11.895
	11.925
	11.97
	15.115
	15.125
	15.145
	15.17
	15.215
	15.225
	15.255
	15.29
	15.34
	15.355
	15.37
	15.38
	15.445
	17.725
	17.735
	17.75
	17.76
	17.77
	17.805
	17.835
	17.865
	17.875
	21.455
	21.51
	21.66
	21.665
	21.72
	21.73
	21.745
	21.75

	MHz
Lisbon	25.69
S Gabriel	6.025
	6.155
	9.615
	9.62
	9.635
	9.74
	11.8
	11.84
	11.875
	11.925
	11.935
	15.125
	15.14
	15.2
	15.34
	17.88
	17.895
	21.495
	21.7
	21.735
Sines	6.01
	6.015
	6.115
	7.285
	7.295
	9.565
	9.585
	9.605
	9.61
	9.615
	9.635
	9.65
	9.66
	9.665
	9.67
	9.675
	9.685
	9.715
	11.825
	11.865
	11.905
	11.915
	15.16
	15.185
	15.245
	15.305
	15.315
	15.32
	17.82

MHz

Portugal – contd

Sines 21.55

QATAR

Doha 9.57

REUNION

S Denis 7.245

RHODESIA

Gwelo 3.307
3.396
4.828
5.016
5.975
6.02
7.175
7.285

ROUMANIA

Bucharest 5.99
6.15
6.155
6.18
6.19
7.12
7.175
7.195
7.225
9.51
9.53
9.54
9.55
9.57
9.59
9.595
9.625
9.64
9.685
9.69
9.75
11.705
11.71
11.725
11.735
11.74
11.775
11.79
11.81
11.83

MHz

Bucharest 11.84
11.86
11.885
11.94
11.95
11.97
11.98
15.25
15.255
15.335
15.34
15.345
15.365
15.38
15.39
17.72
17.73
17.75
17.785
17.79
17.805
17.815
17.825
17.83
17.835
17.84
17.85
17.87
17.89
21.49

RWANDA

Kigali 3.33
6.055
6.16
7.225
9.565
9.7
9.735
11.765
11.785
11.965
15.41
17.765
17.8
21.54

ST THOMAS ISLANDS

Sao Thome 4.807
5.339

MHz

SAUDI ARABIA

Diriyya 6.0
7.22
9.72
11.95

Jeddah 9.67
11.855
15.115

Riyadh 2.739
3.97
5.26
5.39
5.876
6.08
6.085
6.807
7.11
9.605
9.63
9.73
9.887
11.5
11.685
11.715
11.72
11.73
11.775
11.8
11.82
11.845
11.855
11.865
11.87
11.88
11.885
11.89
11.9
11.905
11.91
11.915
11.965
12.113
13.396
15.06
15.11
15.115
15.155
15.175
15.19
15.22

	MHz
Saudi Arabia – contd	
Riyadh	15.23
	15.33
	15.35
	15.365
	15.37
	15.665
	17.755
	17.76
	17.77
	21.48
	21.5
	21.505
	21.52
	21.53
	21.55
	21.555
	21.59
Unknown	10.385

SENEGAL

	MHz
Dakar	4.89
	7.17
	7.21
	11.895
Tambacounda	6.045
	6.08
	6.18
Ziguinchor	3.336
	6.18
	7.21

SEYCHELLES

	MHz
Mahe	6.185
	9.545
	9.615
	9.65
	11.705
	11.725
	11.755
	11.785
	11.8
	11.805
	11.84
	11.855
	11.86
	11.865
	11.945
	15.16
	15.325
Mahe	15.406
	15.43
	17.715

SIERRA LEONE

	MHz
Freetown	3.316
Goderich	5.98
Waterloo	5.98

SINGAPORE

	MHz
Jurong	4.97
	5.01
	5.052
	6.2
Kranji	6.01
	6.05
	6.065
	6.08
	6.195
	7.18
	9.57
	9.58
	9.59
	9.725
	9.74
	11.71
	11.75
	11.85
	11.865
	11.91
	11.92
	11.925
	11.955
	15.27
	15.28
	15.31
	15.435
	15.36
	15.38
	17.715
	17.88
Singapore	6.0
	6.155
	7.17
	7.25
	9.53
	9.635
	11.94
	15.2
Unknown	11.85

SOLOMON ISLANDS

	MHz
Honiara	9.545

SOMALIA

	MHz
Hargeisa	7.16
	11.645
Mogadiscio	6.095
	7.12
	9.585

SOUTH AFRICA

	MHz
Bloemendal	3.25
	3.388
	3.955
	3.965
	3.98
	3.995
	4.88
Bloemfontein	3.32
	4.81
	4.875
Johannesburg	3.23
Meyerton	3.285
	4.835
	4.99
	5.98
	6.01
	7.17
	7.27
	9.525
	9.56
	9.585
	9.61
	9.65
	9.68
	9.71
	11.79
	11.8
	11.885
	11.9
	11.935
	11.955
	15.125
	15.155
	15.2
	15.22
	15.285

	MHz
South Africa – contd	
Meyerton	15.38
	17.78
	17.805
	21.535
	21.555
	21.695
	25.79
SPAIN	
Arganda	5.97
	5.99
	6.0
	6.045
	6.065
	6.07
	6.1
	6.14
	6.16
	7.105
	7.11
	7.155
	7.2
	7.225
	7.245
	7.275
	9.505
	9.57
	9.68
	9.675
	9.685
	11.785
	11.83
	11.855
	11.92
	11.925
	11.93
	15.125
	15.205
	15.22
	15.315
	15.33
	15.335
	15.395
Madrid	9.36
	9.79
Noblejas	6.05
	6.065
	6.075
	6.085

	MHz
Noblejas	6.12
	6.19
	7.105
	7.185
	7.195
	9.505
	9.51
	9.52
	9.53
	9.55
	9.56
	9.57
	9.58
	9.595
	9.6
	9.63
	9.69
	9.695
	9.725
	9.745
	11.715
	11.73
	11.74
	11.775
	11.785
	11.81
	11.815
	11.84
	11.85
	11.88
	11.89
	11.91
	11.92
	11.925
	11.93
	11.945
	11.965
	15.125
	15.185
	15.24
	15.25
	15.29
	15.365
	15.375
	15.39
	15.395
	15.405
	17.72
	17.735
	17.75

	MHz
Noblejas	17.865
	17.87
	17.89
	21.535
	21.62
Playa de Pals	7.155
	7.165
	7.19
	7.22
	7.245
	7.295
	9.52
	9.625
	9.66
	9.68
	9.705
	9.715
	11.875
	11.885
	11.895
	11.915
	11.925
	11.935
	11.945
	11.97
	15.115
	15.13
	15.29
	15.34
	15.37
	15.38
	15.445
	17.725
	17.75
	17.76
	17.77
	17.865
	17.885
	17.895
SUDAN	
Khartoum	5.038
Omdurman	6.15
	7.2
	9.505
	11.835
SURINAM	
Paramaribo	4.85

	MHz
SWAZILAND	
Manzini	5.955
	6.07
	7.215
	7.255
	7.28
	9.52
	9.59
	9.73
	11.71
	11.76
	11.81
	11.91
	11.955
	15.165
	17.745
Mbabane	3.223
	4.76
	4.79
SWEDEN	
Hoerby	6.065
	6.1
	9.59
	9.605
	9.635
	9.66
	9.665
	9.69
	9.695
	9.745
	11.705
	11.79
	11.8
	11.81
	11.845
	11.85
	11.855
	11.905
	11.915
	11.935
	11.955
	15.12
	15.125
	15.13
	15.205
	15.24
	15.39
	15.415
	15.435
Hoerby	17.735
	17.77
	17.79
	17.885
	21.505
	21.61
	21.615
	21.66
	21.69
Karlsborg	6.065
	6.13
	7.145
	9.605
	9.63
	9.66
	9.665
	9.69
	9.695
	11.705
	11.85
	11.905
	15.24
	15.305
	17.735
	17.77
	21.65
	21.69
	21.7
Varberg	11.952
	15.192
	17.777
	17.787
	21.552
	21.557
SWITZERLAND	
Beromunster	7.21
Geneva (UNO)	7.443
Lenk	6.165
Sarnen	9.535
Schwarzenburg	3.985
	5.97
	6.045
	6.135
	7.21
	9.535
	9.56
	9.625
	9.66
	9.685
Schwarzenburg	9.7
	9.725
	9.765
	11.715
	11.72
	11.765
	11.77
	11.775
	11.78
	11.782
	11.85
	11.87
	15.14
	15.17
	15.19
	15.23
	15.235
	15.305
	15.35
	15.385
	15.43
	17.715
	17.735
	17.74
	17.742
	17.77
	17.79
	17.83
	17.84
	21.52
	21.545
	21.547
	21.57
	21.585
	21.63
Sottens	9.765
	11.72
	11.735
	15.305
	15.43
	17.715
	17.77
	17.815
	21.545
	21.57
TAHITI	
Papeete	6.135
	9.75
	11.825
	15.17

TANZANIA

	MHz
Dar-es-Salaam	4.785
	4.825
	5.05
	5.985
	6.105
	7.165
	7.23
	7.28
	9.53
	9.55
	9.595
	9.75
	15.435
Zanzibar	3.339

THAILAND

	MHz
Bangkok	4.832
	6.07
	6.08
	7.115
	9.655
	11.905

TIBET

	MHz
Lhasa	4.035
	5.935
	9.49
	9.655

TIMOR

	MHz
Dili	3.55
	3.85

TOGO

	MHz
Lama Karma	3.222
Lome	5.047
Togblekope	7.265

TRISTAN DA CUNHA

	MHz
Tristan da-Cunha	3.29

TUNIS

	MHz
Sfax	7.225
	7.275
	11.97
	15.225

TURKEY

	MHz
Ankara	6.185
	6.9
	7.17
	7.27
	9.515
	9.665
	11.8
	11.955
	11.965
	15.125
	15.135
	15.145
Hakkari	7.649
Unknown	6.34

UGANDA

	MHz
Kampala	3.334
	4.976
	5.026
	6.03
	9.515
	9.685
	9.73
	15.235
	15.25
	15.325

UKRAINE

	MHz
Ivanofransk	7.115
	9.635
	9.665
	9.76
	11.88
	11.955
	15.33
	17.785
Kharkov	11.76
	11.95
	11.97
Kiev	4.92
	4.94
	5.915
	6.02
	6.085
	6.14
	6.165
	7.165
	7.23
	7.3
Kiev	7.39
	9.58
	9.67
	9.71
	9.715
	9.72
	11.69
	11.77
	11.8
	11.925
	12.07
	15.11
	15.24
	17.82
	17.86
	21.515
Lvov	6.115
	6.125
	7.12
	7.15
	7.16
	7.23
	9.685
	9.705
	9.735
	9.765
	11.72
	11.75
	11.78
	11.9
	15.135
	15.175
	17.76
	17.78
	17.845
	17.89
	21.565
Simferopol	6.03
	6.04
	6.09
	6.13
	7.105
	7.195
	9.515
	9.535
	9.565
	9.675
	9.74
	9.75
	11.705

	MHz
Ukraine – contd	
Simferopol	11.81
	15.13
	15.185
	15.205
	15.245
	15.26
	15.325
	15.425
	17.815
	17.855
	21.515
	21.745
Starobelsk	6.055
	6.145
	6.17
	7.175
	9.58
	9.59
	15.33
Vinnitsa	6.1
	6.175
	7.215
	7.245
	9.53
	9.575
	9.665
	9.72
	11.73
	11.735
	11.85
	15.135
	15.17
	15.435
	17.755
	17.775
	17.835
	17.885
UNITED ARAB EMIRATES	
Abu Dhabi	9.62
Dubai	6.04
UNITED KINGDOM	
London	3.952
	3.97
	3.975
	4.467
	5.872

	MHz
London	5.955
	5.965
	5.975
	5.99
	5.995
	6.005
	6.01
	6.015
	6.03
	6.04
	6.05
	6.06
	6.065
	6.07
	6.08
	6.085
	6.11
	6.115
	6.12
	6.125
	6.13
	6.14
	6.15
	6.16
	6.17
	6.18
	6.185
	6.195
	6.838
	7.105
	7.11
	7.12
	7.14
	7.15
	7.155
	7.17
	7.18
	7.185
	7.2
	7.21
	7.22
	7.23
	7.235
	7.24
	7.245
	7.25
	7.255
	7.26
	7.265
	7.28

	MHz
London	7.285
	7.295
	7.32
	7.325
	7.844
	7.848
	7.893
	7.973
	7.976
	7.991
	9.317
	9.323
	9.41
	9.505
	9.51
	9.515
	9.53
	9.54
	9.555
	9.565
	9.57
	9.58
	9.6
	9.615
	9.62
	9.625
	9.635
	9.64
	9.645
	9.66
	9.68
	9.69
	9.7
	9.705
	9.715
	9.735
	9.745
	9.75
	9.76
	9.77
	9.825
	9.858
	9.915
	11.68
	11.71
	11.715
	11.72
	11.74
	11.75
	11.76

	MHz
United Kingdom — contd	
London	11.775
	11.78
	11.785
	11.79
	11.805
	11.83
	11.835
	11.845
	11.85
	11.855
	11.865
	11.905
	11.91
	11.915
	11.925
	11.935
	11.945
	11.955
	11.96
	12.04
	12.095
	12.127
	12.179
	15.07
	15.105
	15.11
	15.12
	15.14
	15.16
	15.165
	15.18
	15.195
	15.205
	15.215
	15.235
	15.245
	15.25
	15.26
	15.265
	15.27
	15.28
	15.31
	15.315
	15.32
	15.35
	15.355
	15.39
	15.42
	15.425

	MHz
London	15.435
	15.44
	15.445
	15.589
	15.67
	15.849
	15.91
	17.642
	17.695
	17.705
	17.715
	17.74
	17.76
	17.77
	17.78
	17.79
	17.795
	17.8
	17.81
	17.815
	17.855
	17.86
	17.87
	18.08
	19.455
	20.345
	21.47
	21.5
	21.55
	21.57
	21.59
	21.61
	21.63
	21.64
	21.65
	21.695
	21.71
	22.93
	23.191
Rugby Standard Frequency	2.5
	5.0
	10.0
UPPER VOLTA	
Ouagadougou	4.815
	7.23
U.S.A.	
AFRTS Bethany	9.929
	10.869

	MHz
Bethany	16.222
	19.261
	25.88
Boulder Std Freq.	5.0
	10.0
	15.0
	20.0
Cincinnati	6.03
	6.04
	6.125
	6.155
	9.525
	9.685
	9.755
	11.74
	11.79
	11.83
	11.835
	11.895
	15.14
	15.16
	15.19
	15.25
	15.33
	15.43
	17.71
	21.47
	21.485
	21.5
	21.66
Delano	6.01
	6.145
	6.185
	9.545
	9.565
	9.65
	9.66
	9.68
	11.82
	11.83
	11.85
	11.865
	11.9
	11.915
	13.86
	15.205
	15.225
	15.27
	15.345
	15.365

	MHz
U.S.A. – contd	
Delano	17.72
	17.75
	17.78
	17.84
	17.895
	19.912
	21.46
	21.695
	21.745
	25.99
Dixon	5.955
	6.015
	6.055
	6.095
	6.125
	6.185
	9.545
	9.65
	9.66
	9.7
	9.73
	11.74
	11.805
	11.83
	11.85
	11.9
	11.92
	15.25
	15.33
	15.345
	17.72
	17.75
	17.765
	17.82
	17.84
	18.135
	19.48
	21.61
	21.65
	21.745
	25.62
	26.095
Greenville	5.745
	5.995
	6.02
	6.03
	6.055
	6.08
	6.125
Greenville	6.13
	6.135
	6.19
	6.873
	7.651
	7.768
	7.77
	9.51
	9.53
	9.54
	9.565
	9.58
	9.6
	9.62
	9.625
	9.635
	9.64
	9.65
	9.67
	9.68
	9.715
	9.725
	9.745
	9.765
	10.235
	10.38
	10.454
	10.869
	11.715
	11.73
	11.74
	11.75
	11.79
	11.805
	11.83
	11.835
	11.84
	11.845
	11.855
	11.89
	11.9
	11.905
	11.915
	11.92
	11.935
	11.955
	11.96
	11.965
	13.491
	15.12
Greenville	15.125
	15.14
	15.15
	15.155
	15.16
	15.17
	15.195
	15.205
	15.225
	15.235
	15.24
	15.25
	15.26
	15.28
	15.315
	15.35
	15.375
	15.395
	15.4
	15.41
	15.415
	15.425
	15.43
	15.65
	15.752
	15.77
	16.43
	17.73
	17.74
	17.755
	17.775
	17.785
	17.8
	17.815
	17.82
	17.855
	17.86
	17.895
	18.275
	18.782
	19.505
	19.721
	20.06
	20.125
	21.47
	21.5
	21.535
	21.54
	21.55
	21.59

	MHz
U.S.A. – contd	
Greenville	21.61
	21.67
	21.69
	26.04
New York	13.272
Okeechobee	9.525
	9.555
	9.63
	9.66
	9.69
	9.715
	11.705
	11.72
	11.77
	11.8
	11.805
	11.815
	11.845
	11.85
	11.855
	11.875
	11.885
	11.925
	15.11
	15.115
	15.13
	15.16
	15.195
	15.215
	15.44
	17.73
	17.785
	17.805
	17.845
	17.87
	17.875
	21.525
	21.615
Red Lion	11.905
	15.145
	15.175
	15.185
	15.305
	17.72
	17.73
	21.565
Redwood City	5.98
	9.505
	9.53
Redwood City	9.555
	9.6
	9.615
	9.625
	9.685
	9.69
	9.75
	15.28
	15.355
Scituate	5.985
	6.155
	9.525
	9.69
	9.715
	11.78
	11.805
	11.815
	11.85
	11.855
	11.925
	15.11
	15.13
	15.135
	15.16
	15.26
	15.27
	15.35
	15.39
	15.44
	17.73
	17.785
	17.845
	17.865
	17.875
	21.525
	21.61
Unknown	5.23
	5.703
	10.537
U.S.S.R.	
Achkhabad	4.825
	4.895
	6.125
	11.93
	15.195
	17.74
Alma Ata	4.545
	4.99
	5.26
Alma Ata	5.915
	5.96
	5.97
	6.06
	6.135
	6.18
	6.87
	7.14
	7.185
	9.505
	9.61
	11.745
	11.92
	11.95
	15.11
	15.155
	15.21
	15.23
	17.85
Arkhangelsk	5.015
Armavir	5.915
	5.96
	5.965
	6.065
	6.105
	6.12
	6.13
	6.17
	7.155
	7.17
	7.19
	7.205
	7.215
	7.265
	9.52
	9.575
	9.61
	9.62
	9.655
	9.675
	9.765
	11.71
	11.745
	11.775
	11.88
	11.9
	11.93
	11.96
	15.155
	15.18

	MHz
U.S.S.R. – contd	
Armavir	15.21
	15.24
	15.265
	15.28
	15.32
	15.35
	15.375
	15.385
	15.395
	15.405
	17.73
	17.775
	17.84
	17.85
	17.86
	21.51
	21.645
	21.53
	21.555
	21.635
	21.705
	21.745
Baku	4.785
	4.957
	6.11
	6.135
	6.195
	9.65
	11.92
	15.24
	15.26
	15.415
	17.8
	17.825
	21.49
	21.68
Blagovesh-chensk	4.975
Blagoevechtchen	
	5.96
	7.295
	9.52
	9.58
	11.79
	11.94
	15.17
Duchanbe	4.635
	6.105
	6.12

	MHz
Duchanbe	7.115
	7.275
	9.605
	9.735
	9.755
	11.715
	11.825
	11.9
	15.155
	15.22
	17.765
	21.505
	21.585
Dzhambul	4.76
Erevan	4.04
	4.81
	4.99
	6.055
	7.27
	9.515
	9.685
	11.72
	11.76
	11.89
	12.04
	15.15
	15.19
	15.435
	17.735
	17.755
	21.505
Frunze	4.01
	6.08
	7.205
	7.23
	9.53
	9.565
	9.655
	9.695
	9.71
	11.71
	11.72
	11.74
	11.82
	11.85
	11.89
	11.905
	11.97
	15.265
	15.435

	MHz
Frunze	17.745
	17.775
	17.785
	17.815
	17.825
	17.84
	21.515
	21.53
	21.625
	21.745
Gorkii	7.175
	7.185
	11.86
	15.265
	15.385
Irkutsk	5.98
	6.05
	6.09
	6.24
	7.2
	7.265
	9.51
	9.555
	9.565
	9.575
	9.59
	9.67
	9.685
	9.735
	9.755
	11.765
	11.775
	15.175
	15.195
	15.21
	15.285
	15.36
	15.42
	17.73
	17.805
Iujnsakh-klinsk	7.26
	11.86
Jigulevsk	6.17
	7.2
	11.895
	15.115
	15.185
	15.375
Kalatch	15.2
Kalinin	6.105

	MHz
U.S.S.R. – contd	
Kalinin	7.31
	21.635
Kaunas	6.1
	9.635
	9.71
	11.87
Kazan	6.04
	6.065
	6.11
	6.15
	6.175
	7.125
	7.14
	7.16
	7.17
	7.18
	7.2
	7.21
	7.23
	7.285
	9.51
	9.52
	9.58
	9.59
	9.695
	9.7
	9.795
	11.71
	11.735
	11.75
	11.765
	11.805
	11.845
	11.85
	11.92
	11.96
	11.965
	15.175
	15.26
	15.295
	15.32
	15.375
	15.4
	15.415
	17.72
	17.82
Kenga	7.125
	7.255
	7.27

	MHz
Kenga	7.275
	9.505
	9.555
	9.6
	9.7
	9.745
	9.77
	11.73
	11.795
	11.87
	11.89
	11.96
	15.175
	15.255
	15.265
	15.375
	15.425
Khabarovsk	5.985
	6.02
	6.07
	6.115
	6.135
	6.145
	6.175
	7.11
	7.125
	7.175
	7.21
	7.245
	9.545
	9.61
	9.645
	9.7
	11.705
	11.72
	11.73
	11.82
	11.87
	15.115
	15.21
	15.255
	15.28
	15.295
	15.405
	15.42
	17.72
	17.775
	17.805
	17.85
	17.86

	MHz
Khabarovsk	17.87
	21.505
Khanty-Mansiysk	4.52
Kinghisepp	7.175
	9.665
	9.76
Komsomolskamur	5.97
	6.03
	6.06
	6.08
	6.15
	7.215
	7.265
	7.275
	7.28
	9.505
	9.53
	9.71
	9.75
	11.71
	11.715
	11.81
	11.9
	11.91
	11.97
	15.35
	17.865
Konevo	11.85
Krasnoiarsk	5.29
	6.01
	6.06
	6.125
	6.155
	6.175
	7.15
	7.23
	7.245
	7.275
	7.285
	9.56
	9.75
	11.76
	11.78
	11.82
	11.835
	11.86
	11.925
	11.94
	11.96

	MHz
U.S.S.R. – contd	
Krasnoiarsk	15.11
	15.28
	15.42
	17.775
	17.79
	17.835
Kursk	6.1
	15.33
	17.745
	17.86
	21.54
Leningrad	5.95
	6.2
	7.305
	9.59
	9.765
	11.755
	11.765
	15.18
	15.375
	21.6
Magadan	4.03
	5.94
	7.32
	12.24
Minsk	5.945
	7.33
	7.37
	7.42
	9.795
	12.005
Moskva	4.825
	4.85
	4.86
	5.77
	5.9
	5.905
	5.91
	5.94
	5.965
	5.97
	6.0
	6.01
	6.03
	6.045
	6.08
	6.12
	6.13
	6.145
Moskva	6.16
	6.175
	6.2
	6.945
	7.19
	7.205
	7.21
	7.27
	7.28
	7.285
	7.29
	7.295
	7.3
	7.325
	7.34
	7.35
	7.355
	7.36
	7.38
	7.395
	7.4
	7.41
	7.42
	7.43
	7.44
	9.13
	9.15
	9.45
	9.48
	9.49
	9.5
	9.53
	9.54
	9.55
	9.6
	9.62
	9.625
	9.63
	9.64
	9.65
	9.685
	9.74
	9.745
	9.765
	9.775
	9.78
	9.785
	9.79
	9.8
	9.81
Moskva	9.85
	10.12
	10.338
	10.74
	11.34
	11.63
	11.69
	11.75
	11.795
	11.83
	11.845
	11.88
	11.91
	11.92
	11.93
	11.975
	11.98
	11.985
	12.045
	12.055
	12.07
	15.1
	15.21
	15.36
	15.375
	15.41
	15.455
	17.71
	17.835
	17.9
	21.45
	21.575
Murmansk	5.93
Nikolaevvs-kamur	5.97
	5.99
	6.09
	6.155
	6.19
	7.23
	9.725
	9.735
	9.755
	11.79
	15.24
	15.245
	15.415
	17.705
	17.765
	17.845
Novosibirsk	6.045

U.S.S.R. – contd

	MHz
Novosibirsk	6.12
	6.13
	6.185
	7.02
	7.12
	7.145
	7.17
	7.185
	9.645
	9.675
	9.735
	11.74
	11.865
	11.885
	11.92
	15.24
	15.3
	15.31
	15.335
	15.385
	15.4
	15.445
Okhotsk	9.6
Omsk	6.19
	7.11
	7.16
	9.59
	11.77
	15.15
Orcha	11.985
Orenburg	6.09
Palana	4.52
Petropavlovsk	4.055
	4.485
	5.94
Petropavlo Kam	6.1
	6.185
	7.16
	7.29
	9.54
	9.695
	9.71
	9.75
	11.915
	15.14
	15.18
	15.21
	15.425
	17.72
	17.88
Petrozavodsk	4.78
	5.065
Riazan	5.98
	6.185
	7.265
	9.72
	11.79
	11.89
	15.14
	15.275
	15.44
	17.775
	17.87
Riga	5.935
	7.14
	7.235
	15.14
	15.22
	21.615
Serpukhov	5.99
	6.065
	6.095
	6.15
	7.15
	7.165
	9.51
	9.56
	9.605
	9.63
	9.73
	9.77
	11.705
	11.76
	11.78
	11.81
	11.835
	11.845
	11.87
	11.96
	15.15
	15.185
	15.42
	15.45
	17.73
	17.795
	21.59
Sverdlovsk	5.92
	5.96
	6.075
Sverdlovsk	7.21
	7.235
	7.25
	9.75
	11.715
	11.76
	11.785
	11.91
	11.94
	15.305
	15.33
	15.405
	15.505
	17.845
	17.89
Tashkent	4.85
	5.035
	5.925
	5.945
	5.975
	6.025
	7.145
	7.17
	7.22
	9.52
	9.54
	9.55
	9.6
	9.66
	9.75
	11.73
	11.835
	11.925
	11.96
	15.115
	15.255
	15.265
	15.33
	15.395
	17.71
	17.755
	17.86
	17.87
	17.875
	21.585
	21.725
Tallin	6.055
	6.085
	17.805
Tbilisi	5.04

U.S.S.R. – contd

	MHz
Tbilisi	5.93
	5.98
	6.06
	6.085
	7.115
	7.185
	7.255
	9.625
	9.695
	9.76
	11.71
	11.755
	11.78
	11.795
	11.805
	11.82
	11.96
	15.305
	17.805
	17.885
Tchita	4.86
	5.97
	6.0
	6.045
	6.055
	7.115
	7.15
	7.22
	7.315
	9.575
	9.625
	9.665
	9.68
	9.7
	11.755
	11.81
	11.88
	11.89
	11.925
	15.19
	15.385
	15.42
	17.815
	17.89
Tula	5.905
	5.97
	5.995
	6.07
	6.075
Tula	6.18
	7.12
	7.145
	7.16
	7.195
	7.24
	7.25
	7.285
	9.545
	9.555
	9.61
	9.69
	9.7
	9.73
	9.735
	11.72
	11.935
	11.95
	15.17
	15.19
	15.37
	15.435
	17.765
	17.805
	21.46
	21.49
Tyumen	4.895
Ufa	4.485
Ulan Ude	4.795
U.S.S.R. Freq. Standard	14.996
U.S.S.R. Met. Station	3.417
	3.44
	3.47
	3.93
	4.647
	4.656
	4.665
	4.678
	4.685
	4.695
	6.56
	6.596
	6.615
	6.617
	6.73
Vladivostok	3.96
	3.995
	4.04
	4.465
Vladivostok	5.015
	5.96
	6.035
	6.165
	7.2
	7.22
	7.32
	7.49
	9.51
	9.53
	9.62
	9.635
	9.71
	9.75
	9.77
	9.79
	11.755
	15.13
	17.745
	17.755
	17.86
Volgograd	6.075
	7.265
Vologda	5.91
	11.725
	15.35
Voronej	6.005
	6.14
	9.665
	9.675
	11.82
	15.23
	15.295
	15.305
Yakutsk	4.395
Yuzhno Sakhalinsk	4.05
Unknown	3.46
	3.94
	3.99
	4.425
	4.645
	4.654
	4.71
	4.72
	4.73
	4.747
	4.765
	4.815
	4.82

	MHz
U.S.S.R. – contd	
Unknown	4.83
	4.835
	4.84
	4.93
	4.955
	5.255
	5.29
	5.32
	5.455
	5.47
	5.7
	5.794
	5.815
	5.83
	5.885
	5.92
	5.935
	5.95
	6.205
	6.215
	6.23
	6.243
	6.285
	6.39
	6.4
	6.548
	6.554
	6.568
	6.77
	6.808
	6.822
	6.825
	6.852
	6.89
	6.905
	6.92
	6.98
	6.987
	7.01
	7.03
	7.04
	7.045
	7.08
	7.1
	7.105
	7.305
	7.31
	7.335
	7.345
Unknown	7.37
	7.385
	7.437
	7.53
	7.54
	7.55
	7.605
	7.67
	7.74
	7.925
	7.948
	8.125
	8.903
	8.91
	8.917
	8.97
	9.2
	9.21
	9.24
	9.34
	9.345
	9.39
	9.44
	9.46
	9.47
	9.57
	9.595
	9.815
	9.82
	9.83
	10.09
	10.62
	10.66
	10.69
	10.695
	10.855
	11.495
	11.575
	11.67
	11.7
	11.99
	12.0
	12.01
	12.02
	12.025
	12.03
	12.035
	12.05
	12.06
	12.075
Unknown	12.1
	12.14
	12.165
	12.175
	12.205
	12.222
	12.25
	12.252
	12.28
	13.37
	13.38
	13.59
	13.71
	13.76
	13.82
	13.96
	14.29
	14.44
	14.595
	14.85
	15.045
	15.06
	15.46
	15.465
	15.48
	15.49
	15.497
	15.5
	15.515
	15.52
	15.525
	15.53
	15.535
	15.54
	15.545
	15.6
	15.78
	15.85
	15.87
	16.03
	16.14
	16.19
	16.25
	16.33
	16.87
	17.135
	17.56
	17.58
	17.69
	17.7

	MHz
U.S.S.R. – contd	
Unknown	18.015
	18.195
	18.285
	18.31
	18.37
	18.46
	18.653
	18.83
	19.13
	19.21
	19.725
	19.833
	19.845
	20.06
	20.25
	20.605
	21.495
	21.665
	22.205
	22.77
VATICAN	
S M Galeria	5.995
	6.015
	6.19
	7.155
	7.16
	7.25
	7.235
	9.55
	9.605
	9.615
	9.625
	9.645
	9.73
	11.705
	11.715
	11.725
	11.74
	11.765
	11.81
	11.82
	11.83
	11.845
	11.855
	11.875
	15.12
	15.165
	15.2

	MHz
S M Galeria	15.21
	15.4
	17.705
	17.785
	17.825
	17.84
	17.845
	17.895
	21.485
Vatican City	6.105
	6.19
	6.21
	6.221
	6.23
	7.04
	11.7
	17.79
	17.9
VENEZUELA	
Barcelona	3.385
Barquesimeto	4.8
	4.82
	4.88
	4.9
	4.99
Bolivar	4.77
Caracas	3.245
	4.87
	4.89
	4.92
	5.02
	5.03
	5.05
	6.1
	6.17
	9.64
	9.66
	11.75
	15.39
Carora	4.91
Cumana	4.96
El Tigre	3.25
Maracaibo	4.81
	4.86
Merida	3.395
Puerto Cabello	3.285
San Cristobal	4.93
	4.98
San Filipe	4.94

	MHz
Tovar	3.225
	3.25
Trujillo	3.295
Valencia	3.355
	4.78
Valera	4.84
Villa de Cura	4.97
Unknown	4.96
VIETNAM	
Bac Thai	6.884
	7.08
Cao Lang	4.785
	6.26
Gia Lai-Cong Tum	4.762
Ha Bac	4.647
Ha Tuyen	4.82
Hanoi	4.0
	4.86
	4.932
	4.94
	4.945
	4.995
	6.426
	6.435
	7.375
	7.385
	7.415
	7.47
	7.512
	9.84
	9.887
	10.01
	10.04
	10.06
	10.225
	12.033
	15.009
	15.02
Hoang Wen Sun	5.594
Cho Chi Min City	6.165
	9.62
	9.625
Hue	4.68
Lai Chau	5.927
Nghe Tinh	6.28
Phu Kmanh	5.139
Quang Minh	4.692

	MHz
Vietnam – contd	
Son La	4.77
	6.331
Thanh Hoa	4.881
Thu Dau Mot City	4.6
Vinh Phu	4.284
YEMEN	
Sanaa	4.853
	4.975
	6.05
	7.265
	9.77
	9.78
YUGOSLAVIA	
Belgrade	6.1
	6.15
	7.2
	7.24
	9.505
	11.735
	15.24
ZAIRE	
Badundu	7.115
Kananga	6.125
Kinshasa	7.255
	9.66
	9.77
	11.72
	11.795
	15.245
	15.35
Kisangani	6.085
Lubumbashi	4.75
	7.205
	11.865
Mbandaka	5.995
Mbujimayi	7.295
ZAMBIA	
Lusaka	3.295
	3.346
	4.91
	4.965
	6.06
	6.165
	7.22
	7.235
	7.25
	9.58
	11.88
	17.895
ZANZIBAR	
Marhubi	6.005
LOCATIONS UNKNOWN	
Bizam R.	5.914
	6.2
	9.5
	9.585
German Figure Groups	6.453
Kashmir R.	3.915
Mebo II	6.206
National Voice of Iran	6.025
R Echo of Hope	6.348
RFE/R Liberty	4.475
	4.505
	4.565
	4.695
	5.125
	5.295
	5.79
	5.845
	5.945
	5.98
	6.97
	6.975
	6.995
	7.44
	9.09
	9.091
	9.17
	9.25
	10.19
	10.315
	10.42
	11.575
	11.675
	12.246
	13.69
	14.712
	14.715
	15.73
	15.775
	16.065
	16.24
	17.445
	20.215
	20.71
	22.97
	25.35
R Free Portugal	14.44
R Freedom for South Yemen	9.953
R Kulmis	7.235
R of the Patriots	6.1
	9.5
	9.65
	11.7
R. Sandino	7.588
R. Yemen A Arab Rep	6.005
V of Arab Syria	9.51
V of Arabian Peninsula People	8.36
V of Black Man's Resistance	7.2
V of Communist Party of Turkey	6.2
V of Democratic Kampuchea	4.908
	6.09
	6.825
	7.35
	9.47
	9.755
	11.6
	11.725
	11.99
	15.27
	15.6
	17.705
V of Eritrean Revolution	7.24
V of Front for Redemption of Somalia	9.59

	MHz
Locations Unknown – contd	
V of Iraqi Kurdistan	5.95
	5.99
	6.94
	10.055
V of Kawthulay	4.88
V of Lebanon	6.852
V of the One Lebanon	9.51
V of Malayan Revolution	7.305
	9.62
	11.83
	15.79

	MHz
V of the Namibian People	7.18
V of NUFK	4.675
	7.015
	9.985
	10.08
	10.12
	12.006
V of Peace	6.25
V of the People	7.103
V of the People of Burma	5.110

	MHz
V of the People of Thailand	9.42
V of the Rev Party for Reunification	4.557

EUROPEAN
V.H.F. SOUND BROADCASTING STATIONS

This list includes only those transmitters in Europe with an e.r.p. of 100 kW or more, except in the case of the U.K. where all stations are listed. There are in addition more than 3,500 lower powered transmitters in Europe of which over 1,600 are in Italy. The carrier frequencies of the channel numbers in the first column are given on p. 230. Some carrier frequencies are offset from that allocated by up to 150 kHz. Stations transmitting stereophonic programmes are marked with an 'S'.

		kW
AUSTRIA		
6	Lichtenberg (S)	100
7	Schoeckl (S)	100
8	Jauerling (S)	100
9	Pfaender (S)	100
10	Kahlenberg	100
13	Gaisberg (S)	100
14	Schoeckl (S)	100
15	Jauerling	100
16	Kahlenberg (S)	100
19	Dobratsch-Villacher (S)	100
21	Pfaender (S)	100
27	Lichtenberg	100
28	Schoeckl	100
33	Jauerling (S)	100
35	Lichtenberg (S)	100
36	Dobratsch-Villacher	100
36	Kahlenberg	100
37	Pfaender	100
43	Kahlenberg (S)	100
47	Dobratsch-Villacher (S)	100
EIRE		
9	Truskmore (S)	120
22	Mullaghanish (S)	120
24	Maghera (S)	120
26	Mount Leinster (S)	120
FRANCE		
4	Carcassonne	125
6	Lille (S)	150
	Paris	100
7	Le Mans	100
	Reims (S)	150
8	Limoges	150
10	Rennes (S)	100
12	Nantes	200
13	Carcassonne (S)	125
14	Niort	200
17	Rouen (S)	100
19	Le Mans	100
20	Limoges	150
21	Paris (S)	100
22	Rennes	100
23	Rouen	100
24	Nantes	200
26	Lille	150
31	Niort	200
32	Carcassonne	125
	Rouen	100
33	Reims	150
	Le Mans (S)	100
35	Limoges (S)	150
	Paris	100
37	Lille	150
38	Rennes	100
39	Reims	150
40	Nantes (S)	200
41	Niort	200
GERMANY (East)		
9	Karl Marx Stadt	100
11	Inselberg	100
13	Marlow	100
15	Berlin (S)	100
	Brocken	100
16	Sonneberg	100
18	Inselberg (S)	100
23	Leipzig	100
24	Sonneberg	100
27	Berlin	100
	Schwerin	100

Germany (East) – contd	kW
29 Berlin (S)	100
33 Karl Marx Stadt	100
34 Inselberg	100
35 Brocken	100
Berlin	100
38 Schwerin	100
42 Berlin	100

GERMANY (West)

3 Gottelbörner Hoehe (S)	100
4 Bremen (S)	100
6 Gruenten/Allgaeu (S)	100
Heidelberg (S)	100
Langenberg	100
8 Wendelstein (S)	100
10 Harz	100
10 Stuttgart/Degerloch (S)	100
12 Gruenten/Allgaeu (S)	100
Ochsenkopf (S)	100
Teutoburger Wald	100
14 Gottelborner Hoehe (S)	100
17 Brotjacklriegel (S)	100
Harz (S)	100
Stuttgart/Degerloch	100
20 Kreuzberg/Rhoen (S)	100
21 Teutoburger Wald (S)	100
22 Wendelstein (S)	100
23 Bremen (S)	100
Waldenburg (S)	100
25 Brotjacklriegel	100
26 Stuttgart/Degerloch (S)	100
27 Langenberg (S)	100
28 Gottelborner Hoehe (S)	100
29 Gruenten/Allgaeu	100
30 Ochsenkopf (S)	100
31 Kreuzberg/Rhoen	100
32 Brotjacklriegel (S)	100
Waldenburg	100
33 Teutoburger Wald (S)	100
36 Heidelberg (S)	100
37 Harz (S)	100
38 Kreuzberg/Rhoen (S)	100
Wendelstein	100
39 Waldenburg (S)	100

	kW
41 Langenberg (S)	100
Ochsenkopf	100
43 Heidelberg	100
51 Stuttgart-Frauenkapf (S)	100

LUXEMBOURG

6 Hosingen	100
33 Marnach	100

NETHERLANDS

4 Roermond (S)	100
13 Roermond (S)	100
25 Roermond (S)	100

NORWAY

5 Oslo	100
22 Bokn	100

UNITED KINGDOM

4 Ballachulish	15W
Betws-y-Coed	10W
Bressay	10
Ffestiniog	50W
Forfar(S)	10
Llanidloes (S)	5W
4 Lochgilphead (S)	10W
Londonderry	13
North Hessary Tor (S)	60
Sandale (S)	120
Sutton Coldfield (S)	120
Wensleydale (S)	25W
5 Barnstaple (S)	150W
Campbeltown (S)	35W
Carmel (S)	3.2
Douglas	6
Newry	30W
Pontop Pike (S)	60
Rowridge (S)	60
R. Sheffield†	30W
Skriaig	10
Toward (S)	250W
Windermere (S)	20W
6 Ayr (S)	55W
Bath (S)	35W
Belmont (S)	8
Blaen-Plwyf	60
Brecon	10W
Brougher Mountain (S)	2.5
Cambridge	20W

United Kingdom – contd

Channel	Station	kW
6	Carmarthen (S)	10W
	Isles of Scilly (S)	20W
	Kendal (S)	25W
	Kilkeel (S)	25W
	Llangollen	10
	Maddybenny More (S)	30W
	Meldrum	60
	Millburn Muir (S)	25W
	Northampton (S)	60W
	Oban	1.5
	Okehampton (S)	15W
7	Ashkirk (S)	18
	Ballycastle (S)	40W
	Churchdown Hill (S)	25W
	Kingussie	35W
	Larne (S)	15W
	Llandrindod Wells (S)	1.5
	Melvaig	22
	Perth	15W
	Pitlochry	200W
	Rosneath (S)	25W
	Wrotham (S)	120
8	Fort William	1.5
	Haverfordwest	10
	Holme Moss (S)	120
	Kilvey Hill (S)	1
	Machynlleth	60W
	Orkney	20
	Oxford (S)	22
	Penifiler	6W
	Ventnor (S)	20W
9	Grantown	350W
	Hereford (S)	25W
	Kinlochleven	2W
9	Llanddona	12
	Redruth (S)	9
	Rosemarkie	12
	Tacolneston (S)	120
	Weardale (S)	100W
	Whitby (S)	40W
10	Brighton (S)	150W
	Chatton (S)	5.6
	Divis (S)	60
	Dolgellau	15W
	Kirk O'Shotts (S)	120
	Morecambe Bay (S)	4
	Peterborough	20
	Scarborough (S)	25W
	Sheffield (S)	60W
	Swingate (S)	7
10	Thrumster	10
	Wenvoe (S)	120
11	Ballachulish	15W
	Betws-y-Coed	10W
	Campbeltown	35W
	Ffestiniog	50W
	Llanidloes (S)	5W
	North Hessary Tor (S)	60
	Sandale (S)	120
12	Barnstaple (S)	150W
	Bressay	10
	Carmel (S)	3.2
	Douglas	6
	Forfar (S)	10
	Lochgilphead (S)	10W
	Londonderry	13
	Pontop Pike (S)	60
	Rowridge (S)	60
	Skriaig	10
	Sutton Coldfield (S)	120
	Toward (S)	250W
	Wensleydale (S)	25W
13	Ayr (S)	55W
	Bath (S)	35W
	Belmont (S)	8
	Blaen-Plwyf	60
	Campbeltown (S)	35W
	Isles of Scilly (S)	20W
	Kendal (S)	25W
	Kilkeel (S)	25W
	Llangollen	10
	Maddybenny More (S)	30W
	Meldrum	60
	Millburn Muir (S)	25W
	Newry	30W
	Okehampton (S)	15W
	Windermere (S)	20W
14	Ashkirk (S)	18
	Ballycastle (S)	40W
	Brecon (S)	10W
	Brougher Mountain (S)	2.5
	Carmarthen (S)	10
	Cambridge	20W
	Churchdown Hill (S)	25W
	Kingussie	35W
	Larne (S)	15W
	Les Platons	1.5
	Llandrindod Wells (S)	1.5
	Llangollen	10
	Melvaig	22

United Kingdom – contd

		kW
14	Northampton (S)	60W
	Oban	1.5
	Perth	15W
	Wrotham (S)	120
15	Fort William	1.5
	Haverfordwest	10
	Holme Moss (S)	120
	Machynlleth	60W
	Orkney	20
	Pitlochry	200W
	Rosneath (S)	25W
	Ventnor (S)	20W
16	Hereford (S)	25W
	Kilvey Hill (S)	1
	Kinlochleven	2W
	Llanddona	12
	Oxford (S)	22
	Penifiler	6W
	Redruth (S)	9
	Rosemarkie	12
	Tacolneston (S)	120
	Weardale (S)	100W
	Whitby (S)	40W
17	Grantown	350W
	Kirk O'Shotts (S)	120
	Morecambe Bay (S)	4
	Scarborough (S)	25W
	Sheffield (S)	60W
	Wenvoe (S)	120
18	Ballachulish	15W
	Brighton (S)	150W
	Chatton (S)	5.6
	Divis (S)	60
	Dolgellau	15W
	Ffestiniog	50W
	R. Leeds†	5.2
	Llanidloes (S)	5W
	North Hessary Tor	60
	Peterborough	20
	Sandale	120
	Swingate (S)	7
	Thrumster	10
19	Betws-y-Coed	10W
	Bressay	10
	Campbeltown	35W
	Carmel (S)	3.2
	Douglas	6
	Forfar (S)	10
	Lochgilphead	10W
19	Londonderry	13

		kW
19	Sutton Coldfield (S)	120
	Wensleydale (S)	25W
20	Ayr (S)	55W
	Barnstaple (S)	150W
	Belmont (S)	8
	Blaen-Plwyf	60
	Campbeltown	35W
	Carmarthen	10W
	Kendal (S)	25W
	Maddybenny More (S)	30W
	Meldrum	60
	Newry	30W
	Okehampton	15W
	Pontop Pike (S)	60
	Rowridge	60
	Skriaig	10
	Toward (S)	250W
	Windermere (S)	20W
21	Ballycastle (S)	40W
	Bath (S)	35W
	Brecon	10W
	Brougher Mountain (S)	2.5
	Carmarthen (S)	10W
	Cambridge	20W
	Churchdown Hill (S)	25W
	Isles of Scilly	20W
	Kilkeel (S)	25W
	Llangollen	10
	Millburn Muir (S)	25W
	Northampton (S)	60W
	Oban	1.5
	Perth	15W
22	Ashkirk (S)	18
	Fort William	1.5
	Haverfordwest	10
	Holme Moss (S)	120
	Kingussie	35W
	Larne (S)	15W
	Llandrindod Wells (S)	1.5
	Melvaig	22
	Orkney	20
	Pitlochry	200W
	Rosneath (S)	25W
	Wrotham (S)	120
23	Kilvey Hill (S)	1
	Llanddona	12
	Machynlleth	60W
	Oxford (S)	22
	Penifiler	6W
	Rosemarkie	12

United Kingdom – contd

		kW
23	Ventnor	20W
	Whitby (S)	40W
24	R. Derby†	10W
	Grantown	350W
	Hereford (S)	25W
	Kinlochleven	2W
	Kirk O'Shotts (S)	120
?	Redruth	9
	Scarborough (S)	25W
	Sheffield (S)	60W
	Tacolneston	120
	Weardale (S)	100W
	Wenvoe (S)	120
25	Brighton	150W
	Chatton (S)	5.6
	Divis (S)	60
	Dolgellau	15W
	Morecambe Bay (S)	4
	Peterborough	20
	Swingate (S)	7
	Thrumster	10
26	BRMB Radio (S)*	2
	Les Platons	1.5
	R. London†	16.5
	Sandale	120
27	R. Clyde*	3.4
	R. Hallam (S)*	100W
	R. Leicester†	300W
	R. Manchester†	4.2
	R. Oxford †	4.5
	Swansea Sound (S)*	1
	R. Tees (S)*	2
	R. Victory (S)*	200W

		kW
28	R. Brighton†	500W
	R. Bristol†	5
	R. Newcastle†	3.5
	R. Nottingham†	300W
29	R. Birmingham†	5.5
	Capital Radio*	2
	R. Carlisle†	5
	R. Merseyside†	7.5
30	Downtown Radio (S)*	1
	R. Hallam (S)*	50W
	Pennine Radio (S)*	500W
	Plymouth Sound (S)*	1
	R. Solent†	5
	R. Stoke-on-Trent†	2.5
31	R. Blackburn†	1.6
	R. Trent (S)*	300W
32	R. City (S)*	5
	R. Cleveland†	5
	R. Derby†	5.5
	R. Medway†	5.6
33	R. Forth*	500W
	R. Humberside†	4.5
	Metro Radio (S)*	5
	R. Piccadilly (S)*	2
	Thames Valley Broadcasting*	250W
	Wenvoe (S)	120
34	Beacon Radio (S)*	1
	LBC*	2
	Les Platons	1.5
	R. Orwell (S)*	1
35	R. Sheffield†	5.2

*Independent Local Radio
†BBC Local Radio

EUROPEAN
V.H.F. SOUND BROADCASTING CHANNELS

Channel	MHz	Channel	MHz
1	87.2–87.4	31	96.2–96.4
2	87.5–87.7	32	96.5–96.7
3	87.8–88.0	33	96.8–97.0
4	88.1–88.3	34	97.1–97.3
5	88.4–88.6	35	97.4–97.6
6	88.7–88.9	36	97.7–97.9
7	89.0–89.2	37	98.0–98.2
8	89.3–89.5	38	98.3–98.5
9	89.6–89.8	39	98.6–98.8
10	89.9–90.1	40	98.9–99.1
11	90.2–90.4	41	99.2–99.4
12	90.5–90.7	42	99.5–99.7
13	90.8–91.0	43	99.8–100.0
14	91.1–91.3	44	100.1–100.3
15	91.4–91.6	45	100.4–100.6
16	91.7–91.9	46	100.7–100.9
17	92.0–92.2	47	101.0–101.2
18	92.3–92.5	48	101.3–101.5
19	92.6–92.8	49	101.6–101.8
20	92.9–93.1	50	101.9–102.1
21	93.2–93.4	51	102.2–102.4
22	93.5–93.7	52	102.5–102.7
23	93.8–94.0	53	102.8–103.0
24	94.1–94.3	54	103.1–103.3
25	94.4–94.6	55	103.4–103.6
26	94.7–94.9	56	103.7–103.9
27	95.0–95.2	57	104.0–104.2
28	95.3–95.5	58	104.3–104.5
29	95.6–95.8	59	104.6–104.8
30	95.9–96.1	60	104.9–105.1

Praise for *The Four Purposes of Life*

"Dan's new book offers concise and practical guidance for living a meaningful life with a clear sense of purpose."

— Deepak Chopra, author of *The Soul of Leadership*

"This book skillfully delivers on its promise of bringing renewed meaning and purpose — for anyone seeking clear directions at one of life's crossroads."

— Ken Dychtwald, founder of Age Wave
and author of *A New Purpose*

"Dan Millman reminds us of the power of our essential humanity, how to touch it and use it to make this a better world."

— Thom Hartmann, author of
The Last Hours of Ancient Sunlight

"During one of the darkest periods of my life, Dan's writing in *Way of the Peaceful Warrior* extended a hand to me and helped pull me into the light. Dan is the personification of 'love, laugh, live, and give,' as you will see in his new book, *The Four Purposes of Life*. In this book, he does it again as only he can."

— Quincy Jones

"*The Four Purposes of Life* is both wise and transformative, filled with beautiful reminders and empowering encouragement."

— Jack Kornfield, author of *The Path with Heart*

"Leave it to Dan Millman to come up with a way of making purposeful decisions and helping us see deeply into the center of each moment... a graceful and wise book."

— Elizabeth Lesser, cofounder of Omega Institute
and author of *Broken Open*

"Important reading."

— Edgar Mitchell, *Apollo 14* astronaut and author of *The Way of the Explorer*

"Dan Millman, a great teacher, shares important rules to help people everywhere improve their lives, find their passion, do what they love, and live with inspiration."

— Don Miguel Ruiz, author of *The Four Agreements*

"A refreshingly practical yet profound approach to living your sacred calling."

— Peter Russell, author of *From Science to God*

"Dan Millman's new book distills pearls of wisdom with simplicity and power. It's a wake-up call to live our best life. His personal stories highlight stages we all encounter. Reminding us of our own purpose, Dan shows how to integrate all our knowledge in the present moment. It's a book we can revisit many times."

— Amy Smart, actress and environmentalist

"A wonderful book with extremely helpful practices for finding one's authentic purpose. Highly recommended!"

— Ken Wilber, author of *The Integral Vision*

"Once again, Dan Millman provides us with keys to a well-lived life. This is a lovely read, filled with meaningful insight."

— Marianne Williamson, author of *A Return to Love*

THE FOUR PURPOSES OF LIFE

Books by Dan Millman

The Peaceful Warrior Saga

Way of the Peaceful Warrior
Sacred Journey of the Peaceful Warrior
The Journeys of Socrates
Peaceful Warrior: The Graphic Novel

Guidebooks

The Four Purposes of Life
Wisdom of the Peaceful Warrior
Everyday Enlightenment
The Life You Were Born to Live
No Ordinary Moments
The Laws of Spirit
Body Mind Mastery
Living on Purpose

Inspiration

Bridge Between Worlds (with Doug Childers)

Children's Books

Secret of the Peaceful Warrior
Quest for the Crystal Castle

For further information:
www.peacefulwarrior.com

[redacted]

THE FOUR PURPOSES OF LIFE

FINDING MEANING AND DIRECTION IN A CHANGING WORLD

DAN MILLMAN

H J Kramer
published in a joint venture with
New World Library
Novato, California

An H J Kramer Book
published in a joint venture with
New World Library

Editorial office:
H J Kramer Inc.
PO Box 1082
Tiburon, California 94920

Administrative office:
New World Library
14 Pamaron Way
Novato, California 94949

Text design by Tona Pearce Myers

Library of Congress Cataloging-in-Publication Data
Millman, Dan.
The four purposes of life : finding meaning and direction in a changing world / Dan Millman.
p. cm.
Includes bibliographical references.
ISBN 978-1-932073-49-2 (hardcover : alk. paper)
1. Life. 2. Vocation. 3. Meaning (Philosophy) I: Title.
BD431.M623 2011
128—dc22 2011001245

First paperback printing, January 2016
ISBN 978-1-932073-73-7
Printed in Canada on 100% postconsumer-waste recycled paper

New World Library is a proud member of the Green Press Initiative.

10 9 8 7 6 5 4 3 2 1

To my parents,
Herman and Vivian Millman,
who gave me the freedom
to find my own way.

The purpose of life is a life of purpose.

— ROBERT BYRNE

CONTENTS

THE FOURTH PURPOSE:

ATTENDING TO THIS ARISING MOMENT

Prologue

LIVING ON PURPOSE

I think the purpose of life is, above all,
to matter; to count, to stand for something,
to have made some difference
that you lived at all.

— LEO ROSTEN

WITH THE PACE OF LIFE ACCELERATING, in a world of change, it's not easy to maintain our balance and sense of direction. Yet we strive to do so, because a sense of direction, toward a meaningful goal, may be the better part of happiness. In this pursuit, the journey may indeed matter more than the destination — but without a destination to aim for there is no journey; we can only wander.

We humans are goal seekers from infancy, drawn by the objects of our desire. But somewhere along the way, most often in the dilemmas and angst of adolescence, a sense of confusion obscures the simple desires of childhood. What we *want* is muddied by expectations about what we (or others) think we *should* do. We begin to doubt our desires, mistrust our motives, and wonder where we're going and why.

In my first book, *Way of the Peaceful Warrior*, the old service station mechanic I called Socrates suggested that all seeking — for knowledge or achievement, for power or pleasure, for love or wealth or even spiritual experience — is driven by the promise of happiness. But the search only reinforces the sense of dilemma that sent us seeking in the first place. So he advised me to replace the search for future happiness with the *practice* of "unreasonable happiness" in each arising moment.

When my seeking ended and the practice began, I came to understand that what we all need, even more than a happy feeling, is a clear *purpose* — a meaningful goal or mission that connects us with other human beings. As Viktor Frankl wrote in his book *Man's Search for Meaning*, this fundamental need for purpose and direction may be as important to our psychological growth as eating is to our biological survival.

But the duties of our daily lives leave little time to contemplate life's larger questions, except on rare occasions, in the silent hours or in times of transition or trauma when we are compelled to ask: *What do I really want? How would I know if I had it? What would happen if I got it? Is getting what I want going to take me to where I want to be?* And finally, *What is the purpose of my life?*

Maybe you've wondered why you're here on Earth or what you're here to do — what the French call your *raison d'être*, your *reason for being*, an organizing principle and sense of direction that gives shape and meaning to your life. History provides numerous examples of iconic figures like Joan of Arc, Mohandas Gandhi, Nelson Mandela, and the Dalai Lama, whose clarity of purpose drew others to their missions like moths to the light.

This book, which contains elements from my previous works, presenting them in their full context for the first time, was inspired by my own quest for a purpose in life. I once believed that my purpose was all about work, and I searched through my twenties and well into my thirties for a career and calling. It took another decade of exploration and introspection before I understood that career is only one of four primary purposes in life.

But why *four* purposes? Some might argue that our sole (or soul) purpose is learning to love — that whatever the question, love is the answer — or that spiritual awakening or surrender to God is our ultimate aim. Others point out that our primary biological purpose is family — bonding with a mate, and bearing and caring for children. Still others might propose three or five or more purposes, or even suggest that there are as many purposes as there are people. Yet just as we divide all the days of the year into four seasons, and points on a compass into four primary directions, sorting our experience into four fundamental purposes helps us to create a sense of structure to better organize our lives. These four purposes also prepare us for, and point toward, the ultimate or transcendental awakening promised by all the great spiritual traditions.

The first of four purposes we'll explore in this book — *learning life's lessons* — centers around the premise that Earth is a school and daily life is our classroom, and that our daily challenges (in the core arenas of relationship, work and finances, and health) bring learning, growth, and perspective. The value of our life experience resides in what we learn in the process. Difficult days may provide the most important lessons, helping us develop the awareness and self-reflection that lead to higher wisdom.

The second purpose — *finding your career and calling* — underscores the critical importance of self-knowledge, as well as integrating both logic and intuition, in making the wisest possible life decisions. This section also shows how the service you provide in the world can become a meaningful path of personal and spiritual growth.

The third purpose — *discovering your life path* — addresses a hidden calling you're here to explore, a personal path that for most people remains obscure. The information in this section sheds light on the strengths you possess and challenges you face, highlighting a deeper mission you're here to fulfill.

The fourth purpose — *attending to this arising moment* — brings the first three into sharp focus and down to earth, enabling you to integrate all the others with awareness and grace, here and now.

I wrote this book for anyone seeking deeper insight into themselves and their lives, but especially for those at a crossroads, facing a challenge or change, when "business as usual" no longer applies. Join me now as we explore the four key purposes that provide meaning and direction in a changing world.

The First Purpose

LEARNING LIFE'S LESSONS

• *Smarten Up* • *Grow Up* • *Wake Up*

The wise learn from adversity;
the foolish repeat it.

— PROVERB

EARTH IS A PERFECT SCHOOL, and daily life is the classroom. This idea is hardly new, but what follows will help you appreciate the full value of your life experience. And once this central premise penetrates your psyche, you'll stop seeking and start trusting — because you'll confront a higher truth: You aren't here on Earth merely to strive for success; you're here to *learn* — and daily life is *guaranteed* to teach you all you need in order to grow, and evolve, and awaken to your higher purpose here.

You are evolving even now — and there is no way to fail as long as you continue to learn. Like a stone slowly polished by the river's flow, we're shaped over time by the currents of life. The Indian saint Ramakrishna once observed, "If you try to open a walnut when the shell is still green, it is nearly impossible. But when the walnut is ripe, it opens with just a tap." The classroom of daily life serves your ripening process.

SOME YEARS AGO I RECEIVED A LETTER expressing a dilemma shared by many: "Since reading your first book I've had a growing interest in spiritual practice, but how can I find the time when I have a wife and three children and a full-time job?" I wrote back to remind the writer that his wife and children and work were his most important spiritual practices — because true practice is not separate from daily life but rather its very substance. As writer Adair Lara put it, "And some, like me, are just beginning to guess the powerful religion of ordinary life, a spirituality of freshly mopped floors and stacked dishes, and clothing blowing on the line."

Daily life, and the journey it represents, will remain your master teacher. This book serves as a map to help orient you along the way and guide you through the school of this world.

Here are some reminders about how the process of learning works in the classroom of daily life:

Lessons repeat themselves until we learn them. Sometimes we hear a wake-up call but prefer to pull the covers over our heads so we can slumber and dream a little longer. We may ignore, resist, rationalize, or deny reality for a considerable time. That's okay — our schooling is entirely self-paced, but the lessons keep coming until our actions change.

If we don't learn the easier lessons, they get harder. Resistance to learning (or change) reaps more dramatic consequences over time — not to punish us but to get our attention. As Anaïs Nin wrote, "There came a time when the risk to remain tight in a bud was more painful than the risk it took to blossom."

We learn and grow through challenges, and every adversity has hidden gifts. We've all experienced physical, mental, and emotional pain. Yet each challenge has brought a greater measure of strength, wisdom, and perspective. We may not welcome a challenge or embrace unexpected change, loss, or disappointment, but looking back, and over time, we come to appreciate the gifts of adversity.

Oh, please! you may think. *I've heard all this before*: "When God closes a door, God opens a window." "Every cloud has a silver lining." "What doesn't kill us makes us stronger." But I will not spout platitudes or preach "positivity." The fact is, pain and difficulty are no fun — I speak from my own experience.

Some years ago, when a motorcycle crash left me with a shattered thighbone, my recovery was difficult; it also changed the course of my life. I started asking bigger questions and opened to new paths that might have otherwise escaped my notice. And climbing out of that deep hole gave me the strength to scale inner mountains. I certainly don't recommend fractures as a method of personal growth, yet I've learned to view every difficulty as a form of spiritual weight lifting — and we're all in training.

My injury was one of millions of difficulties that people suffer every day on planet Earth, from debilitating illness to childhood abuse to soul-wrenching poverty. Yet the potential for growth remains. In the November 26, 2010, "Letters" section of the *New York Times Magazine*, Betty Rollin wrote:

> 35 years after my first mastectomy and 26 years after the second, I'm feeling oddly cheerful about the whole cancer experience — not that I'd recommend it. But I slowly came to realize that just because something is awful doesn't mean you can't benefit from it. Awful things happen to a lot of us, and it's a happy moment when you start noticing some kind of payoff. Cancer survivors, for example, notice that they're breathing, the way other people don't. And because they're breathing, they're grateful in a way a lot of people aren't, and grateful is a good place to wind up in life. It beats Poor Me.

We all slip into the "poor me" mode at ti
the person who once asked me, "Why does l
be so difficult?" I answered, "You want life t
Then don't get married or have children; don't take on any responsibilities or form any attachments; always do the minimum, never volunteer; live small. Your life will then seem easier. But are we humans here on Earth to live the easiest possible life? Or are we here to grow stronger and wiser?" I believe that St. Augustine knew the answer, which is why he said, "Lord, I ask not for a lighter load, but for stronger shoulders."

Of course we can find easier and more efficient ways to accomplish our goals — no need to seek out adversity or create unnecessary complications. But meanwhile, life will continue to serve up a varied menu of challenges that contribute to our growth and evolution. Events will arise that are neither predictable nor controllable; we can't alter the waves that roll in, but we can learn to ride them. As my old mentor Socrates once observed with a shrug, "Sometimes you get the elevator, and sometimes you get the shaft." Yet, as cancer survivor Betty Rollin and countless others understand, unexpected benefits can emerge out of great adversity, which is why we sometimes volunteer for it.

Voluntary Adversity

Choosing to train in sports or music, preparing for college or graduate school exams, acting, painting, and practicing other arts are all forms of voluntary adversity, as we must confront inner and outer obstacles along the way, facing self-doubt and frustration. The demands of training reveal our weaknesses and develop our strengths. So commitment to any endeavor can become a path of personal growth.

Daily life provides other forms of voluntary adversity as well. For example, maintaining a long-term relationship and raising children, for all their pleasures and joys, present significant challenges; so do responsibilities in the business world and caring for elderly parents. There's something for everyone in the school of daily life.

Many forms of adversity are *involuntary*: a drunk driver slams into us, we lose our job or a loved one, or we get a cancer diagnosis despite healthy eating and exercise. But we do volunteer for many, if not most, of our daily challenges — we sign up and step willingly into the fray. For example, you have chosen (or made choices leading up to) whom you're with right now, where you live, and what you're currently doing. By taking full responsibility for your past choices, you reclaim the power to make new ones. And as you make the best of your life, your life will also make the best of you.

As Socrates reminded me in *Way of the Peaceful Warrior*, "The way itself creates the warrior. We are all peaceful warriors in training, and every life is a hero's journey."

The School Rules

In addition to our core curriculum (which I'll soon present), our Earth school has a clear set of rules, also known as natural or universal laws. These school rules reflect reality — the mechanics of our universe. They also provide guidance about how to function wisely and well in the course of our schooling. So the better we understand and respect these laws, the smoother goes our education.

In science, universal laws are described with mathematical precision, as in the equation $E = mc^2$. In religion, these laws take the form of, for example, the Ten Commandments, the teachings of the Talmud or Koran, or the Golden Rule.

These laws, or school rules, are based not on moral concepts but rather on action and consequence. While some claim that morality comes from above, others suggest that morality is a human invention. Moral precepts differ from one culture or era to another, which is why philosopher Bertrand Russell proposed that "sin is geographical." But the wise among us have observed that actions lead to consequences and lessons. Whether we learn these lessons — or whether they have to repeat themselves — depends on our current readiness and openness.

The natural world reveals and reflects universal laws in their purest form. Observe, for example, how trees need strong roots but flexible branches if they are to withstand powerful winds; how streams flow gracefully around obstacles, finding the path of least resistance; and how the seasons change in their natural order.

The school rules, revealed in the course of daily life, provide a reality-based approach to living that I call the *peaceful warrior's way.* By observing and aligning our lives and actions to these laws of reality, we learn, grow, and evolve with amazing grace.

Humanity's Curriculum

If Earth is a school, what courses do we need to pass in order to graduate? Some years ago, the answer came to me in the form of twelve core subjects — a specific curriculum hidden within (or behind) the activities of everyday life. These subjects include self-worth, discipline, well-being, money, mind, intuition, emotions, courage, self-knowledge, sexuality, love, and service, which I outline below in the catalog of "Required Courses."

In this perfectly designed, fully integrated curriculum, all subjects are equally significant, and all are required. We may be tested in the area of health or money one day and encounter a pop quiz in courage the next. There are no grades, only pass or fail. A pass indicates progress; a fail provides an opportunity to learn and do better the next time. (So there is no permanent failure, only the need to repeat the course and continue with that line of study.) Bear in mind that spiritual life begins on the ground, not up in the air — even Olympians start with the basics. As your skills improve in each of these areas, you'll begin to experience the true meaning of success in the form of higher states of clarity, energy, and action.

Required Courses

As you read the following summary descriptions of the twelve core subjects in life's curriculum, reflect on your current progress in each, as well as on what improvements are still possible on the path to self-mastery.

Foundations of Self-Worth: Getting Out of Your Own Way

Life provides certain opportunities and choices, yet we allow ourselves to receive, achieve, or enjoy such opportunities only to the extent that we believe ourselves deserving or worthy of them. A well-known parable advises, "Ask and you shall receive." The question is, what are you willing to ask for and strive for? As Ramakrishna put it, "An ocean of bliss may rain down from the heavens, but if you hold up only a thimble, that is all you receive." So if we operate on the belief that "beggars can't be choosers," then few options appear. Low self-worth is a primary cause of self-sabotage. Few of us consciously sabotage our relationships or finances, but we may wonder at times, "Why did I say that? Why did I do that?" This first course in the school of life helps us to shift from holding up only a thimble and saying, "Oh, I really couldn't accept that" to opening our arms and saying, "Yes! Thank you!" Realizing your innate worth expands your horizons and opens you to a larger life.

Roots of Will: The Practice of Self-Discipline

Most of us know the value of regular exercise, good diet, kindness, relaxation, and breaking unhealthy habits. Our greatest single challenge in every area of self-improvement is transforming knowledge into action — turning what we know into what we actually do. While some of us act without thinking, too many of us think without acting. Willpower isn't a mysterious force that descends on us from above; rather, it is an innate power within each of us, waiting to be applied. We do so every time we accomplish a task (like taking out the trash or doing the laundry, office work, or schoolwork) despite a lack of motivation. So this subject, which you face daily, calls you to reclaim your will by following through with your goals. As your skills and understanding improve, you better appreciate that your life will be shaped largely by what you do each day — whether or not you feel like doing it.

Well-Being: Approaches to Health and Vitality

Your body forms the foundation of your earthly existence and is the only possession you are guaranteed to keep for a lifetime. Its care and feeding are key to all else. An energized body enhances strength, mental acuity, healing, social interactions, and every other human capacity. While genetics plays a powerful role in health and longevity, you have wiggle room — it's called *lifestyle* or daily choices. The previous two foundation courses in the school of life — self-worth and self-discipline — open the way to an energizing, healthful lifestyle. This course, over time and through your own experience, teaches the wisdom of regular, moderate exercise, a balanced diet, and sufficient rest — amplifying energy for all other courses in the school of life.

Money and Values: Establishing Stability and Sufficiency

Earning, spending, and saving money preoccupy many of us, who spend a good portion of each day working for more of it. But for those of us who have a religious or spiritual interest, money remains suspect — a topic of mixed feelings best expressed by the late boxing champion Joe Louis, who said, "I don't really like money, but it calms my nerves." We eventually learn that money is a form of energy that only makes us more of who we already are — it can bind us or free us, depending on how we manage it. This course cannot guide everyone to great wealth but rather shows the way to create sufficiency and stability, sometimes even leading to the deep satisfaction of philanthropy. As the Arabic proverb goes, "If you have much, give of your wealth; if you have little, give of your heart."

Exploring the Mind: The Nature of Your Inner World

This subject in the core curriculum helps us to understand the illusory nature of our subjective mind. Lao-tzu advised, "As soon as you have a thought, laugh at it," because reality is not what we think. We perceive the world through a window colored by beliefs, interpretations, and associations. We see things not as *they* are but as *we* are. The same brain that enables us to contemplate philosophy, solve math equations, and create poetry also generates a stream of static known as discursive thoughts, which seem to arise at random, bubbling up into our awareness. Such mental noise is a natural phenomenon, no more of a problem than the dreams that appear in the sleep state. Therefore, our schooling aims not to struggle with random thoughts but to transcend them in the present moment, where no thoughts exist, only awareness. Our mind's liberation awaits not in some imagined future but here and now.

Introduction to Intuition: Accessing Subconscious Guidance

Whispering beneath our everyday awareness is a childlike consciousness — a shaman, a mystic, a weaver of dreams and keeper of instincts. Our subconscious mind (or body-wisdom) holds keys to a treasure-house of intuitive guidance and survival skills. It helps us to make fully informed decisions, and may save lives in times of danger. Daily life eventually teaches us the value of trusting intuitive messages and, in the words of Zen sword master Taisen Deshimaru, learning "to think with the whole body." Nearly all scientific discoveries come from creative flashes of intuitive insight (later tested and verified using the scientific method). For this reason Einstein proposed that imagination is more important than knowledge. As we learn to let intuition become reason's trusted adviser, we integrate reason and faith, building a bridge to wisdom and a new way of sensing our world.

The Nature of Feelings: Achieving Emotional Freedom

Emotions play a central role in human life. Behind all seeking is the desire to feel *good* (happy, content, fulfilled, confident) *more of the time* — and to *feel bad* (anxious, sad, angry, fearful) *less of the time*. In the United States and abroad, people spend billions of dollars on self-medication and buy countless books offering to replace bad feelings or thoughts with good ones. Meanwhile, emotions continue to rise and fall like waves on the sea, of their own accord, passing like clouds in the sky. But over time, the school of reality teaches us to accept, value, and learn from our emotions without trying to fix or control them — and without letting them determine our behavior. Life reveals that we have more control over our behavior than we do over transient emotions or thoughts. This realization helps us liberate our lives from reactive or confused soap operas and establish stable, mature, and responsible behavior. We learn to ride the shifting tides of emotion like skillful surfers as we grasp the great truth that we don't need to *feel* compassionate, peaceful, confident, courageous, happy, or kind — we only need to *behave* that way.

Fundamentals of Courage: Confronting Our Fear

In this subject area we confront the primal emotion of fear, which can lead to paralysis or power, and can end a life or save it. Fear can warn us of genuine danger; it can move us to prepare well, take precautions, or avoid a situation. It can also burden our lives with anxiety or self-doubt. How many of us avoid opportunities due to fear of the unknown? We may also fear rejection, failure, and even success. In the classroom of daily life, we learn to listen to fear's counsel when our physical bodies are at risk. But when we experience a more subjective fear of embarrassment, looking foolish, or feeling rejected, we feel the fear but act with courage. Courage is not the absence of fear but the conquering of it. Heroes feel the same fear as cowards; they just respond differently.

Knowing Yourself: Finding Wholeness in the Shadows

In childhood, our power and charm come from authenticity — our actions are undiluted by subterfuge or hidden agendas. But as the years pass, we learn to tell social lies and use pretense to please or placate others while disowning the disapproved-of parts of our psyche. In this way, we create a shadow persona constructed from fragments of our full identity and potential. But as we mature and begin to grasp that we are often the cause of our own difficulties, we begin a process of compassionate self-observation leading to deeper self-knowledge — denial gives way to authenticity as the light of awareness penetrates our shadow. We come to accept ourselves (and others) as we are rather than as we might want ourselves (or them) to be. And as we embrace the full scope of our humanity, we open the way to genuine growth and transformation.

Your Sexual Life: Understanding the Pleasure Principle

Our drive for sexual intimacy is as natural as thundershowers or the changing seasons. But if we suppress or exploit our sexual energy, we create obsessions, compulsions, and guilty secrets. This course of study in the realm of relationship shows us how to observe, accept, enjoy, and channel sexual-creative energy rather than merely indulging or denying it. Whether we are young or old, sexually active or not, an abuse survivor or "perfectly adjusted" — and no matter what our sexual orientation — we all have areas to explore. Sexual issues are rarely about the sexual act. More often than not, our complications stem from unrealistic beliefs, fear, insecurity or jealousy, and issues of identity. Our coursework in life includes, but also goes beyond, our physical drives into areas of awareness, balance, trust, openness, honesty, and the courage to achieve true intimacy.

The Mastery of Love: Awakening the Heart

Despite the efforts of poets and philosophers through the ages, love defies definition. It may appear as sentimental infatuation or as a chemically induced hormonal state, sexual pleasure, familial devotion, or anything that makes us feel good. We know that the word *love* can be either a noun or a verb, and that love is something we can feel and also something we can do. Our life experience reveals the evolving nature of love as it changes from an emotion that happens to us, rising and falling out of our control, to an art we can learn — from something we receive to something we give. As we mature, this practice of love liberates us from dependence on the changing tides of emotion. We gain the capacity to show loving-kindness to others even when we don't feel like it. This is the power of love and the heart's education.

Service and Meaning: Completing the Circle of Life

We all know that service is a good thing — that serving others lifts us out of exclusive preoccupation with the self and turns our attention out into the world. But service is not only something we do for others; it is also an act of self-transcendence — anything we do without regard to self-interest. This final course in our schooling provides an essential link that fulfills all the others. Service is a form of yoga, a catalyst of friendship, an affirmation of our common humanity. Even the smallest gesture of self-sacrifice — giving of our time, energy, or attention — shifts our focus from "What's in it for me?" to "What is for the highest good of all concerned?" Every service we provide for others represents the fruition of our spiritual maturity, completing both our curriculum and the circle of life. In service we find the master key to a meaningful, purposeful life that connects us to one another, and to our world.

BEFORE YOU MOVE ON TO THE SECOND PURPOSE, take a few moments to appreciate what you've already accomplished in each of the twelve subject areas in Earth's core curriculum — self-worth, discipline, well-being, money, mind, intuition, emotions, courage, self-knowledge, sexuality, love, and service — all the tests you've taken and lessons you've learned. But don't get so wrapped up in the twelve subjects we've covered that you lose sight of the larger purpose — *learning life's lessons* — which leads to greater wisdom, perspective, and appreciation for the value and significance of your daily life.

The Second Purpose

FINDING YOUR CAREER AND CALLING

• Choose Satisfying Work

• Earn a Good Living • Provide a Useful Service

Whatever your life work is, do it well.
Do it so well that no one else could do it better.
If it falls on your lot to be a street sweeper,
sweep streets like Michelangelo painted pictures,
like Shakespeare wrote poetry,
like Beethoven composed music;
sweep streets so well that
all the hosts of heaven and earth
will have to pause and say,
"Here lived a great street sweeper."

— MARTIN LUTHER KING JR.

WHETHER YOU LIVE IN A WELL-HEELED SUBURB, inner city, or rural village, you possess twenty-four hours each day, about sixteen of which you spend in more or less of a waking state. And, like most people, you'll spend many of your waking hours engaged in or pursuing activities that provide a sense of purpose as well as some combination of material, psychological, and emotional rewards. Whether you are currently in school, working full- or part-time, between jobs, contemplating a change, or even approaching retirement years, the second purpose gives a fresh look at the form of service you provide (or might like to provide), consistent with your interests, skills, and values.

The Essence of Career and Calling

Because most people pondering their "purpose in life" are thinking about either their career or their calling, let's clarify the meaning of these two terms.

Your *career* refers to a service you perform — trading your time, effort, attention, knowledge, skills, and experience for a salary or other income and benefits. You may refer to it as your employment, work, livelihood, occupation, living, trade, vocation, profession, or "just a job." You may have many reasons to go to work each day — but unless you are independently wealthy, earning an income is a primary career motive.

Your *calling* refers to a personal interest, attraction, inclination, drive, or passion that is usually (but not always) of a higher order. It isn't just something you want to do, but rather something you *need* to do, something that captures your imagination, touches you deeply and absorbs you, whether or not you can explain why. A calling may (or may not) earn an income or become a career.

A calling can take the form of an art, craft, or other creative endeavor, such as writing, painting, or playing a musical instrument. Or it may involve volunteer service, such as teaching, working with children or the elderly, or charitable work. Some people, wanting to make a difference in their community or the larger world, are called to a religious order, others to military service, politics, or environmental (or other) causes. Parenthood — minding the home front and raising children — may be one of the highest and most fundamental of callings.

Since a true calling is often associated with serving others, personal leisure activities such as golf or bowling, hunting or fishing, sewing or reading, knitting or building miniature ships fall into the realm of hobbies or avocations. But if we end up performing or teaching that hobby, sharing it with others, then our avocation may become both calling and career — an absorbing professional path of learning and growth.

One child's interest in magic tricks may fade with the arrival of young love or some other passion; another child may end up a professional stage magician. One teen's love of video games may wane with age; another young person may become a gifted game designer. Similarly, many people have an interest in yoga, martial arts, or another fitness practice — but for some pioneers and popular teachers, their exercise immersion evolved into a lifetime calling and career.

The primary difference between a career and a calling is that we pursue a career primarily for income and a calling primarily for innate satisfaction. But if you love your career so much that you'd do it for free (if you could afford to do so), then it has likely become a calling as well. And if a calling begins to produce a good income, then it has also become a career.

Why are such distinctions important? Because many of us cling to a calling yet struggle financially because we ignore or resist the practical need for an income-producing day job, insisting, "I must be free to follow my heart and devote my life to my art." Others among us focus so much on climbing a career ladder to success that we abandon a life-affirming calling that might bring even more joy and meaning to our life.

For some of us, career and calling have merged into one; for others, they remain separate and distinct. One way is not necessarily better than another. We each have our own unique process.

Stories of Career and Calling

The following stories highlight different approaches to career and calling. Although I've changed the names and a few identifying details in the interest of privacy, these are true stories about real people — a tiny sample to highlight the variety of experiences people meet on the winding path to their career and calling.

I met Megan Caldwell at Oberlin College, where I coached the men's and women's springboard diving team. I'd rarely encountered a student-athlete who so loved her sport, or had more fun at it, than Megan, who was dedicated to perfecting her diving skills. (I was only vaguely aware that she was majoring in mathematics and physics.) That first year she won the Women's Conference Championships on the three-meter board.

After that, I moved on to another phase of my life and we were out of touch for two decades. During that time, Megan's life moved from reverse dives to inverse problems of acoustics and electronics. She earned a PhD from Indiana and did postdoctoral work at Stanford, eventually working on advanced radar technology, in which she transformed numbers representing voltages to coherent radar images for national defense.

Megan had established a strong career — but for nearly twenty years, except for occasional practice sessions, her athletic calling faded to fond memories. Then, in 1997, when she was in her early forties and doing research at the University of Minnesota, she found a course in beginning springboard diving and signed up. The instructor, seeing her skills and form, suggested that she compete in Masters Diving in her age group. She did so, and won several national titles and a Masters World Championship on the three-meter board.

Even as Professor Megan Caldwell was dedicated to her career, she remained devoted to her calling — springboard diving. Over a hot meal on a cold autumn evening in Berkeley, California, where she was a visiting professor at a math institute, I learned that she made time — three days every week — for a two-hour commute each way, by bus and train and on foot, from Berkeley to Stanford to practice with other Masters divers — on sunny days and in pouring rain (thank goodness for the hot-tub respites between dives!). She was then in the 55–59 age group, but her form, and love of her calling, remained that of the young woman I had first met all those years before.

Megan's life was blessed, but those blessings came from years of work and application. Her hard-earned career and calling were worlds apart, yet each has brought a special kind of satisfaction.

The following story makes quite a different point as it relates how a late bloomer found his calling and transformed it into a career with a surprising twist.

Kevin Kohler found his calling early on but showed little tolerance for paycheck-driven work. Kevin's passion during high school and college was the game of Ultimate Frisbee. His many hours throwing the flying disk led to a certain expertise, but his pastime showed little promise as a profession.

Eventually, Kevin's parents suggested that he move out of his childhood bedroom and into his own apartment

— after all, he was by this time thirty-two years old. Soon after, while he was taking a hot shower, an idea popped into Kevin's mind. Thrilled by his revelation, he quickly dried off, dressed, and made a call to the Wham-O Corporation, which manufactured the Frisbee, and finally got through to a decision maker in their marketing department. "Here's my idea," said Kevin. "I'd like you to give me five hundred free Frisbees with the words *World Peace* written in both English and the Russian Cyrillic alphabet. Then I'd like you to pay my way to Russia and put me up for a month. What I'll do for you is to become a Frisbee goodwill ambassador — I'll go to Red Square every day, once we get permission, and I'll teach people to throw Frisbees. It will be a great cultural exchange and help open up a market for you."

This was back in the 1960s, during the Cold War. The company agreed, since it was not a big investment and might do some good. Kevin traveled to Russia (then part of the USSR), learned to speak the language, and ended up leading numerous Frisbee goodwill tours there. He even married a Russian woman.

Since Kevin couldn't find work that suited him, he did what he loved and got someone to pay him for it. His calling, for some years, became his career.

Not many of us will materialize a career (and calling) based on an idea that appears in a flash, but Kevin's life testifies to what's possible.

The story of Stuart Anders represents yet another approach to career and calling.

I met Stuart Anders for the first time when I began my four-year tenure as head gymnastics coach at Stanford University. With enthusiasm and zeal, I prepared to do whatever it would take to transform what at that time was a weak team into top-caliber athletes. I was twenty-two years old and, like many athletes my age, thought I was bulletproof. The year before, I had co-captained the UC Berkeley gymnastics team to their first National Collegiate Championships, so my standards and expectations were high.

Before my first day meeting the team, the athletic director drew me aside and explained, "Dan, for the past ten years, a man named Stuart Anders has been showing up regularly as a volunteer assistant coach. I realize that you don't know him, but he's a good guy and totally reliable. He only did a little gymnastics years ago, but he loves the sport. It's your call, of course, but it would be a nice gesture if you'd let him come in and help in any way he can." I said I'd be glad to meet with Stuart and see how it went.

As it turned out, he was a relaxed, easygoing guy with a likable personality, who did show up on time, every day. We didn't have much time to talk personally, since we were both focused on training. But it seemed that we'd get on fine.

Then one day a month or two later, Stuart arrived about an hour late. He apologized, explaining that he'd been out flying and a complication had kept him from arriving on time. Curious, and a little surprised that Stuart had a pilot's license,

I asked, "What kind of plane were you flying — a Cessna or Piper Cub?"

"It's a larger craft," Stuart answered. "Boeing's newest, called a 747. I was checking its glide path." It turned out that Stuart was an aeronautical engineer and test pilot for NASA who worked at Moffett Federal Airfield in nearby Mountain View. I also learned that he restored old Porsches as a hobby and was building a one-man experimental jet, rivet by rivet, in his garage.

I quickly lost my twenty-two-year-old sense of self-importance, thinking about how my volunteer assistant had apologized for being a little late to practice. I had known Stuart only through one of his callings — helping young gymnasts hone their craft — even as he devoted most of his day to another calling and professional career, testing aircraft on the cutting edge of flight technology.

Now we turn to someone who chose a different path, following her heart's wisdom and a call to service.

Julia Marsala is an intelligent and industrious woman who loves to read and stay well informed. In high school she maintained close to a straight-A average, but in college, like many students in liberal arts, she lacked a clear career objective or path and changed majors several times. Despite her lack of career direction or ambition, Julia demonstrated a strong work ethic and an inclination for service, which would become a lifelong calling. As a child she helped out every day after school in a small market; in high school and college she worked at a variety of jobs out of the practical necessity of earning money. The type of work was less important than the idea of being useful. This service orientation is where Julia derived satisfaction, making her labors, however humble, feel worthwhile and meaningful.

After graduating from college Julia volunteered to teach at a school while earning money as a waitress. As the years passed, she remained indifferent to the idea of finding a profession. In fact, she never had to do a career search because work always seemed to find her. When Julia married, she worked as a bookkeeper and helped support her husband as he searched for his own career and calling.

With the birth of her first child, Julia's calling shifted naturally into the role of mother. For a time she consciously chose a life as a homemaker, while her husband focused on producing income. He was the breadwinner; she was the bread baker, providing a stable emotional foundation for her family. She embodied the saying "There's no such thing as a nonworking mother." Julia's life was not defined by or confined to the walls of her home. When her children started school, she worked part-time until her husband's income grew. And throughout her children's school years prior to college, she volunteered at school functions. Her life was not only full; it was a full-time job.

Over the years, Julia developed an interest, and then some expertise, in mythology. She took numerous courses, read many books on the subject, and did her own research. She ended up speaking to a few small groups on mythology. But her studies remained a calling, not a career — one facet in a wide variety of roles, as she continued her higher calling of service to her husband, her children, and the world around her.

Not everyone chooses to have children, but those who do so may take a break in career pursuits to raise children as their primary service. Some may later return to a career outside the home; others find a lifelong calling as parents. Still others, like Julia Marsala, live full and meaningful lives without having to define themselves by a career, serving in whatever circumstances life presents.

The following narrative describes the multiple — and passionate — careers and callings of one of the most intriguing people I've known.

I was a college professor teaching various forms of physical training when Charles Edwards enrolled in my trampoline course. He struck me as an unlikely candidate for any gymnastics practice, since he was built more like a football tackle or grizzly bear. He was, in fact, a football and a field hockey player and a member of the college aikido club who worked part-time as a motorcycle mechanic. It turned out that Charles was also a bright young man. Near the end of his senior year, he applied to a number of medical schools.

Charles and I lost touch for nearly a decade. Then I learned that he had become not only a doctor but a top pediatric heart surgeon, performing some of the early transplants. He married Marie, also a physician and researcher, and eventually they had three children. Charles had achieved a third-degree black belt in aikido and was at the top of his profession when he decided to move from the fast track of the big city to a smaller practice in a rustic location.

Life went smoothly for Charles and his family for some years. Then, as he was working in his garage with a circular saw to make a birdhouse for one of his sons, he stepped on a spot of oil. His leg slipped out from under him and he fell. When Charles sat up, he discovered that two fingers from his right (dominant) hand were missing. As he recovered from the acute injury, he practiced intensely using his left hand to tie sutures and use instruments, but it became apparent that he would never perform major surgery again.

After some soul-searching, Charles decided to start law school. Perhaps it isn't surprising that he edited his school's law review and graduated with honors. Specializing in medical law, for years he served in hospital administration.

Charles had become a respected physician, attorney, and hospital policy maker as well as a high-ranking aikido sensei. As he and his wife watch their grown children move into their own careers and callings, Charles is now considering a new career and calling — divinity school. He exemplifies not only a life well lived in the service of others but also a life reinvented in response to changing circumstances.

I once asked Charles, "Where do you find the time to accomplish all that you do?" His reply might be a reminder to us all: "I never *found* the time, so I made some." Charles's life shows what is possible when a passionate engagement with life combines with a focused application of energy. His every endeavor has become his calling.

Despite the saying that "life begins at forty," we can sometimes experience a rebirth decades later. Retirees who have completed their career arc may find a new calling. Take the case of Bud Gardner.

Bud Gardner, former college English teacher, writer, and writing coach, prepared to follow his heart and play golf into his retirement years. Then he read a study about how playing a musical instrument late in life was good for aging brains (and spirits). So he surprised himself by buying a harmonica.

It wasn't entirely out of the blue; he had played old favorites on the mouth organ for sixty years, ever since his dad had taught him. Soon bored playing the same old three songs, he placed an ad in a local paper hoping to find someone to teach him more. After twenty people showed up at their first meeting, the "Harmonicoots" group — the Coots, for short — was born. For the seven years since then, the Coots — sixty men and women over fifty-five — have met weekly with three goals: having fun, learning new songs, and playing together.

They have since played more than 250 gigs in retirement homes, hospitals, parades, elementary schools, and churches, often bringing tears to grateful listeners. The Coots now have a mission to "entice the world" to the joys of playing, inspiring and exciting young and old to a lifetime of musical enjoyment. They've helped hospital residents improve their breathing capacity, and they've played carols on harmonica over the holidays. Some of the members have traveled worldwide.

What began as a postretirement whim turned into a new calling — and could have also become a late-blooming career, except that Bud and the Coots use any income they make to purchase harmonicas that they donate to elementary school students. Thanks to the Coots, these students enjoy a fresh breath of life.

This final story relates how a young man followed a call to accomplish something great against all odds.

In 2001, during a severe drought in his village in Malawi, fourteen-year-old William Kamkwamba was forced to drop out of school because his family couldn't afford the tuition. It was all they could do to sustain themselves on one meal a day from their meager farm income.

Young William spent his time in a nearby library, fascinated by a book on windmills. Not knowing any better, he believed that he could build a windmill for his village, assembled from old car batteries, bike parts, tractor fans, and plastic pipes. Spent wood from local blue-gum trees would serve as a tower. His parents and everyone else thought he'd lost his reason, but their doubt only increased this young man's determination.

Three months later, William illuminated his family's home with a lightbulb powered by his first windmill. He later built four more in his village, including one at a local school where he taught others how to build windmills. This resulted in electricity for the village, which enabled them to pump in their own water — a gift that became a village treasure.

William's story and the stories that preceded it are but a tiny sampling among millions of stories of career and calling, as interesting and varied as the people on our planet. Yet the most significant story is your own. Your personal memories are your treasures — and each story, each memory, can provide teachable moments.

It doesn't ultimately matter whether your career and calling are united or are two separate parts of your life. In an ideal world, career and calling might merge — we would feel drawn, as if from above, to do the work we do each day. But this is the real world, where not every calling becomes a career or every career a calling. Most of us go to work, put in our time, enjoy aspects of our job, then look forward to doing what we do for love alone during our discretionary time.

There are, after all, benefits in *not* centering your life around your career. When your work is "just a job" that you leave behind each evening, you aren't as likely to get overly stressed or define your worth by the work you do, even as you strive to do your work well. Your family may also benefit from the extra time, attention, and energy you have to spend with them.

The balance between career, calling, and family will naturally change over time, so reevaluating and fine-tuning this balance can help transform midlife crises into midcourse corrections and create a space for refueling and recharging. Maintaining such balance involves a process of self-examination and insight that ripens over time.

Self-Knowledge and Career Decisions

Thousands of years ago, the Athenian lawgiver Solon offered two words of guidance that have echoed through the ages: *Know thyself*. These words were inscribed in the forecourt of the Temple of Apollo at Delphi and also attributed to numerous sages, including Heraclites, Pythagoras, and Socrates (who proposed that "the unexamined life is not worth living"). Why would such sages recommend self-awareness above other human qualities? Perhaps it's because self-knowledge informs all our choices and decisions (in education, careers, and relationships) and thereby shapes the quality of our lives.

In high school and college we begin, by necessity, to learn something of our interests and capacities. We begin to sort out whether we're more of a workhorse or an innovator, and whether we lean toward science or liberal arts. But it can take time to discern what we really *want* to do as opposed to what we think we *should* do (to please parents or to garner status or respect).

If self-knowledge were easy or automatic, then most of us would make clear and wise decisions from our early years about schooling, career, mate, and so on. But all too often, through our twenties, our choices are ill-informed — a mixture of hope and guesswork. Jobs and relationships become a series of learning experiences as life doesn't work out as we might have hoped or planned. (If only we had known then what we know now...)

The majority of twenty-somethings leave college with limited self-knowledge, a bachelor's degree, and a big question: *Now what?* Throughout the "trying twenties" — when we try this, then try that — we strive to understand ourselves enough to find our place in the world.

Deciding on a Career

Decision making is a fundamental life skill, and (as with choosing a life partner) your career choice is one of life's major decisions. Decisions demand that you come to know yourself. Otherwise, you may make the right choice for the wrong person — the one you hoped or believed you were rather than the one you are.

There is no best career — only the best one for you at a given time of your life. There are both satisfied and dissatisfied people in medicine, plumbing, hairstyling, law, sales, teaching, and every other line of work. Life is an experiment, a laboratory of self-exploration. So, until you find a truly satisfying vocation, just get a job. Choose the best option available now. Meanwhile, stay open to new opportunities, until you find a career or calling you are ready to commit to for a significant period of time. Today, it's not unusual to go through several careers in the course of our working lives.

My Own Search

In my own search for a career and calling, I traveled a winding path. After college graduation, I sold life insurance in Los Angeles for two months, until I was offered a head coaching position at Stanford University; after four successful years, I was invited to join the faculty at Oberlin College. Professionally, it felt as if I was leading a charmed life — a head coach at twenty-two, then a college professor — until I decided to return to Berkeley to teach at a spiritual institute for six months.

Then my life changed considerably. I taught gymnastics part-time at the Berkeley YMCA (three months), went on unemployment (two months), nearly joined the Navy but ended up in a spiritual community (on and off, eight years), during which time I worked as a typist-secretary (was fired after four months), taught at a small gymnastics studio in San Francisco (six months), moved to Atlanta to coach a girls gymnastics team (nine months), attended graduate school in psychology in

California (four months), started court reporting school in San Francisco (six months), taught a martial arts/acrobatics class at the Berkeley YMCA (three months), served as the women's gymnastics coach at UC Berkeley (two years), and returned to Oberlin College (two years), where my wife, Joy, served as dorm director and where I wrote the final draft of *Way of the Peaceful Warrior*. Soon after publication, the book went out of print — and I returned to the San Francisco Bay Area, where I found temporary work doing data entry for Bank of America, then spent the next eleven months word processing for a management firm before moving on to two typing jobs — one from 5 to 8:30 AM and another from 9 to 5. The long workdays lasted eight months, until I found an administrative position, which lasted a year. Then, in 1983, I started my own personal fitness coaching business.

It wasn't until 1984, at age thirty-eight, when *Way of the Peaceful Warrior* was republished and word of mouth began to build, that I created some Peaceful Warrior audio programs and began teaching workshops. At last, I had found my career and calling — writing, teaching, and speaking.

They say it takes ten years to become an "overnight success." In my case it took sixteen years of experimentation and uncertainty — treading water until I learned to swim. So the observations I share are not abstract theories. I've been in the trenches; I've done the soul-searching, experienced the disorientation and doubt. Those years of searching were, in retrospect, a necessary period of introspection and self-inquiry.

On the quest to find our career and calling, we're like people driving in the dark — we can see only as far as our headlights illuminate. Meanwhile, the better we understand ourselves, the better we steer our course. That is why self-knowledge may be the heart of wisdom — and wisdom is earned over time.

Talents, Interests, and Values

The path to career and calling is like a treasure hunt — and at a crossroads where three key qualities intersect, you'll find a signpost pointing toward work you are most likely to find fulfilling. Those qualities can be expressed by three questions, the answers to which require time and experience to fully grasp: *What are my talents? What are my interests? What are my values?*

Talents

During my freshman year in college, I was preoccupied with women who didn't seem to know that I existed. I asked a few out until I finally got the message that they just weren't interested. Meanwhile, I tended to discount women who were friendly toward me. When I shared my situation with a teammate, he said something I never forgot: "Dan, some girls aren't going to like you no matter how much you try, and other girls are going to like you just as you are. Why waste your time on the ones who aren't that into you?"

That story and advice also apply to your career search. For any number of reasons, many people pursue work that doesn't match their talents, hoping to improve as they go — a little like buying a suit, then trying to tailor yourself to fit the clothing. There's nothing wrong with experimenting and trying different sorts of work, but why not pursue a career consistent with your innate talents and established abilities?

Some of us are stronger in sports or mathematics than in English or art, or the opposite may be true. As Will Rogers said, "We're all ignorant, only on different subjects." In searching for a career, we naturally want to learn all we can about the career, online and in the field — even interviewing those in a profession. But even more important, we have to "research ourselves." And one important part of self-knowledge follows from a realistic assessment of our strengths and weaknesses, revealed over time in the school of experience. This reminds me of a story:

I was about to catch the team bus to the gymnastics conference championships, where I would compete the next day, when I dropped by the old Texaco station. As soon as Socrates saw me, he beckoned me to come into the garage, where he handed me a heavy case of engine oil. I was asking him where he wanted me to put it when he said, "Just hold on to it for a minute" — and he placed a second case on top of the first one. I could barely hold both cases. Ignoring my strained expression, he then started piling individual cans of oil on top of the two cases in my shaking arms. "Socrates," I said, panting, "I can't —"

My arms collapsed, and the boxes and cans crashed to the floor at my feet. I stared down at them, then looked up at Socrates, whose expression revealed nothing. "Okay," I said. "This was some kind of a test, right? And I failed..."

When Socrates finished laughing at my sullen expression, he said, "Life is not about success or failure, Dan — it's about stretching yourself. How will you ever know your limits until you've tested them? And how do you test them unless you're willing to fail brilliantly?"

That competition the next day was one of the best of my life, because I tested my limits and discovered the truth of Soc's words. As playwright and orator George

Bernard Shaw wrote, "I learned to speak as I learned to skate or cycle — by doggedly making a fool of myself until I got used to it."

Never assume limits until you've tested them. You may not lack talent; you may only lack experience. The impressions we form of our level of talent — even early on, as children — can be based on our far too limited exposure, or even a single incident. Years ago I watched as my daughter took a children's gymnastics class taught by a new instructor. He had the youngsters line up and told them it was time to learn cartwheels. He told them to hold up their arms, put one leg forward, then "kick over the top" and attempt a cartwheel. The children attempted one after another. My daughter, like some of the other children, did the cartwheels easily and well, because she had previous experience. But some of the children had never before attempted this movement and, as beginners, naturally fell down once, and again. This is where a skilled instructor would have given more helpful feedback for these beginners, but he could only tell them to "keep your head up and your arms straight." They fell again. As a parent, it was not my place to take over the class. I could only observe the frustration and disappointment on their little faces as the beliefs took shape that they weren't as talented as their classmates.

You've also formed assumptions about your talent

(or lack of talent) in math or singing, cooking or sports, art or reading comprehension. Maybe those assumptions are correct — or maybe they're not. But if you're willing to apply yourself, even modest talent combined with effort over time can reap significant improvement. Still, you're more likely to invest your time and effort in an area of abiding interest.

In every field there are those with abundant talent who have little interest in a given arena — and others who have great interest but little talent. Of the two, interest in (or passion for) a given vocation may be the more important factor.

Interests

On the face of it, you'd think interest would be easy to determine. After all, you either have an interest in something or you don't. But your level of interest is often complicated by beliefs in your capacity in a given area — by your self-concept. For example, I became "less interested" in painting because I formed a mistaken belief that I wasn't much good at it. It happened like this:

I started kindergarten a week or two late due to a flu. On my first day I was given paint, a brush, and an easel and told to paint a tree. My tree looked like a green lollipop. Still, I was moderately pleased with my creation — until I compared it to those of the children around me, whose trees had branches and individual leaves. Unaware that the other children had been practicing painting trees nearly every day, I mistakenly concluded that I wasn't as good as the other students. So the next day, when I was given a choice to go paint pictures or to climb the monkey bars, you can guess which activity I chose. My interest in climbing grew along with my skills, and painting was all but abandoned. So due to a simple misunderstanding, I turned away from the path of art even as a future in gymnastics beckoned.

Added to this basic complication of (often distorted) self-concept influencing our interests, we spend much of our teens and twenties sorting out what we (or our parents or peers) think we *should* or *could* do from what we actually *like* doing. We can only guess at the number of misguided decisions that have turned someone away from an interest because they were told that it had "no future."

Many high achievers have been pushed to the top of their professional ladder by social pressures or parental expectations, only to discover later that the ladder was "leaning against the wrong wall." For example, I know two former attorneys who switched careers midstream — one began a long and successful career as the head swimming coach at Stanford University; the other became a respected sports chiropractor, whose clients include a major city ballet company and a professional sports team.

So consider carefully the kinds of areas or activities you genuinely enjoy — since interest is a clear guidepost at any career crossroads — which is why I advised my daughters, "Do what you love, then get someone to pay you for it."

Values

We humans cling to a variety of divergent values. Some of us, for example, have more puritan values (such as self-denial and delayed gratification) — we go with the rules. Others of us value a more hedonist lifestyle (valuing self-indulgence and immediate gratification) — we go with our impulses. Some lean toward the sentimental; others have a tougher, more rigid character. Still others embody the lone wolf or the social butterfly, the believer or the skeptic. These values, along with our interests and talents, all propel us along our career and life paths.

Your unique combination of values — what seems

right, ethical, and worthwhile to you — is revealed in your choices and behavior. I don't mean to suggest that one set of values is better than another. Remember the rules in the school of life — it's not about right or wrong, but about choices and consequences.

Your Core Value

On the path to self-knowledge, in the arena of career and calling it's important to know which of three criteria, or values, is most important to you: location, relationship, or career. Most of us appreciate that all three are important — but if you had to pick a core value and subordinate (or sacrifice) the other two, which might that be? Reflecting on your life and choices, you may realize that one has taken precedence over other areas on your personal scale of values. Let's consider a few brief examples.

Location: Rebecca's star is hitched to Boulder, Colorado, a city that captures her imagination. Centered on that location, she's set on moving there. Once she settles in, she'll then meet people and find whatever work she can, but for her, the people and work are secondary priorities.

Relationship: Jason is a carpenter and can work anywhere. His partner, just completing an internship, is applying to a number of medical residency programs. Jason is committed to the love of his life, around which all else revolves. Home for him is where the heart is, wherever his love may go. He's committed to a person more than any given place, and he'll find work wherever he goes.

Career: Elizabeth is devoted to her work, the central value in her life, where she derives the strongest sense of purpose and meaning. She will go to whatever location best serves her career and develop a circle of friends there in the free time her work allows.

We each have different values and make our choices accordingly. Some years ago, a friend and fellow professor at a college in the Midwest lived on campus, while her husband worked as a government attorney on the West Coast. They loved each other dearly. She flew to visit him every few weeks, and he did the same for her. But their work (and necessary locations) took priority over living together — a mutual choice based on their values and circumstances.

Our values of place, people, or work are not necessarily fixed for life; they can change over time. If, for example, Boulder-loving Rebecca meets the love of her life and that person is offered a wonderful career opportunity in another state, then Rebecca (and her new partner) may both need to reexamine their core values.

In literature, in film, and in life, *character and values are revealed by choices people make under pressure.* Self-knowledge — in this case, knowing your central value — helps you to make clear decisions, even as the decisions you make also add to your inventory of self-knowledge. As E. M. Forster wrote, "How do I know what I think until I see what I do?"

Career Criteria: The Trinity of Needs

Knowing your core value is only one facet of self-knowledge and career choice. You also need to measure any potential career against three essential key criteria: *Do I find the work satisfying? Can I make good money? Does it provide a useful service?* These questions point to the three essential components of a fulfilling work life. Two out of the three may seem sufficient for a while, but all three elements are usually required in a satisfying long-term career. And (along with unfair or incompetent superiors or repressive company policies) the lack of one or more of these three basic elements is the primary cause of dissatisfaction and complaints in the workplace.

Satisfying Work

In America and elsewhere during the Great Depression of the 1930s, hungry men and women would travel if necessary and do virtually any work they could (preferably legal) to earn a few dollars. My father grew into adulthood during those difficult years. When I once asked him whether his work made him happy, he looked puzzled for a moment, then replied, "It never occurred to me." For him, happiness was not part of the career equation. His goal was supporting his family in the best way he knew how.

Many immigrants today also do whatever work is available — they don't likely ponder whether the work fits their aesthetic sensibilities. Yet they derive a sense of meaning and purpose from being able to support their families and seek better opportunities for their children. The same is true for many people in the world today. But thanks to the labors of your parents and grandparents, and your education and skills, you have a wider array of work options and opportunities; you are able to consider the issue of "right livelihood" and to seek a career appropriate to your interests, temperament, and needs.

Some kinds of work primarily entail physical exertion and skill; other kinds involve more mental exertion.

Some careers involve contact with the public; others call for solitude. Some professions confer higher status or social recognition than others. There's no "best" career, only the most suitable one for you at a given time of your life. Different people enjoy every sort of vocation. Zen and Taoist literature is full of enlightened butchers and wise craftspeople, and long ago, there was that carpenter from Nazareth who also found a higher calling.

What you do matters, but *how* you do it matters even more. *Fulfillment lies not in the work itself but in the quality and care you bring to it.* Those who take pride in the quality of their work, no matter how humble, experience a meditative absorption that makes time fly, and that in itself can be fulfilling.

In Japan years ago, while waiting on a train platform for the bullet train, I observed a railroad employee cleaning one of hundreds of silver posts supporting a long railing that ran the entire length of the platform. He polished the posts, devoting about thirty seconds to each, making sure it was shining and spotless before moving on to the next. The man showed the concentration of a Zen master as he moved gracefully through his task, absorbed in his work. Few of us would consider polishing posts creative or satisfying work, and perhaps it wasn't what this man had dreamed of doing as a child. But from what I observed, the pride he took made his efforts a work of art. In a sense he was also polishing his spirit.

So bear in mind that you don't only *find* a fulfilling career; you create it — like that Zen master on the railway platform.

Good Money

What you or I may consider a "good income" will differ, depending on our skills, experience, and circumstances. But in each case, the compensation must feel fair, appropriate to the work, and sufficient to meet our needs. Otherwise, unless you are either a masochist or a saint, you're unlikely to tolerate underpayment over time.

There are, of course, other forms of compensation besides monetary income. You may accept work offering minimum pay but exceedingly pleasant surroundings or medical or other benefits, or because you believe in your employer's mission.

A woman I'll call Jean came to work for me part-time some years ago. She was decompressing from a high-paying, high-pressure job with a software start-up company in Silicon Valley. On the verge of burnout after working weeknights and weekends, she'd read our small announcement for part-time work and contacted us. Jean had read a few of my books and was thrilled to be able to work in a relaxed, informal, and positive atmosphere; reclaim some personal space; rebalance her life; and reset her priorities. Her delight with the work never wavered, but after about six months, she had to move on again — in her words, to "find better-paying, full-time work more suitable to her 'market value.' " We wished her well in finding the middle ground she sought, a position that would meet her financial needs without undue pressure or demands.

Useful Service

When I began trampoline training, a "call to service" was the farthest thing from my mind. The same is probably true for others who strive for success in amateur or professional sports or in nonathletic games such as poker or chess. Self-interest, the desire for success and mastery, and personal tastes provide the initial drive and motive for most professional undertakings. Yet the satisfaction derived from service — doing something that benefits others — moves some athletes (or chess or poker players) to serve as mentors and role models in their chosen fields.

Those who work in service-oriented professions likely sleep more peacefully than drug dealers, thieves, or other criminals whose work is ultimately destructive

— and who, even if they make good money and find their work satisfying, may feel disconnected from the light of a larger world and higher possibilities. Whenever we uplift or improve the lives of others, we experience a quality of fulfillment unattainable through self-serving pursuits.

Not everyone is going to become a life-saving neurosurgeon or great philanthropist or celebrated master artist. Most of us do our jobs without fanfare or wide recognition. We assist and labor in stores and small businesses, in corporate buildings and markets and shops and gardens, doing work no less important than that of the rich and famous. But how many of us take ourselves and our skills for granted? How few of us appreciate the important service we provide?

Most people think of "service" as volunteering for the needy, helping with charities, or delivering food to the elderly or homebound. Such acts are certainly commendable. Yet someone who helps assemble furnaces or automobiles or televisions, making sure that every part fits, that every screw is tight, and that the construction is top quality, provides a significant service on whose skills others depend.

It's easy to devalue the worth of our work if we provide a less-creative service like doing data entry or answering phones in a small office or large corporation.

But such work connects us with others and helps our world function. We are all links in the chain, and every link counts — even the criminal and the poker player, who also eventually learn what they need in life's continuing education.

THESE THREE ESSENTIAL CRITERIA — doing satisfying work, making a good income, and offering a useful service — provide another facet of self-knowledge that leads to better-informed decisions, and point us toward a fulfilling and sustainable career.

Career Decisions: Analysis and Imagination

Most of us have grown up in a world that places the highest value on logic, analysis, and the scientific method. Yet *nearly all scientific discoveries and breakthroughs arise from our imaginative and intuitive faculties* (only to be later tested and validated through experiment and analysis).

Few of us find a career through logic, however — more often we find work through personal or secondary connections and sometimes through sheer coincidence. Here's a personal example: I left Berkeley and moved back to Los Angeles with my pregnant wife and an undergraduate degree in psychology. Since we had less than $500 remaining in our account, I searched through the want ads and found work as a trainee at a local office of New England Life Insurance. I got the job, bought a suit, and did my best to sell life insurance products. After two months I knew two things: First, that selling insurance might be perfect for some people, but it wasn't for me; second, I needed to move out of Los Angeles and back to the San Francisco Bay Area. So I quit the job, drove up to Sacramento, where I dropped off my wife to visit with her parents for a few days, and from there drove the two hours down to Berkeley to explore my options, thinking that with my background I might find work as a recreation director or something similar. I stopped by the office of my old coach to say hello and let him know I was looking for a job. He said, "Good timing, Dan — I just heard this morning that the Stanford [gymnastics] coaching job opened up yesterday — why don't you go speak with the athletic director?" I did so — and the next day I was hired as the head gymnastics coach at Stanford University.

I may have found that job through timing and luck — but it was preceded by years of intensive training in the field, confirming that luck happens when preparation meets opportunity. I hadn't been looking for a college coaching job, and I didn't find it through reason or analysis. In fact, logic is not well suited for making career decisions, because our analytical brain can only weigh variables — pros and cons, benefits and liabilities — trying to sort out a reasonable course of action. Doing so can generate for many people a state of "paralysis by analysis."

Have you ever made a rational decision, then awakened in the night or the next morning feeling uneasy about that choice without knowing why? It may be wise to pay attention to your intuitive sense regarding any important decision. Ideally, you want to use both imagination and reason in any problem-solving task — just not necessarily at the same time. Reason reflects an intellectual skill of the conscious mind; intuition reflects the deeper grasp of the subconscious mind, which sends messages to the conscious mind in various ways, including dreams.

During my tenure as the Stanford coach, one of the gymnasts on our team shared an assignment he was given that brought home the power and value of the subconscious mind in problem solving. Ted told me about a strange riddle he'd been assigned earlier that day in Dr. William Dement's Sleep and Dreaming course. The entire riddle consisted of eight letters: H-I-J-K-L-M-N-O. Ted didn't have a clue what it was about. His assignment was to ruminate on the riddle before going to sleep, with the intention of solving it — even if he didn't understand it — then to write down his dreams when he awoke the next day. Ted came to the gym the next morning and told me he'd written down his dreams but still had no clue about the riddle's meaning or solution. He showed me his dream notes — images of an old sailing ship in a stormy sea, with waves crashing and rain pouring down in torrents. It made no sense at all — until Ted learned the answer to the riddle and realized that his subconscious mind had indeed solved it beautifully, all without any conscious clue about what was going on. The riddle, H-I-J-K-L-M-N-O, was composed of the letters H to O, or H_2O, the chemical symbol for water, images of which had filled Ted's dreams.

Years later, as a writer, I experienced a similar phenomenon again and again. After I had written myself into a corner and the plot line was stuck, or I didn't know where to go next with a particular train of thought, I'd sleep on it — and, as often as not, the solution would pop into my head upon waking the next morning.

Time Travel with Your Subconscious Mind

You're about to experience a form of "time travel" that engages your imaginative capacities — both your conscious and subconscious mind — to help you make better-informed decisions about career or any other facet of life. With practice, this process becomes simple and straightforward, and it requires an investment of only ten to fifteen minutes. Sending your imagination forward through time can help you choose the best of two, three, four, or more options. But for the sake of explaining the process, let's say someone's choosing between A and B.

Examples: Kristen is deciding between two colleges; Mahmoud is weighing whether to look for work in his current city, where his girlfriend lives, or to accept a good job overseas; and Maliha is considering the offers of two admirers who have both proposed to her. Or there's Caleb and his start-up software company, which is beginning to take off, and he can either expand or accept a buyout from a software giant; Samantha can accept a high-paying corporate accounting job or start her own firm; Curtis has to choose whether to undergo expensive, high-risk surgery, chemotherapy, and radiation or settle his affairs and let go; and Sarah must decide whether to take an early-retirement package and minimum Social Security payments or to push on for a few more years so she'll receive maximum Social Security benefits.

The decision you face may be similar to or quite different from the examples above, but the process is the same. So here's the time-travel exercise.

1. Choose A. Settle into the idea that you have *fully* committed to A. This step is essential.

2. Ask yourself this three-part question:

   Having committed to A, What will I be doing, what will I be feeling, and what will I look like *one hour from now*?

   Ideally, write down the answers to the three parts of the question. Your answers may be brief, but write them down. At the very least, imagine the answers vividly. See what comes up. It's not likely to be too difficult to imagine one hour from now.

3. Ask yourself the same three-part question, but instead of *one hour from now*, imagine *one day from now*. Again, write down the responses that come up.

4. Same three-part question, but imagine *one month from now*.

5. Same three-part question, but imagine *one year from now*.

6. Same three-part question, but imagine *ten years from now*. At this point, you may feel like saying, "Wait a minute. How can I possibly know what

> I'll be doing, be feeling, and look like (having committed to A) a decade hence? That's just my imagination." Right. It's your imagination — a bridge to intuitive sight. After all, why do you imagine one thing and not another?

When this process is complete, repeat it for option B. That is, say to yourself, "What was I thinking? Clearly, B is the best choice!" And commit to that. Feel into it. You have chosen B. Now go through all the steps just mentioned.

When you've written down (and/or vividly imagined) all your responses for both A and B (and C, if there's a third option), you will have transformed tunnel vision into a more expansive, intuitive depth and breadth of vision. This doesn't mean you will now have absolute certainty, but you'll be better prepared to make a firm decision.

Deciding to Decide

Most of us view a decision as a mental process and conclusion — but in fact, no decision becomes real until one acts on it. This became clear to me some years ago, while I was on a solo hike in the high Sierra. I was heading back to camp as the sun sank toward the crest of a peak, when I came to a fork in the road. I wasn't sure which path to take. I peered ahead, examined the terrain, and weighed my options. First the right fork looked more promising; then I changed my mind and leaned toward the left, until my intuition urged me again to the right. This indecision continued as azure sky turned a darker shade. I decided left; I decided right — but I made no progress until I took action and started down one path.

Decide to decide — and when you act, do so with the full force of your being. Second-guessing yourself is a form of self-abuse. As Zen master Ummon reminded his students, "When you sit, sit; when you stand, stand — just don't wobble between the two!"

Remain committed to whatever course you've chosen unless you get new and compelling reasons to change course. If you've committed to cross a rushing mountain stream only to realize a third of the way across that the current is too swift and deep for a safe crossing, a change of plans may be not only appropriate but wise. The same is true in your larger life. If you're staying in an abusive relationship or punishing job, it may be time to reconsider and recommit to a more life-affirming direction, taking into account your boundaries, rights, and worth. (There's a difference between commitment and masochism.)

However, for most decisions we make, even as doubts and difficulties arise, commitment means marching forward, persisting through challenges, and keeping faith in yourself and in your course of action. *Faith is the courage to live as if everything that happens, and every choice we make, is for our highest good and learning.* Faith is also nourished by the higher understanding that every decision eventually leads to wisdom.

Career Notes: From Entry Level to Leadership

Any insight into our second purpose — career and calling — must include a look at the roles we play in the evolving cycles of our lives at work. I've experienced entry level and also played a leadership role; so have you — at home, at work, or in the larger world. Every time we face some new task or responsibility, we experience the entry-level challenges, the beginner's mind. And leadership turns out to be more about a personal set of qualities than a position of power, corner office, or company title. You will soon discover how it's possible to demonstrate qualities of leadership even at entry level, or any stage of your personal career arc.

Now it's time to address key facets upon which our work experience will turn.

Committing to Excellence: You may have a number of careers or callings in the course of your life. But no matter what your tasks and duties appear to be, committing to excellence and quality adds meaning and value to your days. For example, some years ago, I was supporting my family by working two jobs: overload typing at a law firm early in the mornings, followed by data entry at a real estate firm the rest of the day. Not the most creative of duties, nor my long-term choice of careers, but the best employment I could find at the time. So I aimed to be an excellent data-entry person. I look back on that period without regret (and with some pride), because I provided high-quality work for my employer until it was time to move on.

Rungs on the Career Ladder: Why is it that some of us move up the ladder, while others are content with lower-level positions, and still others work for themselves? We might as well ask why some athletes become Olympians and others enjoy the junior varsity — or why some people work behind the camera and others in front of it. If we did ask why, we might come up with factors including genetics, birth order, life experience, beliefs and self-concept, educational opportunities, socioeconomic status, role models, friends and networks, family support, energy level, drive, talent, intelligence, exposure, location, the state of the economy, and a measure of chance or destiny.

Your own unique combination of such factors may direct you to a subordinate position or drive you upward to the rarified air and higher pressures of a top-floor corner office.

Those who ascend to the peaks may, or may not, be happier or more satisfied than those who live down in the valleys. The main thing is to find the most suitable work you can (keeping in mind the criteria we've already covered) while staying open to new opportunities. Your career is not a marriage; you don't have to make a lifetime commitment to your employer (any more than your employer makes such a commitment to you). So there's nothing wrong with staying apprised of opportunities in your field or another that may better fit your talents, values, and interests.

Two Secrets of Success for the Self-Employed: I sometimes tell people that I have an extremely demanding boss — and I'm self-employed. If you're in a similar situation, note the two key factors that will enhance your chances of success: 1. *Be good at what you do*; and 2. *be good at promoting what you do* (or hire someone who will do that for you). If you have good skills but fall short at marketing yourself, keep in mind that your business can only serve people who know that you exist.

Your Leadership Qualities

I learned about leadership skills and responsibilities in the field, while coaching college athletes. I needed to set high standards, provide feedback, motivate and support team members, troubleshoot when necessary, and create a productive environment that encouraged innovation — the same responsibilities taken on by CEOs, high-level managers, politicians, parents, coaches, teachers, and leaders in any walk of life.

No matter what role you play at work or at home, you influence people around you — you teach and lead by example — because people notice what you do. Therefore, leadership is a universal calling that has nothing to do with one's status or level of authority in an organization. Tyrants, dictators, and despots may be powerful rulers yet poor leaders. One can serve as chairman of the board but lack essential leadership skills, while someone who works in the mail room inspires others to quality work. Many of us play a leadership role in our circle of family or friends.

So what are the key qualities of an effective leader? As you read the following, consider how you (might) demonstrate these qualities among your friends, loved ones, and colleagues.

As a leader you inspire by example. Humanitarian-physician Albert Schweitzer once said, "In influencing other people, example is not the main thing; it is the only thing." You embody the qualities that you'd like to see in others.

As a leader you offer support where needed. People don't care how much you know unless they also know how much you care. Support means setting the bar high, then asking, "How can I help you accomplish this goal?"

As a leader you motivate by pointing out a higher purpose. You transform a job into a service-oriented mission, incorporating meaningful, big-picture goals into the work at hand. You present tasks not as an end in themselves but as a means of personal growth, lifting the game to another level.

As a leader you appreciate the value of mistakes. You understand that mistakes can lead to breakthroughs, to new insight and innovation. So true leaders encourage risk and accept new and creative errors in the process of learning and growth.

As a leader you encourage collaboration rather than competition. Competition for external rewards, perks, and bonuses results in workers who feel like rats running for food pellets — but "the problem with the rat race," quipped actress Lily Tomlin, "is that even if you win you're still a rat." And highly competitive environments are often described as "soul-killing" by those caught in the maze. The competitive spirit that stimulates short-term drive among high achievers can also lead to burnout and resentment. In contrast, a spirit of collaboration — when one person's contribution (supported by others) reflects well on the entire team — generates a connected, productive, and friendly workplace where open systems and information sharing replace self-protective strategies.

As a leader you empower others. You become a bridge over which others may cross. You remind your team that *no one is smarter than all of us* — and you focus on what's right, not who's right. Using not only the brains you have but all you can borrow, you shift the focus from "What's in it for me?" to "What serves the highest good of all concerned?"

How Will You Spend Your Life?

Contemplate for a moment the mysterious aspects of how you ended up where you are now. Consider the open road ahead, which, like any good novel, may bring surprising twists and turns. As you grow stronger and wiser through the hurdles you meet along the way, hold to the faith that your life is unfolding as it should — that every bump in the road is a part of your unique process. Treasure your career and calling, no matter how humble; trust the changing rhythms of your life, shaped by the talents, interests, and values you hold dear.

Our discussion of the second purpose of life — *finding your career and calling* — opened with a reminder about the hours we are given each day. Now we return to this central issue and close with some big-picture perspectives about your life and work.

The hours of your life are the most valuable currency you will ever have. How will you spend them? Imagine winning a huge lottery — you'd never again need to work for money. What then would you do with your time? How would you spend your life? Would you give of your time and talent in creative service to the world? Or would you pursue amusements, pleasures, traveling, and celebrity, collecting sensory experiences and possessions, playing with power and status and influence? Let's suppose that "billionaire you" bought everything you've ever wanted. Then you traveled around the world on the finest cruises and stayed at luxury hotels. You acquired beautifully furnished homes with staff in numerous prime locations. If you became bored or restless in one place, you'd take your private jet to another — and buy (or have your personal assistant purchase) all the cars and toys and clothing you could want and the best schools for your children.

Then what? Would you preview the newest movies on the biggest screens? Play video games day after day? Arrange for one sensory pleasure after another? Be seen with the rich and famous? And *then* what? Entertainments and amusements can distract almost anyone from a sense of emptiness — at least for a while. But when our life nears its end, what do we want to look back on? So the question remains for us each to answer for ourselves: Can a self-pleasuring life lead to genuine fulfillment? The following tale addresses this perennial question.

A thief named Willy, making his escape from a high-rise penthouse, falls from the fire escape toward certain death thirty-five floors below. A moment later, it seems, he finds himself in a fancy pool hall, in perfect physical shape. Willy concludes that he has died and gone to heaven — because pool is his all-time favorite hobby. With a sense of relief and joy, he racks up and discovers he has somehow become a master at the game. His only regret is that no one else can see his triumph in this empty establishment as he sinks one shot after another.

Later that day as he explores his new environment,

he visits a swimming pool and finds himself performing Olympic-caliber dives from the three-meter board, then cutting through the water in what must be record time. At a nearby casino, where he meets taciturn croupiers and silent dealers, he starts with a few dollars in his pocket and emerges a millionaire — the luckiest man alive. He finds a hotel and pays in advance for a month in the executive suite. The next day is a blissful repeat of the day before. He races cars, wins another fortune at video poker, and beds six gorgeous women, having discovered a profound new virility.

It's an amazing week, and a pretty good month, but as time passes, Willy begins to tire of the constant winning and his mastery of everything and everyone. Within six months he feels like he's going mad. He undertakes a search for St. Peter or whoever's in charge, so he can express his dissatisfaction. After an exhausting search he finally reaches the main administration hall. An efficient secretary waves him into a grand office, where he finds a well-dressed man sitting behind a massive mahogany desk. "May I help you?" asks the official.

With that, Willy begins a litany of complaints: the terrible boredom, the effortless success and victory at every turn, ending with, "I never thought that heaven would be anything like this."

The official stands slowly, and his eyes reflect a reddish glow. "What," he asks, "makes you think that this is heaven?"

The wealthiest and most privileged among us may have special appreciation for Willy's dilemma, because no matter how many moments of self-gratification money can buy, fulfillment eludes us until we find meaning, purpose, and connection. Indeed, a growing number of billionaires today find that meaning by donating a significant share of their wealth in service of others, and celebrities find worthy causes, using their fame for a higher purpose.

Years ago I met a woman I'll call Doris who had hundreds of millions of dollars from a family trust. When I first met her, she fit my image of a "spoiled rich girl." She seemed to possess a sense of self-importance but not a lot of personality. She made no effort to be kind or even courteous. Perhaps from her view she didn't need to observe social niceties. She ignored the live-in staff as they cleaned and gardened. She deigned to speak with me only because, being an avid reader, she had enjoyed some of my books.

Doris was not an easy person to like. At that time I had no knowledge that she was desperately unhappy in her relationship and dissatisfied with her life. Only years later did I find out that several months after our meeting, and after a difficult divorce, she had made two significant changes in her life: First, Doris set up a philanthropic foundation to help others in need; second, she found regular part-time work in a small flower shop. She had always loved flowers and now had a way to provide a direct, person-to-person service. Doris had created new meaning and purpose in her life, and a heartfelt connection with the larger world.

Unlike Doris, many millions of people on this planet are born into poverty. Their day-to-day lives are fully occupied with finding food and shelter. As Mohandas Gandhi put it, "To a starving man, God is bread." Yet despite, and even because of, their struggle for the bare necessities, many of the poorest among us have a clear sense of meaning, purpose, and connection as they cling together for the common good, doing their best to care for their families and communities.

The same call to service applies to retirees with discretionary funds and time, who may at first pursue leisure activities, luxuriating in their new freedom. But for many, unlimited leisure becomes an endless "long weekend." So, echoing the search of young college graduates (and even the independently wealthy), retirees also confront the need to reexamine their current values, interests, and skills. Some go back to part-time work — not only for income but as a means to feel useful, reestablish personal or professional contacts, reaffirm their sense of relevance, or join a larger mission. And like younger people who may change jobs several times, retirees may also explore different ways to best serve, whether mentoring, consulting in a field of expertise, or working retail a few hours a day (as Doris did). So the paths of poverty, wealth, and even retirement can all lead to service.

The film *Groundhog Day* relates the story of Phil, a cynical television weather forecaster caught in an existential time loop while on a location shoot in Punxsutawney, Pennsylvania — he awakens every morning to the same day in the same town. He is the only character aware that he's living the same day over and over. At first he tries exploiting his knowledge, taking advantage of people and circumstance. As the endless repetition of days continues, he grows distraught and commits suicide again and again in a variety of ways, hoping for escape, but immediately awakens to yet another same day. Eventually, having explored every experience and exhausted all other options, Phil stops fighting and begins to improve himself — to study and practice piano, medicine, and other skills — until he finally realizes the ultimate lesson: There's nothing left to do but to serve, to become a useful force in the world. With this recognition, he finds his purpose (and love and salvation).

Groundhog Day represents our own evolutionary journey of awakening. As we come to appreciate how the world has supported us, fed us, taught us, and tested us — whether or not we earned these blessings or seem to deserve such grace — the call to service awakens. As my friend Lou, who sells life insurance and provides financial advice, told me, "I just use my work as an excuse to connect with people, to offer some kind and helpful words. Maybe we do some business, and maybe we don't. I hope their lives are richer for it; I know that mine is, for the wealth of friends I've made." Lou had realized what most of us are discovering — that we can't serve others without also elevating ourselves.

You have a sacred calling;
The question is, will you
take the time to heed that call?
Will you blaze your own path?
You are the author of your own life. . . .
Don't let others define it for you.
Real power comes by doing
what you are meant to be doing,
and doing it well.

— OPRAH WINFREY

As you consider the arc of your career and the challenges of moving through the low points and difficulties that come with the territory, recall the first of the four purposes — learning life's lessons. You've learned much so far, and will continue to adapt and grow, mature and evolve. And during those times you feel stuck, stagnant, or even that you're slipping backward, you may actually be backing up to get a running start. So persist. Stand up for yourself. Lead by example. Practice the quiet courage of everyday life.

Your work — the service you provide, no matter how humble it may be — connects you to the larger community. So never take your work, or yourself, for granted. Author Leo Buscaglia reminded us all that "small acts of caring can turn a life around." One life you will turn around, through career and calling, is your own.

The Third Purpose

DISCOVERING YOUR LIFE PATH

- *Understand Your Hidden Calling*
- *Follow Your Higher Potential*

The shoe that fits one person pinches another;
there is no recipe for living that suits all cases.
Each of us carries our own life-plan,
which cannot be superseded by any other.

— C. G. JUNG

THUS FAR WE'VE EXPLORED the geography of the school of life, then cast off and sailed into the changing currents of career and calling — key areas on the surface of our lives. Now it's time to submerge into the depths. According to philosopher Søren Kierkegaard, "At birth we set sail into the world with sealed orders." In this, the third purpose of life, you'll gain access to those "sealed orders" and delve into your hidden calling, or life path.

The third purpose begins with the precept that we are each climbing a mountain toward our highest potential but by differing paths. Because few of us can see the drives, challenges, and gifts on our own life path from the valley of ordinary awareness, I present core elements of the Life-Purpose System — a method first detailed in my book *The Life You Were Born to Live*. This system provides a numeric key to the third purpose you are here to fulfill.

Hidden Calling, Higher Potential

Comprehending the third purpose requires an open mind and curious heart, because it introduces a system, drawn from an ancient mystical tradition, that doesn't easily lend itself to scientific explanation. Yet its accuracy points to unsolved mysteries and unexplored territory awaiting further study.

How I Discovered the Life-Purpose System

Despite nearly twenty years spent exploring the human psyche, various insight traditions, and metaphysical models of reality, I remain an empiricist at heart. I value the scientific method, which utilizes well-designed (double-blind, controlled) experiments to test whether a theory or hypothesis is in fact true — whether, for example, a medication, herb, or other substance works better than a placebo (sugar pill). The methods of science helped to pull humanity out of the dark ages of superstition. So over the years I've been skeptical of magical or wishful thinking and untested notions.

I hadn't even considered the existence of a deeper

calling, or anything like a "life path," until 1984, when I met an unusual mentor whom I'll refer to as MB. He had read my first book, *Way of the Peaceful Warrior*, and had decided to take me under his wing.

Not long after our initial encounter, MB sat me down and gave me a "reading," which changed the course of my life. In that session, he revealed in-depth information that clarified my past, present, and potential future, including details that rang true and proved amazingly accurate. It was as if his words removed a veil that had previously obscured my vision, and awakened in me the first clear recognition of the life I was born to live.

I was astonished that he could have such insight into my life but had no idea how he gained access to such information. I was at that time well versed in the "cold-read" techniques and tricks used by so-called psychics. MB claimed no psychic abilities, however, stating only that he had been trained to know "where to look" for such information. He would say no more about it at that time.

In the months that followed that session, I began to learn life's lessons with greater ease and openness, and to engage with life in a more assertive way. Understanding what I was here to do, I set out to do it. My family's financial situation improved correspondingly as I refined an approach to living that I called the *peaceful warrior's way*.

Meanwhile, I remained fascinated with MB's apparent intuitive abilities. So when he announced an advanced training he was offering in Hawaii, where he would teach, among other things, the basic elements of the system that gave him such uncanny insight into people's lives, I was the first to sign up. I could hardly believe that I might learn to do for other people what he had done for me.

At the training, I sat with about twenty other participants as MB began a series of lectures on this mysterious system. The first thing he revealed was an objective method of adding up the digits of anyone's date of birth, and then deriving meanings that provided insight into their lives.

I was initially disappointed by this revelation, as when a magician shows how a wondrous illusion is accomplished with mirrors and sleight of hand; the "magic" vanished. Besides, using people's birth dates sounded like numerology, an occult art that had never attracted me. It made no sense that adding the digits of someone's date of birth could yield accurate information about central issues in that person's life.

MB went on to explain that such methods had been

passed down in various cultures over the centuries, but that they differed in interpretations and degrees of accuracy. He added, "Once you learn more about this approach, you can determine for yourself the validity of the system." He then spent several evenings presenting information that pinpointed key issues in each of our lives.

The life path, or hidden calling, that you are about to explore points to what you are *really* here to fulfill — the innate drives, challenges, and gifts that for most people remain unseen or obscure.

I took careful notes from MB's lectures, outlining in about twenty pages the fundamental elements of the system. As soon as I returned home, I began to give free life-purpose readings to family and friends, using the basic information in my notes. Within a few weeks I had internalized the information and no longer needed any notes. Eventually, after working with many hundreds of people, my insight deepened and expanded.

Eight years later, I trained a small number of therapists, health professionals, and life coaches in the fundamentals of what I came to call the Life-Purpose System, and eventually I went on to write *The Life You Were Born to Live*.

Here, in the context of the four purposes of life, I present core elements of this system as a means to help you clarify your life path and hidden calling — the third purpose of life.

A final note prior to addressing the specifics of this system: If your sense of logic and reality leave you reluctant or resistant to applying a numerological system that uses one's date of birth as a means to access core issues of life, you have the option of ignoring the following calculations. Instead, you can simply review the nine life paths I'll soon describe, and contemplate, as objectively as you can, which of these life paths may stand out the most for you. You always have the option of later checking your own conclusions against those that your birth number points to, in order to determine for yourself the validity and utility of the system as a tool for focused insight.

Determining Your Birth Number

You can choose either of two methods to determine your birth number, the key to your life path. The first is the quicker and more accurate.

Method One: Use the Life-Purpose Calculator Online

1. Go to www.peacefulwarrior.com using your computer, smart phone, or other electronic device with access to the Internet.

2. Click on the Life Purpose link to access the Life-Purpose Calculator.

3. Input your year, month, and day of birth. You will then see your birth number, along with a paragraph of summary information about your life path. Make note of your birth number before returning to this book.

Method Two: Do the Math

If you don't have access to the Internet, turn to the Appendix at the end of this book (page 151) for instructions on doing the math yourself. Once you have your birth number, make note of it and read on.

Your Birth Number and the Third Purpose

Like most people living today, you have a three-digit or four-digit birth number (or, like a small percentage of people born on certain dates after the year 2000, you might have a single-digit birth number).

You will soon learn the meanings and significance of the digits 1 through 9 as they relate to the challenges on your life path. Every digit in your birth number provides information and insight into your life path. However, please make note of the following.

- If you have a *three-digit* birth number (e.g., 27/9), then the final digit, to the right of the slash, represents your hidden calling — the third purpose you're here to fulfill.
- If you have a *four-digit* birth number (e.g., 29/11), then each of the two middle digits of your birth number (on either side of the slash) represents key elements of your third purpose.
- If you have a rare one-digit birth number, then that single digit alone represents your third purpose.
- A zero in your birth number (for example 20/2 or 28/10) indicates potential spiritual gifts, such as empathy, strength, or intuitive discernment.
- As you mature along your life path, the qualities associated with your birth number tend to evolve over time from more negative (less mature) aspects into more positive (constructive) forms.

To make sense of all these numbers and principles — bringing them down to earth and into your life — we now turn to the primary meanings and issues associated with each digit (1 through 9).

The Nine Life Paths

Study each of the following descriptions through the filter of your own history and experience, paying special attention to the digits of your birth number. And remember that each of the following life paths or callings (associated with the digits 1 through 9) represents primary drives and challenges.

1: Creativity

The 1 centers around creative energy — in the arts (writing, visual media, acting, music), or in education, business, family and childrearing, or any other field where creativity gives birth to new directions, solutions, or approaches. The 1 also indicates a knack for inventiveness or reinvention. When those with 1 (or double 1) in their birth number tackle their innate insecurity — an underlying sense of inferiority and driving need to prove themselves — and stop trying to fit in, then their creativity flows like a surging river. But creative energy is a double-edged sword; it can manifest in positive, productive, constructive forms or in negative or destructive ways (such as abuse or addiction). Physical exercise helps those working 1 energy to stay grounded and balanced. Applied in the positive, their energy charms, attracts, and inspires others to their own creative endeavors.

2: Cooperation

The 2 represents a drive to, and issues around, cooperative service but also a tendency to overcooperate, or start a relationship by overhelping, sometimes to the point of servitude — giving and giving (whether or not anyone asked) — then later resenting and resisting, withdrawing (physically or emotionally), feeling taken for granted. Those with 2 in their birth number (like those with 4) can feel overly responsible and take on responsibility to solve problems ("I'll just do it myself!"), and need to delegate at times. A key for those with 2 involves establishing a balance (and boundaries) between the needs (and responsibilities) of others and their own. Those working 2 can be strong and supportive worker bees, nurturers, and caregivers of the world. While fully capable of leading, their deepest and most satisfying calling is bringing loyal, capable support to a person, organization, or mission in service of a higher good.

3: Expression

The 3 points to an emotionally needy and sensitive soul (whether or not outwardly visible to others) — a romantic whose acute sensitivity makes criticism especially painful but also helps them intuitively tune in to and connect with others. The 3 centers around a powerful drive for self-expression — through speaking, writing, music, art, public relations, coaching, or teaching. If the expressive drive of those with 3 is blocked, stifled, or obstructed, they can get moody, depressed, or manipulative (hinting at rather than clearly expressing what they need). In the negative, 3 energy takes the form of complaining, sarcasm, insults, gossip, or cutting wit; in the positive, those with 3 in their birth number bring creative, constructive, encouraging, uplifting, enthusiastic expression. The biggest 3 hurdle is an innate feeling of self-doubt (a sense of being unprepared or not capable enough). The courage to push past this hurdle enables those with 3 to bring uplifting expression into the world.

4: Stability

The 4 represents a deep call to structure and stability, but with tendencies to do the opposite. Those with 4 in their birth number are ambitious and gregarious problem solvers with a natural analytical ability; thus, they make excellent managers and organizers. But 4 indicates a tendency to overanalyze that can lead to a sense of confusion, so writing down options and listening to gut feelings can help. Those with 4 have innate strength that can turn to stubbornness, because they view themselves, others, and the world as they "should" be. Like those with 2, those with 4 are so responsible they need to resign as "general managers of the universe." Those with 4 benefit immensely from using a clear, step-by-step process to accomplish their goals. Practicing anything over time, or engaging in a construction project, can be immensely helpful. When they balance strength with flexibility, and analysis with intuition, and temper impulsiveness with patience and persistence, they establish stable foundations in their lives and relationships, and can accomplish great things.

5: Freedom

The 5 represents freedom — a drive to experience many facets of life. Those working 5 are vivacious, quick-minded, multifaceted individuals who can see life from many angles, but their wide variety of interests can leave them scattered or burned out. Easily bored, they seek drama and excitement — or they create it. Most of these freedom-loving people favor new experience or adventure (direct or vicarious) over security or money in the bank. They have a powerful a drive to rescue others and to fight for the underdog. A notable challenge for those with 5 is a tendency to swing from dependence to independence — so they benefit from establishing interdependent relationships and healthy self-reliance in their own lives. Their key to freedom is discipline in the form of clear priorities and focus — digging one well one hundred feet deep rather than ten wells ten feet deep. Ultimately, in helping to liberate others, they find their own freedom.

6: Vision

The 6 points to high ideals and standards of beauty, purity, justice, fairness, and authenticity. But idealism can lead to disappointment, since few people or situations live up to the initial expectations of those working 6. *Disillusion* is a big word for them, both in the negative sense of disappointment and in the positive sense of freedom from illusions — maturing into realism and a sense of perspective. In a group-learning environment, those with 6 tend to compare themselves to the most accomplished — and even if they do twenty things well, they tend to focus on a single mistake. Once they replace perfectionism with perspective, and learn to accept life as it is, they learn to celebrate their progress rather than obsess about so-called mistakes in the past. As those working 6 come to appreciate the innate perfection of life unfolding as it is, in its own good time, their strong sense of duty and commitment contributes to their vision of a better world.

7: Trust

The 7 centers around issues of trust — of self, others, and life unfolding. The 7 is an introspective, insightful energy, and those working it possess incisive minds that can read between the lines. Natural scholars and researchers (or hermits, despite a social persona), those with 7 favor and need solitude. Their private, suspicious, individualistic nature — and tendency to overshare or undershare what they're thinking or feeling — can lead to misunderstandings, which they may view as betrayals. Those working 7 are here to learn to trust their own bodies, instincts, and sense of discernment rather than relying mostly on the guidance of books, experts, and the theories of others (including conspiracy theories). Otherwise clear and capable, they can refresh and revitalize their spirits out in the natural world, whether garden, beach, desert, or mountains. Once they learn to trust themselves, they finally establish trust with others through clear agreements — and achieve a relaxed openness in a world they can rely on.

8: Recognition

The 8 involves drives, struggles, mixed feelings, and deep satisfaction in the realm of material success — work, money, influence, self-control, and authority. Those with 8 may deny (even to themselves) an innate drive for recognition and may resent displays of wealth, power, or authority in others. Whatever their grades in school, those with 8 have powerful, logical, and strategic minds that see the path ahead and the writing on the wall, but (like those with 4) they need to follow a diligent process of effort over time to reach their goals. Because those with 8 like to make the rules rather than follow them, they sometimes need strong lessons to break through their resistance to feedback. Accepting their place in the material realm puts them in touch with their personal power to manifest their goals if only they step forward and follow through. They evolve and fulfill their calling through productive enterprise — sometimes philanthropic or humanitarian causes — achieving material success in service of a higher good.

9: Integrity

The 9 represents the one who leads the way — the wise man or woman whose life and example draw others to follow. However, this example of integrity and wisdom is achieved only after those working 9 overcome hurdles, learn the lessons of experience, and "get real." They may be slow to gain wisdom, but when they do so, they can be a great resource for others. Due to their depth and charisma, those working 9 find themselves in positions of leadership. They will either become role models of integrity or will suffer the consequences. They may have friends and family yet feel lonely due to a tendency to live in their own world of mental concepts and ideas at odds with reality. They have strong opinions and are also acutely sensitive to the opinions of others. Only when they free themselves from the "god of opinion" can they hear the god or goddess of their heart and find the wisdom that is their birthright and destiny, demonstrating the true value of leadership as they help light the way for others.

Putting It Together

This may be a good time to take a deep breath and let this information settle in. If you haven't read *The Life You Were Born to Live*, it may take a little while to absorb these highlighted issues and qualities associated with the primary "paths of potential" 1 through 9. Give yourself some time to explore how the descriptions may (or may not) apply to your past or present experience. All that follows is designed to further clarify the life-path information as it relates to the third purpose of life — your hidden calling and higher potential.

At this juncture it's entirely possible that you relate to issues in most or all of the nine categories: 1. We've all done something creative or felt insecure at times; 2. we've supported someone who took us for granted; 3. we've felt self-doubt or had trouble expressing ourselves; 4. we've felt impatient, stubborn, or confused or have acted impulsively; 5. we've desired independence or rooted for the underdog; 6. we've idealized someone who disappointed us or failed to meet our expectations; 7. we've needed alone time or felt betrayed; 8. we've wished to influence others or struggled with money; and 9. we've assumed a leadership role, had a slip of integrity, or needed a reality check.

I recognize many facets of the primary numbers in my own life as well, because all these issues are a part of the human experience. However, if each birth number were a suit of clothing, you might come to appreciate — after trying on different birth numbers (and related issues) — that your own birth number is indeed a better fit in expressing those strengths and challenges you have faced in the theater of your history and everyday life. For example:

- If you have 1 in your birth number, you will feel (or have felt) a need to prove yourself, and feel most on track when your creative juice is flowing.
- If you have 2 in your birth number, you will struggle (or have struggled) to balance giving with receiving, and feel most on track when your support is accepted and appreciated.
- If you have 3 in your birth number, you will feel (or have felt) self-doubt and emotional sensitivity, and feel most on track when you find a receptive listener.
- If you have 4 in your birth number, you will feel (or have felt) confused due to overanalysis, and feel most on track among family or other foundations in life.
- If you have 5 in your birth number, you will feel (or have felt) both dependent and independent, and feel most on track in the adventure of service to others.
- If you have 6 in your birth number, you will feel (or have felt) that you could have done better, and will feel most on track as you accept yourself and the world as it is.

- If you have 7 in your birth number, you feel (or have felt) betrayed, and will feel most on track as you learn to trust your body, your mind, and your life.
- If you have 8 in your birth number, you feel (or have felt) mixed emotions about money, fame, and authority, and will feel most on track as you gain recognition for excellent work in service of others.
- If you have 9 in your birth number, you feel (or have felt) clear messages about integrity, and feel most on track when leading and guiding by your example.

Individual Differences, Shared Patterns

No two individuals are exactly alike, even if they share the same life path. Even identical twins do not remain completely identical; they grow into unique personalities with their own interests. Factors including birth order, gender, family dynamics, and childhood experiences influence our choices and responses and help to shape our lives. No single tree on the planet is exactly like any other, yet we can accurately describe differences between redwoods and pines and birches.

Despite our individual differences, we humans also

fall into recognizable patterns or life paths, each with common issues, potentials, and challenges. You'll work through those shared issues in your own way, in your own time. And your life may look radically different from the life of someone else on the same path if one of you is working through the issues in a more positive, constructive way, while the other animates the negative or destructive aspect.

Even though there are forty-five possible birth numbers shared among many millions of people born from 1900 to the present, you can speak of your birth number as a personal path — because you will experience and respond to those issues in your own unique way. There's ultimately no better or worse way to experience your life path. There's only living and learning.

You may recognize both negative and positive elements in your life as you reflect, with patience and compassion, upon your struggles and progress. Note where you've been, where you are at the present time, and where you're going — that is, who you're becoming. As you do so, bear in mind the often-repeated story of a conversation between an elder Native American and his grandson after the boy had gotten into some trouble: The grandfather says, "Within each of us live two wolves — one kind, productive, and peaceful; the other cruel, destructive, and hurtful."

"Which one is stronger inside me, Grandfather?" asks the boy.

"The one you feed," answers the elder.

This is the freedom and power within you: You can choose whether to animate the negative possibilities of your life path or to feed those more positive aspects as you mature and evolve through time in the school of life. This power of choice means it's never too late to follow your life path where it leads — never too late to fully live the life you were born to live.

Fundamentals on Your Life Path

To deepen and reinforce your understanding, here are some of the basic premises of the Life-Purpose System that relate to your life path.

First, your birth number(s) represent(s) the path you are here to ascend — it's an uphill climb — so what you are here to do isn't necessarily what comes easiest. In fact, your path brings special challenges as well as special potential. As you overcome the hurdles, you experience the benefits. This principle applies to every primary digit in anyone's birth number.

For example, those with 1 in their birth number — especially those with a double 1 (such as 19/10 or 29/11) — have higher creative potential but also feel the challenge of insecurity more acutely. So they won't always or immediately demonstrate creative flair or artistry; they may in fact show less creativity at first (or create in negative ways) until they overcome their insecurity and show the courage to be different, original, and to stand apart from the crowd.

An opposite example is also worth noting: My own birth number (26/8) does not contain the digit 1. This doesn't mean I necessarily lack creativity; rather, it means that I don't have the same sensitivities, challenges, or potential in this arena; that is, neither creativity nor insecurity is as central in my life as would be the case for people with 1 or double 1 in their birth number.

Second, as already noted, any life path has both negative and positive poles — and, more than any other factor, whether you work more in positive or negative ways will shape the course of your life. (The reasons one person chooses to create a software application and a second person comes up with a software virus could fill another book.)

Third, no life path is better (or worse) or easier (or more difficult) than any other; they are only different, each presenting distinct strengths and challenges.

Having understood these critical points about the life path numbers and how they work, you may want to review the nine life paths (or at least those related to primary digits in your birth number) in the context of what you now understand.

Consider the words of novelist Anya Seton: "There are many trails up the mountain, but in time they all reach the top." You were born with the challenges and gifts of your life path; what you do with them is up to you.

Life Paths, Career Paths

In their enthusiasm, those who have newly discovered the Life-Purpose System often hope that the information and insights they gain will help solve all the riddles in their lives, such as revealing the specific occupations or careers they should pursue, or predicting talent or success in a particular area.

For example, since those with an 8 as a predominant number have an innate drive to influence others, should they all become attorneys, politicians, or clerics? Should all those with a 1 (or double 1) go into creative arts of acting, music, or painting? Should those with a life path of 7, naturally inclined to solitude and scholarship, all become researchers? Should businesses groom all charismatic individuals with a 9 life path for leadership positions? Should those with a life path of 5, drawn toward saving others, go into search-and-rescue or social work?

The answer is no, because we are more (complicated) than our birth numbers. As much as I might like to present the Life-Purpose System as a solution to all career searches, your birth number is not intended to determine what kind of work you do in the world, but rather illuminates *how* you approach whatever work you do and the qualities you may bring to it.

Each individual, and each life path, brings different talents, qualities, and challenges to a given type of work. For example, in the field of acting, Meryl Streep, with a birth number of 33/6, demonstrates not only extraordinary sensitivity and expressive ability as a master of dialect (the double 3) but the high standards (of the 6) that impelled her to learn to play the violin in order to prepare for a role as a music teacher. Robert Downey Jr. (29/11), with the 9 and double 1, has both charisma and abundant creative energy that enable him to do each take of a scene with inspiring originality.

Streep and Downey each do great work but in qualitatively different ways reflecting their respective life paths. We find other respected actors (and people in every field) bringing their own unique approaches to their work — approaches that reflect not only the issues of their life paths but the genetic factors, life experience, timing, luck, and effort over time that all contribute to success in any given field. So no single path, and no particular number, is predictive of success or failure. We each bring our own unique combination of resources to our chosen career, calling, or craft as we climb a path to our potential.

Well-known individuals, in widely divergent fields, have throughout human history struggled in different ways with issues on their life paths — some working in the positive and others in the negative. For example: actor Paul Newman, singer-actress Barbra Streisand, astronaut Neil Armstrong, humanitarian César Chávez, inventor Alexander Graham Bell, comedian Groucho Marx, polio vaccine creator Dr. Jonas Salk, Olympian Jesse Owens, artist Michelangelo, and several popes all share the 26/8 life path — and so does Osama bin Laden. And inspiring role model Helen Keller, visionary poet William Blake, humanitarian Coretta Scott King, Facebook founder Mark Zuckerberg, medical pioneer Dr. Karl Menninger, U.S. president Theodore Roosevelt, and philosopher Bertrand Russell all share the 32/5 life path — and so does Adolf Hitler. This sampling of quite different individuals known in a variety of fields all share similar strengths and challenges. How did they have such different impacts on the world? Most of these people overcame their challenges and applied their energy in positive, constructive, creative ways — while a few applied theirs in negative and destructive ways.

Never make the mistake of thinking that your birth number defines you as a person or limits your choices in the world. This system is intended to enhance, clarify, and focus insight and compassion (for self or others) chronically late," or, "Whenever she acts that way I just blow my top," I want to respond, "Now that you know this, *what are you going to do about it?*"

Insight isn't enough. At best it provides a map, but we still have to make the journey. As I said earlier, just *thinking* about doing something is the same as not doing it. We know we've learned a lesson in the school of life only when our *actions* change.

When we explored the first purpose (learning life's lessons), I introduced the school rules — the laws of reality. Certain laws have greater relevance than others, exerting specific leverage to help you overcome the hurdles on your life path — enabling you to *do* something, to take action, to change your behavior. Anyone interested in learning more about the laws most relevant to each life path will find them in the source book for the third purpose, *The Life You Were Born to Live*; they're also presented for more general application in *The Laws of Spirit.*

Keeping the Faith: Your Promise and Potential

Use the Life-Purpose System, combined with honest soul-searching and self-observation, to better understand yourself and your hidden calling — what you are truly here to confront and fulfill on the path to your full potential. What you do with that potential will reveal itself in the unfolding of your life and the fullness of time.

The extent to which you fulfill your hidden calling depends on how you respond to the challenges you meet along the way. No matter where you live, and whether you are born into poverty or privilege, you still have the responsibility, opportunity, and capacity to reach toward the highest light within you. But take to heart the higher truth that your life is unfolding as it should. We cannot predict, or force, where our path may lead — but we can bring a measure of grace to the journey with each small step we take.

Know that your path will guide you — you cannot lose your way. Because wherever you step, the path will appear beneath your feet. The way may twist and turn, and obstacles will appear as you continue onward and upward. But as my old mentor reminded me more than once, and I again remind you, "The way creates the warrior" — the climb itself develops the capacity to complete the journey. And the higher you climb, the better the view.

IF KNOWLEDGE IS POWER, you now wield the force of a deeper understanding of the first three purposes of life: learning essential lessons in the school of life; wisely choosing a satisfying career; and understanding your hidden or higher calling as you climb toward the peaks of your potential.

The Fourth Purpose

ATTENDING TO THIS ARISING MOMENT

• *Pay Close Attention* • *Make Each Moment Count*

There is surely nothing other
than the single purpose of the present moment.
A person's life is a succession of moment after moment.
When one fully understands the present moment,
there will be nothing else to do
and nothing else to pursue.

— YAMAMOTO TSUNETOMO,
HAGAKURE (THE BOOK OF THE SAMURAI)

Consider for a moment the amount of territory you've covered in exploring the first three purposes of life: first, learning life's lessons, the school rules, and our human curriculum; second, applying tools for greater self-knowledge to help you choose a fulfilling career and calling; third, using an unusual method for determining, and focusing on, your life path (or deeper calling).

All this information can feel overwhelming. How can you organize it, distill it, and apply it in everyday life? To that end, it's time to bring it all home with the last piece of the puzzle — the fourth purpose, which integrates the first three purposes and brings them all into the simplicity and power of this present moment.

No matter what life brings — joys and disappointments, burdens and delights — amid all the complications, duties, and responsibilities, you can always handle this arising moment. *Now* is where you gain entry to the fourth purpose of life.

The Man Who Had No Purpose

Over the years, in addition to writing, speaking, and teaching, I've also offered one-on-one coaching — sometimes by phone and sometimes in person at my home in Northern California. One day a man I'll call Peter came to my home for a session. We greeted each other, he paid the agreed-upon fee in advance, and then I began with a key question: "What's your purpose for coming here? What would you like to accomplish?"

When I asked Peter this question, he just sat there, looking down at the floor, before he answered, "I don't know — I have no purpose."

"Well, if you did know your purpose, what might it be?"

Peter shook his head and repeated, "I have no purpose. None at all."

I thought about this, then explained, "This session is based entirely on what you want to accomplish, Peter, and since you're certain you have no purpose, I can only thank you for coming by. Nice meeting you. Now could you close the door on your way out?"

Surprised and a little disoriented, Peter stood up and

started moving toward the door, but then he turned and said, "Wait a minute! I traveled quite a distance to get here, and I just paid you some good money for ninety minutes of your time!"

Smiling, I said, "You know what? I think you've discovered a purpose. Please sit down."

This bit of Zen trickery on my part helped Peter to see that he did indeed have a purpose — to spend ninety minutes with me. And when our session was over, I promised Peter, he would discover another purpose — finding his way back to his car, opening the door, getting in, putting on his seat belt, starting the car, and driving to wherever he was going (without hitting any people or objects on the way). And then, when he arrived at his destination, another purpose would appear.

I then told Peter a brief conversation that took place at the funeral of a great Hasidic master. When a visitor, paying her respects, asked a longtime disciple, "What was the most important thing to your teacher?" he responded, "Whatever he happened to be doing at the moment."

I reminded Peter that the purpose of today is today. The same is true of each moment.

Peter had convinced himself that he had no purpose because he hadn't yet grasped a larger mission for his life. But he had failed to notice what may be the most important purpose of all — the one that appears before each of us, moment to moment. Such moments are the building blocks that form the foundation on which we build our lives.

Attention to this arising moment has been advocated by many different sages, at different times, in a variety of cultures. Yale professor William Phelps said, "I strive to live each day as if it were the first one I had ever experienced, and the last one I would ever live." Zen sword master Taisen Deshimaru reminded his students, "Be happy here and now or you never will be." And writer Margaret Bonanno points out, "It's only possible to live happily ever after on a moment-by-moment basis."

We learn life's lessons (the first purpose), we choose our career and calling (the second purpose), and we fulfill our life path and hidden calling (the third purpose) all *in this present moment*. Welcome to the fourth purpose of life. No matter where you go, and no matter what the clock-face reads, you are always right here, right now.

What Snood Taught Me about Life

Snood is not an Eastern guru or relative of my mentor Socrates. Snood is not even a person, but rather an online game, and an addictive one at that. Like most addictive games, it seduces through simplicity: You shoot a little, round colored bubble, aiming to hit groups of same-colored bubbles; if your aim is true, they vanish, and the goal is to make as many bubbles as possible disappear. Other than developing a measure of aim (so presumably you can dissolve any other colored bubbles you may encounter in the real world), Snood provides a few moments (or many nonproductive hours) of immersion and diversion; it's fair to consider it a shadow form of meditation, a sport without exercise, an insular activity.

But there's one beautiful, largely hidden facet of Snood that may reveal a transcendental teaching about life — and about living in the moment freshly, innocently, and without judgment or expectation. Does this wondrous revelation I'm about to share justify the hours I spend manipulating my mouse in hopes of topping my last score? Probably not. Or maybe so, depending on what you do with what I am about to share. Perhaps my wasted hours were not so wasted after all.

So here's my revelation: Every five shots or so, the entire configuration of all those bubbles shifts. What was, is no longer the same. It all changes, just like life. Any plan or strategy I might have formed for the next shot is now useless. It's a whole new ball (or bubble) game. There's no point in resistance or regret. You have to shift your mental course instantly, look again, freshly, and address what is here and *now*, in this arising moment.

Everything changes — again and again. Each time the demand is the same: What now? What's my purpose in this moment? Snood taught me to linger not on what almost was, what could have been, or what might have or should have been but rather on what is, right now. As Marcus Aurelius wrote, "Time is a river of passing events. No sooner does one thing appear in sight than it is washed away, and another appears in its place. And this, too, shall be washed away." Each moment unfolds like waves breaking on the shore. If you're knocked down by one wave, rise again and ready yourself for the next one. Wave by wave, moment by moment.

Life as a Series of Moments

The entirety of our lives consists of a series of moments; as poet Emily Dickinson wrote, "Forever is composed of moments." No one is intelligent or kind or cruel or neurotic or saintly or enlightened or funny or sad all the time. But some of us have more intelligent, kind, cruel, neurotic, saintly, enlightened, funny, or sad moments. One realistic aim might be to increase the number of kind, enlightened, intelligent, or happy moments.

While we live, we're doing something in every

moment. That doing may be sleeping and dreaming, or laughing or playing, or sitting very still in meditation, or writing or stretching. Life consists of action (or stillness) moment to moment. So even if you tend to dream big, start small; then connect the dots. Those dots are made of moments.

What are you doing in this moment? Are you sitting on a chair or a sofa, treadmill or stationary bike? Maybe you're lying in bed, sitting at the table or under an umbrella at a resort or on a beach. Whatever your location, reading is likely your current purpose. Maybe you'll stop for a few moments, look up and around, take a breath, sip a drink, or bite into a snack — which then becomes your purpose in the moments that follow.

Each moment is a gift. And as the proverb goes, "One moment can change a day, one day can change a life, one life can change the world."

There's No Such Thing as a Future Decision

All decisions are made in the moment, for the moment. Trying to make a permanent decision is like trying to eat once and for all. Waves keep rolling in; circumstances change. Each new moment is fresh. Want to know what you've decided? Observe what you do.

Begin your training in making decisions (and living

in the moment) by noting the following guidelines and observations.

All you have to decide is what you will do right now. That is the only decision you can make or will ever make. Decisions only matter in the moment: Resolutions are at best good intentions, so make them if you wish — then face the next moment, ride the next wave. What will you do now?

You only need to make a decision in the moment you need to make it. Deciding which college you're going to attend, which job you'll take, or which man or woman you'll date or marry isn't necessary until it's fish-or-cut-bait time. Meantime, don't agonize over it. Just do what you're naturally doing in the moment.

We only need to manage (or change) our lives in relevant moments. One of the primary sayings in Alcoholics Anonymous is "Take it one day at a time." In reality, all we have to do is take it one *moment* at a time. Those who experience a powerful compulsion to drink do not feel compelled all the time. There's no need for them to decide not to take a drink except when the impulse to do so arises. In that moment, they can decide not to drink and instead call a sponsor or attend a meeting.

Decisions get clearer the closer you are to having to make them. Making a decision about what you will do next year or when you retire is extraordinarily difficult and usually ends up being inaccurate — guesswork at best. Again, it's only a wish or intention. Deciding what to do in this moment arises naturally from circumstances at hand.

This moment is all you need to manage. *This* is your moment of power.

We Cannot Grasp the Now

Any physicist will remind you that what we normally think of as "the present moment" doesn't have any objective reality — it doesn't actually exist. If you say the word *now* as quickly as possible, between the instant you begin to sound out "n—" and the time you reach "—ow," millions of nanoseconds have passed. As scientist and philosopher William James put it, "Where is it, this present moment? It has slipped from our grasp, fled ere we could touch it, vanished in the instant of becoming." This is the fleeting quality of the present moment.

Yet, paradoxically, all we have is the eternal present — we float downstream in the river of time, flowing with the current, thus remaining in perfect stillness. Just so, we move through time yet remain within the stillness of this moment. Past and future are both happening now. There is only now. *Wherever we go is now*.

Past is memory; future is imagination. Nice places

to visit, but you don't want to live there. Purposes tied to past or future have no reality; they are phantom concepts conjured by the mind. Handle just what is in front of you. By attending to the fourth purpose — this arising moment — you find the simple life. Reality is where you are now — moving in stillness, floating with the current in the river of time, resting in the eternal present.

Refrain from comparing this moment with memories of the past or an imagined future, and you find contentment — here and now. As a Zen sage said, "When I'm hungry, I eat; when I'm thirsty, I drink; when I'm tired, I rest."

The Challenge of Attending to the Present

The challenge of attending to this arising moment can be demonstrated with a single question: How often do you consciously eat a meal? Consider this question carefully. When was the last time you only ate — not ate while reading or watching television or listening to the radio or talking or checking your phone or computer, but the last time you sat and ate, only ate — seeing the food, smelling, tasting, and savoring it, aware you were chewing and swallowing? When was the last time that eating was enough to engage your full attention? (This task is probably no easier for me than it may be for you.)

Much of the time, we humans are bored with the present. So the prime difficulty of learning to live in and attend to the present moment is that *we don't really want to*. It doesn't seem enough for us, because we're not really paying attention. Isn't it fascinating that we can immerse ourselves in a video game or movie for minutes or hours, yet we can hardly pay full attention to dinner for more than a few seconds? Our restless minds flit about from past to future, remembering, judging, regretting, replaying, anticipating, planning, expecting. And none of that has anything to do with this moment. As author Alan Watts wrote, "The power of memories and expectations is such that for most human beings the past and the future are . . . *more* real than the present."

One night in the gymnasium, after Socrates watched me fly from the high bar, perform a difficult aerial maneuver, and stick my landing, I raised my fists in elation and declared my workout over. I pulled off my sweatshirt and stuffed it into my workout bag, and we left the gym. As we headed down the hallway, he said, "You know, Dan, that last move you did was pretty sloppy."

"What are you talking about?" I asked. "That was the best dismount I've done in a long time."

"I'm not talking about the dismount," he responded. "I'm talking about the way you took off your sweatshirt and stuffed it into your bag."

I realized I had treated one moment, doing gymnastics, as special — and another moment as ordinary. Socrates reminded me, once again, that there are no ordinary moments. This realization, this challenge, lies at the very crux of your life and will determine the quality of your every moment: Can you, will you, learn to love the present moment? Can you develop the capacity to show each arising moment the same attention that you might give to a cherished friend or a lover — or at least to an online game? Will you wake up to the inestimable value of each passing moment as it slips, like a grain of sand, through the hourglass of your remaining time here? To the extent that you pay attention to the moment, you will have fulfilled the promise of the fourth purpose of life.

The Gift of Life: A Matter of Perspective

You appeared from nowhere, born onto the material realm on this blue-green speck suspended, spinning, in a vast galaxy in the cosmos. You may ruminate on how or why you came to be, or contemplate where space ends, but you find no answers, because all great questions lead to wonder and awe — and perhaps a measure of gratitude for the miracle of being invited to the feast. In counting your blessings, you may experience a desire to give something back — thus begins the path of service, the fruit of your journey, the doorway to joy.

Imagine for a moment that, through a strange twist of fate, you wake up one morning and find yourself imprisoned — scheduled for execution at midnight. You gaze out through prison bars and see the first rays of your last sunrise. A rooster crowing in the distance sounds poignantly sweet. You feel greedy for every sight, sound, taste, and aroma that day, which passes so quickly! You eat your last meal as the shadows grow long outside. And as the sun sets, you say your final farewell to the light of day, for you will never see another dawn. Each passing minute draws you closer to your final good-bye, your last prayer, your last breath.

Such a final day awaits every one of us. We may know in advance that our end is near, or we may have only a few seconds' notice, or none at all. But when the executioner raises the ax, how many of us might want to cry out, "Please, another moment! Let me take one more sweet breath! Give me one more sight, one more sound, one more touch of my loved ones!"

Now would be a perfect time to look up and around, to listen, to breathe deeply, to spend some time in nature — whether mountain, beach, or garden — to touch your loved ones, and to bring your best to life while you have the life to spend. How much time, how many more moments, that may be, no one can say.

Life itself is a near-death experience — ephemeral and brief. One among billions, you go on, loving, serving, grieving, and celebrating, seeking meaning and fulfillment. Every life represents a hero's journey. Every moment counts.

The Highest Practice of All

We *do* things all the time — we do the dishes, do the laundry, do the shopping. We go to work and dig or type or lift or speak. But few of us *practice*. The key difference between doing and practicing is this: When you practice an action, you aren't merely repeating by rote but rather are striving to improve or to refine whatever you are doing — whether it's signing your name, opening a door, carrying groceries, merging with traffic, or folding laundry.

We're familiar with practicing a sport or a dance or a game or a musical instrument. Naturally, in these formal training activities, we understand that we want to improve our stroke or leap or swing or strum. We treat these activities as somehow special and separate from the practice of everyday life, as if they were more deserving of our full attention. Which is why Socrates once said to me, "The difference between us, Dan, is that you practice gymnastics; I practice everything."

Think of it! What would it be like to practice every moment and everything we do? To attend to how we lift a forkful of food to our mouths, and breathe, and chew — to attend to the words we speak and how we speak them. I grant you, in the hands of some of our more obsessive or neurotic friends, this idea of practicing everything might take the shape of a compulsive, never-ending self-improvement program. But that isn't at all the point.

The practice I'm recommending creates an art of living. Your life becomes an art form. Your performance is unique — unlike any other — because no one else can live your life as you can. When you *practice* all that you do, your attention naturally returns to each arising moment. You step into the flow. You enter the zone. You become a cloud drifting through the skies, neither racing ahead of the wind nor dragging behind, but moving naturally, once again in harmony with what the Chinese sages called the Tao.

This is my practice. I remain something of a beginner, but a devoted one. I practice with the faith in this principle — that we improve over time. The first step is to form the sincere intention to respect each passing moment and to treat each action as if you were performing it before an audience of thousands. Please remember that *how you do anything reflects how you do everything* — which is why the Zen masters say, "If you can serve tea properly, you can do anything well." In mastering one thing, you master yourself. So whatever you may be doing, ask yourself, in random moments: Am I breathing? Am I relaxed? Am I moving with grace?

The concept of a Way is a foundation stone of the Eastern spiritual traditions, which gave us the arts of flower arranging, calligraphy, the tea ceremony, and many of the martial disciplines, as well as the art of sitting or walking meditation — not as ends in themselves, but to light the way, to refresh our hearts and minds, to provide a template in the art of living and a doorway to the kind of ego dissolution in moments of absorption that we enjoyed as young children. What we have lost, we can find again. This is how we "become like little children" to enter the "kingdom of heaven."

Practice whatever you do. Notice the small things — the subtle shift of posture that relieves stress, the relaxing breath, the act of remembering to smile for the pleasure of it. Six words: Here and now, breathe and relax. Practice that in all you do, and your life will change for the better.

Despite what you may have heard about "just being"

— or the notion that "it's not about what you do but about who you are" — I beg to differ. I propose that in terms of our influence on and existence in the world around us, you are (largely) what you do. So changing your behavior changes your identity, moment to moment.

You don't have to think positive thoughts. You don't have to feel peaceful or confident or compassionate or happy or loving — you only need to behave that way, as much as you can, whenever you remember to do so. This practice animates and celebrates the best of your humanity and spirit. This practice reflects the heart of the *peaceful warrior's way*.

The fourth purpose of life, which integrates and distills the other three purposes into one manageable moment at a time, is a lifelong practice. Calling forth all you have learned on your journey, you can transform everyday life into a path of personal evolution and infuse each moment with new meaning and purpose.

Attend to this moment with each breath. And in random moments, silently ask yourself, "What is my purpose in this arising moment?" Then do whatever needs to be done, in a wondrous and changing parade of purposes that shape the story of your life, and of all our lives.

Epilogue

OUR SPIRITUAL PURPOSE

If Earth is a school, then what does graduation feel like? What's waiting for you on the mountaintop? Could your life be more mysterious than you can see from the valley of conventional awareness? Since ancient times, mystics and shamans have consulted with oracles and sought to communicate with the spirit world, while countless others have simply gazed up into the night sky or looked deep within to find answers. It seems fitting here to complete our exploration of the four purposes of life by addressing the larger spiritual purpose of our existence.

I relate the story of my own spiritual journey in *Way of the Peaceful Warrior*. Many readers have told me that the book changed their lives. Few are able to articulate exactly what changed, but I believe it comes down to this: The story describes my shift from a narrow focus on personal success to an understanding of life's bigger picture — and readers share that experience.

In the film adaptation of my first book, the full scope of the quest gives way to a smaller story about a young gymnast who learns nuggets of wisdom, culminating in a deeper understanding of the power of the present moment. In other words, the film ends with the fourth purpose of life.

But this epilogue, like my first book, takes us beyond self-improvement to what may be humanity's ultimate purpose — transcendence, usually defined as rising above or beyond. Any dictionary definitions fall short, because the experience, by its very nature, goes beyond conventional language and defies description, which is why Lao-tzu said, "Those who know do not speak [of it], and those who speak [of it] do not know."

Many of the sages and saints who have seen the light, or truth or reality as it is, have felt moved to provide clues, methods, metaphors, and paths pointing toward that which defies description. For example, Zen masters assign their students *koans* — riddles that can't be solved by the rational mind (such as "What was your original face before you were born?"). The *koan* stretches the student's awareness to leap beyond conventional mind to experience *kensho*, a breakthrough insight similar to, yet transcending, the "aha" experience of solving a more conventional riddle.

Zen *koans* are but one way to facilitate an awakening. Seeking to achieve a state of liberation or bliss, Sufi masters twirl and dance, while other seekers go on vision quests or ingest psychotropic substances; still others pray, contemplate, meditate, chant, or do practices involving breath and visualization. All for a glimpse of the transcendent.

Why do they bother? Isn't this conventional world, and everyday life, enough? Perhaps transcendence doesn't really exist except in the minds of true believers. Maybe ideas about the divine, and soul and spirit, and heavenly realms during or after physical death, are fanciful notions or flights of imagination. Our souls may or may not ascend to a heavenly realm, and even if we have past or future lives, we're not likely to remember any of them with certainty, but we certainly do have to live *this* life.

Yet, in moments of awe and wonder — or more often when tragedy strikes, in times of great dislocation, distress, or when dealing with mortality — we are moved to ask the existential questions, such as "Why live if we must die?" and "What is my purpose here?" At times like these, the possibility of a divine intelligence, a prime mover or creative source and substance of the universe, can seem not only appealing but somehow self-evident. (As one religious scholar put it, "There's God, then there's not paying attention.") After all, innumerable sages, mystics, and clerics in holy orders, who have devoted their lives to the quest for Spirit, God, or Truth, report how the "gates of heaven" appeared as they broke through the illusions of conventional mind to glimpse the truth that sets us free — a vision of the eternal and absolute Consciousness at the heart of our existence. They call us to turn our eyes upward toward our spiritual purpose even as we live our day-to-day lives; to merge heaven and earth with a leap of faith; to live with our heads in the clouds and our feet on the ground.

Most of the time conventional reality monopolizes our attention. We have daily tasks, duties, pleasures, and problems. We play out our dramas in the theater of gain and loss, desire and satisfaction. We pursue fulfillment, pleasure, and success, trying to make life work out according to our hopes and wishes. No matter what our degree of success or status, we experience attachment, anxiety, and disappointment in this world of change. This realization can set in motion a sincere and open-minded quest to realize a higher or transcendent truth.

One way to understand the difference between conventional and transcendental awareness is to imagine the experience of lying on your belly down in a valley and examining stones, weeds, and grass versus standing on a mountain peak and surveying the panorama below. Both experiences have value, but only one provides the bigger picture. To enjoy the panorama, we need to make the climb. But in daily life, this shift of awareness can take place in an instant. When we make this shift — remembering life's bigger picture even as we function in everyday life — the conventional and transcendental worlds become one.

Seeking the transcendent does not mean rejecting our conventional world but rather embracing it fully, releasing our resistance, attachments, and expectations. As we do, we experience a lighthearted wisdom; we take ourselves and our dramas less seriously. A momentary glimpse of the transcendent can restore our humor and refresh our spirits, which is why many awakening individuals consider our spiritual purpose the ultimate quest even as we do the laundry, care for children, and work for a better world.

Freedom and joy do not reside elsewhere; they are right here, right now, right before our eyes. All paths, and the four purposes I've shared in this book, lead toward an awakening where theories and concepts, models and maps all dissolve into the eternal present, into reality as it is.

The four purposes not only lend meaning to our lives but also provide the foundation for this great spiritual purpose: the quest for illumination. Once we come full circle and complete this "journey without distance," we recognize that we've been here before. We have always been right here, precisely where we stand, in this arising moment of absolute mystery and divine perfection.

In quiet moments, many of us are touched by the wonder, miracle, and mystery at the heart of our existence. As expressed in the epigraph below, from my novel *The Journeys of Socrates*, my old mentor, after an arduous odyssey, comes to the following conclusion, which speaks to a realization awaiting us all.

When I was young, I believed that life
might unfold in an orderly way,
according to my hopes and expectations.
But now I understand that the Way winds like a river,
always changing, ever onward, following God's gravity
toward the Great Sea of Being.
My journeys revealed that the Way itself creates the warrior;
that in the fullness of time,
every path leads to peace, and every choice to wisdom.
And that life has always been, and will always be,
arising in Mystery.

— DAN MILLMAN, *THE JOURNEYS OF SOCRATES*

Acknowledgments

Each book has its own story to tell, and its own cast of characters. Most, like this one, begin as a solitary enterprise — an idea in the author's mind, an impulse that becomes a mission. Then the lonely hours of isolation, confronting and filling empty page after page in the first creative outpouring. But sooner or later, the author needs others to provide feedback, perspective, and editorial guidance.

Many helped in this process: My wife, Joy, my daughter Sierra, and my friend Doug Childers volunteered to hack and hew their way through the thicket of verbiage that made up the first rough draft, looking for paragraphs worth rescuing, helping me clear away the overgrowth so that the second draft resembled something more like a landscaped garden, with ideas deserving further cultivation laid out in orderly rows.

I was then ready to show the manuscript to publisher Linda Kramer of H J Kramer Inc., which publishes in collaboration with New World Library. She immediately saw its potential and made an offer on the spot. I now had a committed publishing house for this work in progress.

Then a group of readers, including Martin Adams, Holly Deme, Alex Deme, Peter Ingraham, David Moyer, and Beth Wilson offered helpful comments and perspectives.

After additional rewriting, I sent the manuscript to freelance editor Nancy Carleton, whose line edit generated additional rewriting before I handed the completed manuscript to the editorial and production team at New World Library. There the manuscript was polished by editors Kristen Cashman and Jason Gardner. Then Alan Hebel and Ian Shimkoviak of *the*BookDesigners designed the cover, Tona Pearce Myers typeset the book, Karen Stough proofread it, and Munro Magruder and Monique Muhlenkamp planned the publicity.

That's when Suezen Stone, H J Kramer's hardworking director of foreign rights, contacted publishers overseas to invite offers on the new book.

It takes a village not just to raise a child but also to help bring a book to full fruition and send it out into the world. I'm grateful to each and all.

Further Resources

SOURCE MATERIAL

If you would like to explore the following topics more deeply, I refer you to the source books, from which I've drawn material to present within the context of *The Four Purposes of Life*.

Life as a School

No Ordinary Moments: A Peaceful Warrior's Guide to Daily Life
Dan Millman (H J Kramer/New World Library)

As indicated by the subtitle, this book presents fundamental practices of the peaceful warrior's way and the school of daily life, and contains sections dealing with issues such as universal addictions and energy management not found in my other books.

Required Courses

Everyday Enlightenment: The Twelve Gateways to Personal Growth
Dan Millman (Grand Central Publishing)

The course catalog and curriculum presents a summary introduction of each subject in the school of life. *Everyday Enlightenment* offers an in-depth treatment of these twelve arenas of life.

School Rules

The Laws of Spirit: A Tale of Transformation
Dan Millman (H J Kramer/New World Library)

This book relates an adventure in the mountains with a mysterious woman sage who reveals specific laws of spirit, including balance, choices, process, action, surrender, and unity, and describes how they apply in everyday life.

The Nine Life Paths

The Life You Were Born to Live: A Guide to Finding Your Life Purpose
Dan Millman (H J Kramer/New World Library)

The summary descriptions of each of the nine life-path numbers provide a brief introduction to a far more comprehensive treatment of the Life-Purpose System presented in this book. (Also see the Life-Purpose App at the iTunes store.)

Tales of Transcendence

Bridge Between Worlds: Extraordinary Experiences That Changed Lives
Dan Millman and Doug Childers (H J Kramer/New World Library)

This book speaks to readers who appreciate documented stories that stretch our imaginations and remind us of the mysterious nature of life — and those who enjoy dramatic reminders of life's intriguing possibilities.

Final Epigraph
(Permission to reprint courtesy of HarperOne Publishers)

The Journeys of Socrates: A Novel
Dan Millman (HarperOne Publishers)

This Russian odyssey, beginning with a death and birth in 1872, relates the life, the ordeals, and the evolution of the man I would years later name Socrates. It's a story about how a boy became a man, how a man became a warrior, and how a warrior found peace.

Appendix

DETERMINING YOUR LIFE-PATH NUMBER BY DOING THE MATH

If you can add a few numbers together (such as 2 + 4 + 3 + 6), you can compute your birth number in three simple steps:

1. Write out an accurate date of birth numerically: For example: April 20, 1975, becomes 4-20-1975. *Note:* Be sure to include all four digits of the year, as in 1975 (not '75) or 2010 (not '10).
2. Add up every single digit to form a sum. In this example: 4 + 2 + 0 + 1 + 9 + 7 + 5 = 28. *Note:* Be sure to separate all digits. So for April 20, you would add 4 + 2 + 0 (*not* 4 + 20).
3. Once you have a sum — in this example, 28 — you then "resolve" these two digits by adding them together to make a final sum, so 28 becomes

2 + 8 = 10 — and the final *birth number* is written (in this example) as 28/10.

Other examples:

- October 20, 1988: 10-20-1988 — 1 + 0 + 2 + 0 + 1 + 9 + 8 + 8 = 29, resolved to 29/11 (a four-digit birth number).
- March 3, 1960: 3-3-1960 — 3 + 3 + 1 + 9 + 6 + 0 = 22, resolved to 22/4 (a three-digit birth number).
- February 20, 2003: 2-20-2003 — 2 + 2 + 0 + 2 + 0 + 0 + 3 = 9 (a single-digit birth number).

Be sure to check your math; even though it is simple addition, you want to make sure your calculations are correct. (I encourage you to check your results via the life-purpose calculator at www.peacefulwarrior.com, which also provides additional information.)

Further Information

You can find more details on birth numbers and the Life-Purpose System, including answers to other common questions, in *The Life You Were Born to Live* (see Further Resources) or by downloading the Life-Purpose App (available for the iPhone, Android, and other mobile devices).

About the Author

DAN MILLMAN — a former world champion athlete, coach, martial arts instructor, and college professor — is the author of numerous books read by millions of people in twenty-nine languages. He teaches worldwide, and for three decades has influenced people from all walks of life, including leaders in the fields of health, psychology, education, business, politics, sports, entertainment, and the arts.

For details: www.peacefulwarrior.com

H J Kramer and New World Library are dedicated to publishing books and other media that inspire and challenge us to improve the quality of our lives and the world.

We are a socially and environmentally aware company, and we strive to embody the ideals presented in our publications. We recognize that we have an ethical responsibility to our customers, our staff members, and our planet.

We serve our customers by creating the finest publications possible on personal growth, creativity, spirituality, wellness, and other areas of emerging importance. We serve New World Library employees with generous benefits, significant profit sharing, and constant encouragement to pursue their most expansive dreams.

As a member of the Green Press Initiative, we print an increasing number of books with soy-based ink on 100 percent postconsumer-waste recycled paper. Also, we power our offices with solar energy and contribute to nonprofit organizations working to make the world a better place for us all.